GLOBAL CLIMATE CHANGE

GLOBAL CLIMATE CHANGE
Pressing Dangers

E.B. Homes
Gina Jeffords
John Kelly

Kruger Brentt
Publishers

2023

Kruger Brentt Publishers UK. LTD.
Company Number 9728962

Regd. Office: 68 St Margarets Road, Edgware, Middlesex HA8 9UU

ISBN: 9781787150492

Disclaimer:

Every possible effort has been made to ensure that the information contained in this book is accurate at the time of going to press, and the publisher and author cannot accept responsibility for any errors or omissions, however caused. No responsibility for loss or damage occasioned to any person acting, or refraining from action, as a result of the material in this publication can be accepted by the editor, the publisher or the author. The Publisher is not associated with any product or vendor mentioned in the book. The contents of this work are intended to further general scientific research, understanding and discussion only. Readers should consult with a specialist where appropriate.

Every effort has been made to trace the owners of copyright material used in this book, if any. The author and the publisher will be grateful for any omission brought to their notice for acknowledgement in the future editions of the book.

For information on all our publications visit our website at http://krugerbrentt.com/

PREFACE

Climate change is a major global challenge today, and the world is becoming more vulnerable to this change. Climate change refers to the changes in Earth's climate condition. It describes the changes in the atmosphere which have taken place over a period, ranging from decades to millions of years. The recent report from the United Nations predicted that average global temperature could increase by 6° Celsius at the end of the century. Climate change has an adverse effect on the environment and ecosystem. The Earth's climate has always changed and evolved. Some of these changes have been due to natural causes such as volcanic eruptions, floods, forest fires etc., but quite a few of them are due to human activities. Human activities such as deforestation, burning fossil fuels, farming livestock etc., generate an enormous amount of greenhouse gases. This results in greenhouse effect and global warming which are the major causes for climate change. If the current situation of climate change continues in a similar manner, then it will impact all forms of life on the earth. The earth temperature will rise, the monsoon patterns will change, sea levels will rise, and storms, volcanic eruptions and natural disasters will occur frequently. The biological and ecological balance of the earth will get disturbed. The environment will get polluted and humans will not be able to get fresh air to breath and fresh water to drink. Life on earth will come to an end.

The Government of India has taken many measures to improve the dire situation of Climate Change. The Ministry of Environment and Forests is the nodal agency for climate change issues in India. It has initiated several climate-friendly measures, particularly in the area of renewable energy. India took several steps and policy initiatives to create awareness about climate change, and help capacity building for adaptation measures. It has initiated a "Green India" programme under which various trees are planted to make the forest land greener and more fertile. We need to follow the path of sustainable development to effectively address the concerns of climate change. We need to minimise the use of fossil fuels, which is the major cause for global warming. We must adopt alternative sources of energy, such as hydropower, solar and wind energy to make a progressive transition to clean energy. Mahatma Gandhi said that, "Earth provides enough to satisfy every man's need, but not any man's greed". With this view, we must remodel our outlook and achieve the goal of sustainable development. By adoption of clean technologies, equitable distribution of resources and addressing the issues of equity and justice, we can make our developmental process more harmonious with nature.

The present book contains twenty one chapters covering all related disciplines. These chapters include Global Climate Change: An Introduction, Impact of Climate Change Projections -2030 in India, Climate Change: A 'Glocal' Problem Requiring 'Glocal' Action, Modelling the Climate, The Impact of Climate Change on the Agricultural Sector, Global Warming and Climate Change, land value changes and carbon sequestration: as an ecosystem, climate change and mental health, global environmental change and health, climate change: twenty-first century and beyond, major causes, effects and solutions of global warming, climate change: causes, effects and mitigationmeasures, globalization, climate change, and human health, the global environmental change and human security (gechs), climate changes and agriculture, impacts of climate changes on food production and safety, managing weather and climate risks in agriculture, impacts climate changes on pest and disease prevalence, effects of climate change on animal production and livestock, climate changes and land use, the toxicology of climate changes, Related terminology is given at the end for ready reference. It addresses holistic issues related to climate change and its contribution in triggering the temperature increase with a multitude of impacts on natural processes. It is organized into information on the past, present, and future of climate change to lead to a more complete understanding and therefore effective solutions. It will prove to be invaluable to a wide range of readers, from students in the life sciences who need a brief overview of the basics of climate science, to atmospheric science, geography, geoscience and environmental science students. It is also a valuable reference text for those involved in environmental monitoring, conservation and policy making.

We are grateful to all those persons as well as various books, manuals, periodicals, magazines, journals etc. that helped in the preparation of this book. In spite of the best efforts, it is possible that some errors may have occurred into the compilation and editing of the book. Further queries, constructive suggestions and criticisms for the improvement of the book are always welcome and shall be thankfully acknowledged.

E.B. Homes

Gina Jeffords

John Kelly

CONTENTS

1

GLOBAL CLIMATE CHANGE AN INTRODUCTION

1.1 INTRODUCTION

Many changes have been observed in global climate over the past century. The nature and causes of these changes have been comprehensively chronicled in a variety of recent reports, such as those by the Intergovernmental Panel on Climate Change (IPCC) and the U.S. Climate Change Science Program (CCSP). This section does not intend to duplicate these comprehensive efforts, but rather to provide a brief synthesis, and to integrate more recent work with the assessments of the IPCC, CCSP, and others. Human activities have led to large increases in heat-trapping gases over the past century (Fig. 1).

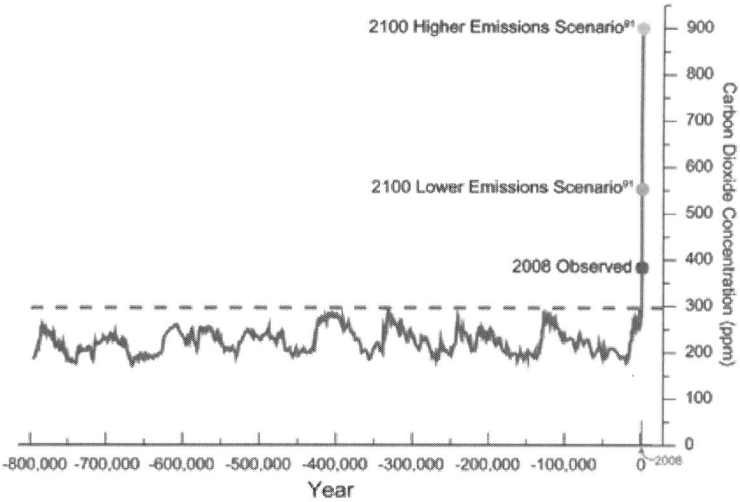

Fig.1: Analysis of air bubbles trapped in an Antarctic ice core and the Earth's changing carbon dioxide concentration

Analysis of air bubbles trapped in an Antarctic ice core extending back 800,000 years documents the Earth's changing carbon dioxide concentration. Over this long period, natural factors have caused the atmospheric carbon dioxide concentration to vary within a range of about 170 to 300 parts per million (ppm). Temperature-related data make clear that these variations have played a central role in determining the global climate. As a result of human activities, the present carbon dioxide concentration of about 385 ppm is about 30 percent above its highest level over at least the last 800,000 years. In the absence of strong control measures, emissions projected for this century would result in the carbon dioxide concentration increasing to a level that is roughly 2 to 3 times the highest level occurring over the glacial-interglacial era that spans the last 800,000 or more years.

The Earth's climate depends on the functioning of a natural "greenhouse effect." This effect is the result of heat-trapping gases (also known as greenhouse gases) like water vapor, carbon dioxide, ozone, methane, and nitrous oxide, which absorb heat radiated from the Earth's surface and lower atmosphere and then radiate much of the energy back toward the surface. Without this natural greenhouse effect, the average surface temperature of the Earth would be about 60°F colder. However, human activities have been releasing additional heat-trapping gases, intensifying the natural greenhouse effect, thereby changing the Earth's climate. Climate is influenced by a variety of factors, both human-induced and natural. The increase in the carbon dioxide concentration has been the principal factor causing warming over the past 50 years. Its concentration has been building up in the Earth's atmosphere since the beginning of the industrial era in the mid-1700s, primarily due to the burning of fossil fuels (coal, oil, and natural gas) and the clearing of forests. Human activities have also increased the emissions of other greenhouse gases, such as methane, nitrous oxide, and halocarbons (Fig. 2).

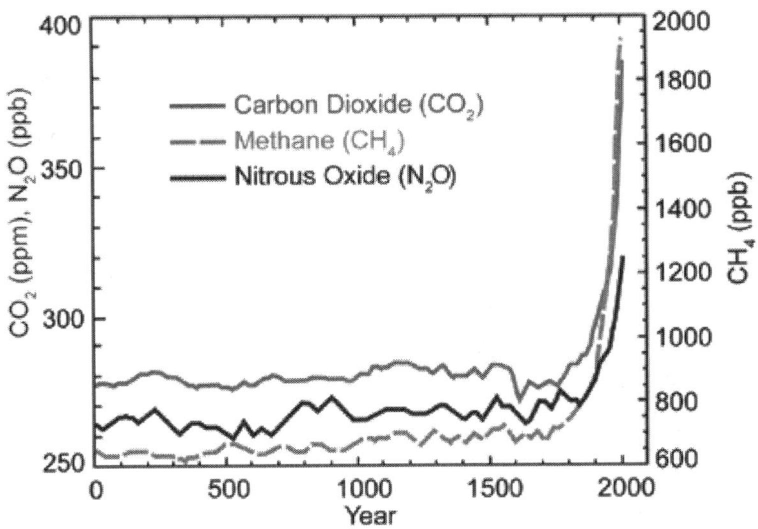

Fig. 2 Human activities have also increased the emissions of other greenhouse gases.

Increases in concentrations of these gases since 1750 are due to human activities in the industrial era. Concentration units are parts per million (ppm) or parts per billion (ppb), indicating the number of molecules of the greenhouse gas per million or billion molecules of air.

These emissions are thickening the blanket of heat-trapping gases in Earth's atmosphere, causing surface temperatures to rise.

1.2 HEAT-TRAPPING GASES

Carbon dioxide concentration has increased due to the use of fossil fuels in electricity generation, transportation, and industrial and household uses. It is also produced as a by-product during the manufacturing of cement. Deforestation provides a source of carbon dioxide and reduces its uptake by trees and other plants. Globally, over the past several decades, about 80 percent of human-induced carbon dioxide emissions came from the burning of fossil fuels, while about 20 percent resulted from deforestation and associated agricultural practices.

The concentration of carbon dioxide in the atmosphere has increased by roughly 35 percent since the start of the industrial revolution. Methane concentration has increased mainly as a result of agriculture; raising livestock (which produce methane in their digestive tracts); mining, transportation, and use of certain fossil fuels; sewage; and decomposing garbage in landfills. About 70 percent of the emissions of atmospheric methane are now related to human activities.4 Nitrous oxide concentration is increasing as a result of fertilizer use and fossil fuel burning. Halocarbon emissions come from the release of certain manufactured chemicals to the atmosphere. Examples include chlorofluorocarbons (CFCs), which were used extensively in refrigeration and for other industrial processes before their presence in the atmosphere was found to cause stratospheric ozone depletion. The abundance of these gases in the atmosphere is now decreasing as a result of international regulations designed to protect the ozone layer. Continued decreases in ozone-depleting halocarbon emissions are expected to reduce their relative influence on climate change in the future. Many halocarbon replacements, however, are potent greenhouse gases, and their concentrations are increasing.

Ozone is a greenhouse gas, and is continually produced and destroyed in the atmosphere by chemical reactions. In the troposphere, the lowest 5 to 10 miles of the atmosphere near the surface, human activities have increased the ozone concentration through the release of gases such as carbon monoxide, hydrocarbons, and nitrogen oxides. These gases undergo chemical reactions to produce ozone in the presence of sunlight. In addition to trapping heat, excess ozone in the troposphere causes respiratory illnesses and other human health problems.

In the stratosphere, the layer above the troposphere, ozone exists naturally and protects life on Earth from exposure to excessive ultraviolet radiation from the Sun. As mentioned previously, halocarbons released by human activities destroy ozone in the stratosphere and have caused the ozone hole over Antarctica. Changes in the stratospheric ozone layer have contributed to changes in wind patterns and regional climates in Antarctica. Water vapor is the most important and abundant greenhouse gas in the atmosphere. Human activities produce only a very small increase in water vapor through irrigation and combustion processes. However, the surface warming caused by human- produced increases in other greenhouse gases leads to an increase in atmospheric water vapor, since a warmer climate increases evaporation and allows the atmosphere to hold more moisture. This creates an amplifying "feedback loop," leading to more warming.

1.3 OTHER HUMAN INFLUENCES

In addition to the global-scale climate effects of heat-trapping gases, human activities also produce additional local and regional effects. Some of these activities partially offset the warming caused by greenhouse gases, while others increase the warming.

One such influence on climate is caused by tiny particles called "aerosols" (not to be confused with aerosol spray cans)". For example, the burning of coal produces emissions of sulfur-containing compounds. These compounds form "sulfate aerosol" particles, which reflect some of the incoming sunlight away from the Earth, causing a cooling influence at the surface. Sulfate aerosols also tend to make clouds more efficient at reflecting sunlight, causing an additional indirect cooling effect.

Another type of aerosol, often referred to as soot or black carbon, absorbs incoming sunlight and traps heat in the atmosphere. Thus, depending on their type, aerosols can either mask or increase the warming caused by increased levels of greenhouse gases. On a globally averaged basis, the sum of these aerosol effects offsets some of the warming caused by heat-trapping gases.

The effects of various greenhouse gases and aerosol particles on Earth's climate depend in part on how long these gases and particles remain in the atmosphere. After emission, the atmospheric concentration of carbon dioxide remains elevated for thousands of years, and that of methane for decades, while the elevated concentrations of aerosols only persist for days to weeks. The climate effects of reductions in emissions of carbon dioxide and other long-lived gases do not become apparent for at least several decades. In contrast, reductions in emissions of short-lived compounds can have a rapid, but complex effect since the geographic patterns of their climatic influence and the resulting surface temperature responses are quite different. One modeling study found that while the greatest emissions of short-lived pollutants in summertime by late this century are projected to come from Asia, the strongest climate response is projected to be over the central United States.

Human activities have also changed the land surface in ways that alter how much heat is reflected or absorbed by the surface. Such changes include the cutting and burning of forests, the replacement of other areas of natural vegetation with agriculture and cities, and large-scale irrigation. These transformations of the land surface can cause local (and even regional) warming or cooling. Globally, the net effect of these changes has probably been a slight cooling of the Earth's surface over the past100 years.

Natural influences

Two important natural factors also influence climate: the Sun and volcanic eruptions. Over the past three decades, human influences on climate have become increasingly obvious, and global temperatures have risen sharply. During the same period, the Sun's energy output (as measured by satellites since 1979) has followed its historical 11-year cycle of small ups and downs, but with no net increase (see figure page 20).16 The two major volcanic eruptions of the past 30 years have had short-term cooling effects on climate, lasting 2 to 3 years. Thus, these natural factors cannot explain the warming of recent decades; in fact, their net effect on climate has probably been a slight cooling influence over this period. Slow changes in Earth's orbit around the Sun and its tilt toward or away from the Sun are also a purely natural influence on climate, but are only important on timescales from thousands to many tens of thousands of years. The climate changes that have occurred over the last century are not solely caused by the human and natural factors described above. In addition to these influences, there are also fluctuations in climate that occur even in the absence of changes in human activities, the Sun, or volcanoes. One example is the *El Niño* phenomenon, which has important influences on many aspects of regional and global climate. Many other modes of variability have been identified by climate scientists and their effects on climate occur at the same time as the effects of human activities, the Sun, and volcanoes (Fig. 3).

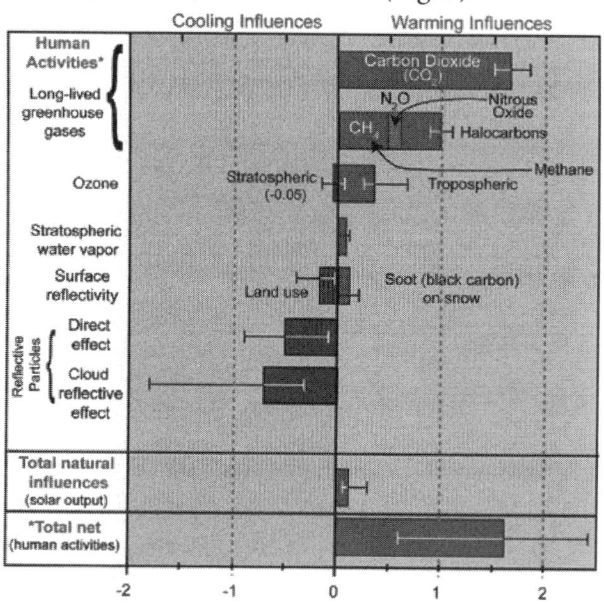

Fig. 3: The amount of warming influence (red bars) or cooling influence (blue bars) that different factors have had on Earth's climate over the industrial age .

The figure above shows the amount of warming influence (red bars) or cooling influence (blue bars) that different factors have had on Earth's climate over the industrial age (from about 1750 to the present). Results are in watts per square meter. The longer the bar, the greater the influence on climate. The top part of the box includes all the major human-induced factors, while the second part of the box includes the Sun, the only major natural factor with a long-term effect on climate. The cooling effect of individual volcanoes is also natural, but is relatively short-lived (2 to 3 years), thus their influence is not included in this figure. The bottom part of the box shows that the total net effect (warming influences minus cooling influences) of human activities is a strong warming influence. The thin lines on each bar provide an estimate of the range of uncertainty.

1.4 CARBON RELEASE AND UPTAKE

Once carbon dioxide is emitted to the atmosphere, some of it is absorbed by the oceans and taken up by vegetation, although this storage may be temporary. About 45 percent of the carbon dioxide emitted by human activities in the last 50 years is now stored in the oceans and vegetation. The rest has remained in the air, increasing the atmospheric concentration. It is thus important to understand not only how much carbon dioxide is emitted, but also how much is taken up, over what time scales, and how these sources and "sinks" of carbon dioxide might change as climate continues to warm.

For example, it is known from long records of Earth's climate history that under warmer conditions, carbon tends to be released, for instance, from thawing permafrost, initiating a feedback loop in which more carbon release leads to more warming which leads to further release, and so on. Global emissions of carbon dioxide have been accelerating. The growth rate increased from 1.3 percent per year in the 1990s to 3.3 percent per year between 2000 and 2006. The increasing emissions of carbon dioxide are the primary cause of the increased concentration of carbon dioxide observed in the atmosphere. There is also evidence that a smaller fraction of the annual human induced emissions is now being taken up than in the past, leading to a greater fraction remaining in the atmosphere and an accelerating rate of increase in the carbon dioxide concentration.

1.5 OCEAN ACIDIFICATION

As the ocean absorbs carbon dioxide from the atmosphere, seawater is becoming less alkaline (its pH is decreasing) through a process generally referred to as ocean acidification. The pH of seawater has decreased significantly since 1750, and is projected to drop much more dramatically by the end of the century if carbon dioxide concentrations continue to increase. Such ocean acidification is essentially irreversible over a time scale of centuries. As discussed in the Ecosystems sector and Coasts region, ocean acidification affects the process of calcification by which living things create shells and skeletons, with substantial negative consequences for coral reefs, mollusks, and some plankton species important to ocean food chains.

Global average temperature and sea level have increased, and precipitation patterns have changed.

Temperatures are rising

Global average surface air temperature has increased substantially since 1970.26 The estimated change in the average temperature of Earth's surface is based on measurements from thousands of weather stations, ships, and buoys around the world, as well as from satellites. These measurements are independently compiled, analyzed, and processed by different research groups. There are a number of important steps in the data processing. These include identifying and adjusting for the effects of changes in the instruments used to measure temperature, the measurement times and locations, the local environment around the measuring site, and such factors as satellite orbital drift. For instance, the growth of cities can cause localized "urban heat island" effects.

A number of research groups around the world have produced estimates of global-scale changes in surface temperature. The warming trend that is apparent in all of these temperature records is confirmed by other independent observations, such as the melting of Arctic sea ice, the retreat of mountain glaciers on every continent, reductions in the extent of snow cover, earlier blooming of plants in spring, and increased melting of the Greenland and Antarctic ice sheets. Because snow and ice reflect the Sun's heat, this melting causes more heat to be absorbed, which causes more melting, resulting in another feedback loop (Fig. 4).

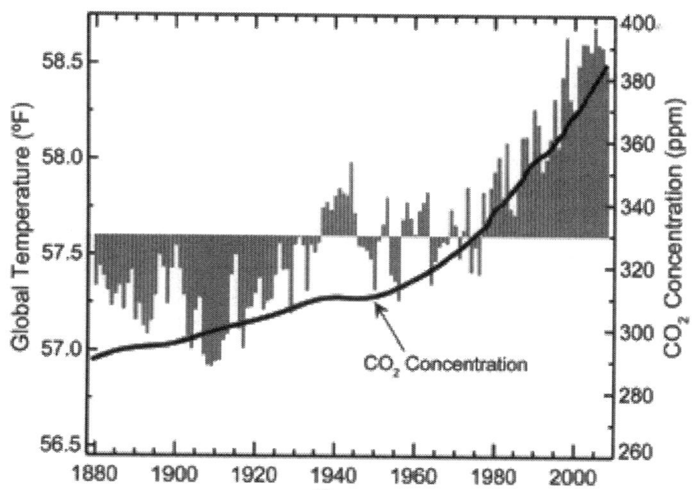

Fig. 4: Global annual average temperature (as measured over both land and oceans).

Global annual average temperature (as measured over both land and oceans). Red bars indicate temperatures above and blue bars indicate temperatures below the average temperature for the period 1901-2000. The black line shows atmospheric carbon dioxide (CO_2) concentration in parts per million (ppm). While there is a clear long- term global warming trend, each individual year does not show a temperature increase relative to the previous year, and some years show greater changes than others. These year-to-year fluctuations in temperature are due to natural processes, such as the effects of *El Niños, La Niñas,* and the eruption of large volcanoes.

Additionally, temperature measurements above the surface have been made by weather balloons since the late 1940s, and from satellites since 1979. These measurements show warming of the troposphere, consistent with the surface warming. They also reveal cooling in the stratosphere. This pattern of tropospheric warming and stratospheric cooling agrees with our understanding of how atmospheric temperature would be expected to change in response to increasing greenhouse gas concentrations and the observed depletion of stratospheric ozone.

Precipitation patterns are changing Precipitation is not distributed evenly over the globe. Its average distribution is governed primarily by atmospheric circulation patterns, the availability of moisture, and surface terrain effects. The first two of these factors are influenced by temperature. Thus, human-caused changes in temperature are expected to alter precipitation patterns. Observations show that such shifts are occur- ring. Changes have been observed in the amount, intensity, frequency, and type of precipitation. Pronounced increases in precipitation over the past 100 years have been observed in eastern North America, southern South America, and northern Europe. Decreases have been seen in the Mediterranean, most of Africa, and southern Asia. Changes in the geographical distribution of droughts and flooding have been complex. In some regions, there have been increases in the occurrences of both droughts and floods. As the world warms, northern regions and mountainous areas are experiencing more precipitation falling as rain rather than snow. Widespread increases in heavy precipitation events have occurred, even in places where total rain amounts have decreased. These changes are associated with the fact that warmer air holds more water vapor evaporating from the world's oceans and land surface. This increase in atmospheric water vapor has been observed from satellites, and is primarily due to human influences.

1.6 SEA LEVEL IS RISING

After at least 2,000 years of little change, sea level rose by roughly 8 inches over the past century.

Satellite data available over the past 15 years show sea level rising at a rate roughly double the rate observed over the past century. There are two principal ways in which global warming causes sea level to rise. First, ocean water expands as it warms, and therefore takes up more space. Warming has been observed in each of the world's major ocean basins, and has been directly linked to human influences.

Second, warming leads to the melting of glaciers and ice sheets, which raises sea level by adding water to the oceans. Glaciers have been retreating worldwide for at least the last century, and the rate of retreat has increased in the past decade. Only a few glaciers are actually advancing (in locations that were well below freezing, and where increased precipitation has outpaced melting). The total volume of glaciers on Earth is declining sharply. The progressive disappearance of glaciers has implications not only for the rise in global sea level, but also for water supplies in certain densely populated regions of Asia and South America. The Earth has major ice sheets on Greenland and Antarctica. These ice sheets are currently losing ice volume by increased melting and calving of icebergs, contributing to sea-level rise. The Green- land Ice Sheet has also been experiencing record amounts of surface melting, and a large increase in the rate of mass loss in the past decade.41 If the en- tire Greenland Ice Sheet melted, it would raise sea level by about 20 feet. The Antarctic Ice Sheet consists of two portions, the West Antarctic Ice Sheet and the East Antarctic Ice Sheet. The West Antarctic Ice Sheet, the more vulnerable to melting of the two, contains enough water to raise global sea levels by about 16 to 20 feet. If the East Antarctic Ice Sheet melted entirely, it would raise global sea level by about 200 feet. Complete melting of these ice sheets over this century or the next is thought to be virtually impossible, although past climate records provide precedent for very significant decreases in ice volume, and therefore increases in sea level (Fig. 5).

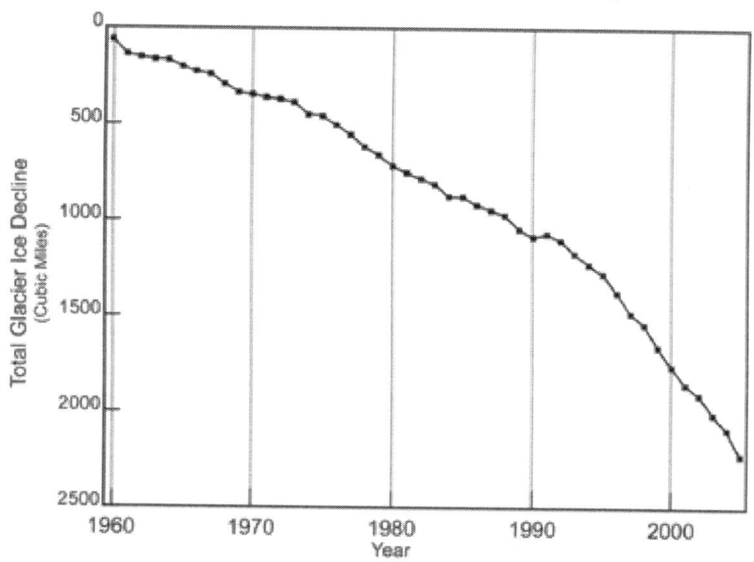

Fig. 5: As temperatures have risen, glaciers around the world have shrunk. The graph shows the cumulative decline in glacier ice worldwide.

The global warming of the past 50 years is due primarily to human-induced increases in heat-trapping gases. Human "fingerprints" also have been identified in many other aspects of the climate system, including changes in ocean heat content, precipitation, atmospheric moisture, and Arctic sea ice. In 1996, the IPCC Second Assessment Report cautiously concluded that "the balance of evidence suggests a discernible human influence on global climate." Since then, a number of national and international assessments have come to much stronger conclusions about the reality of human effects on climate. Recent scientific assessments find that most of the warming of the Earth's surface over the past 50 years has been caused by human activities.

This conclusion rests on multiple lines of evidence. Like the warming "signal" that has gradually emerged from the "noise" of natural climate variability, the scientific evidence for a human influence on global climate has accumulated over the past several decades, from many hundreds of studies. No single study is a "smoking gun." Nor has any single study or combination of studies undermined the large body of evidence supporting the conclusion that human activity is the primary driver of recent warming.

The first line of evidence is our basic physical understanding of how greenhouse gases trap heat, how the climate system responds to increases in greenhouse gases, and how other human and natural factors influence climate. The second line of evidence is from indirect estimates of climate changes over the last 1,000 to 2,000 years. These records are obtained from living things and their remains (like tree rings and corals) and from physical quantities (like the ratio between lighter and heavier isotopes of oxygen in ice cores) which change in measurable ways as climate changes. The lesson from these data is that global surface temperatures over the last several decades are clearly unusual, in that they were higher than at any time during at least the past 400 years.47 For the Northern Hemisphere, the recent temperature rise is clearly unusual in at least the last 1,000 years.

The third line of evidence is based on the broad, qualitative consistency between observed changes in climate and the computer model simulations of how climate would be expected to change in response to human activities. For example, when climate models are run with historical increases in greenhouse gases, they show gradual warming of the Earth and ocean surface, increases in ocean heat content and the temperature of the lower atmosphere, a rise in global sea level, retreat of sea ice and snow cover, cooling of the stratosphere, an increase in the amount of atmospheric water vapor, and changes in large-scale precipitation and pressure patterns. These and other aspects of modeled climate change are in agreement with observations.

Finally, there is extensive statistical evidence from so-called "fingerprint" studies. Each factor that affects climate produces a unique pattern of climate response, much as each person has a unique fingerprint. Fingerprint studies exploit these unique signatures, and allow detailed comparisons of modeled and observed climate change patterns. Scientists rely on such studies to attribute observed changes in climate to a particular cause or set of causes. In the real world, the climate changes that have occurred since the start of the Industrial Revolution are due to a complex mixture of human and natural causes. The importance of each individual influence in this mixture changes over time. Of course, there are not multiple Earths, which would allow an experimenter to change one factor at a time on each Earth, thus helping to isolate different fingerprints. Therefore, climate models are used to study how individual factors affect climate. For example, a single factor (like greenhouse gases) or a set of factors can be varied, and the response of the modeled climate system to these individual or combined changes can thus be studied.

For example, when climate model simulations of the last century include all of the major influences on climate, both human-induced and natural, they can reproduce many important features of observed climate change patterns. When human influences are removed from the model experiments, results suggest that the surface of the Earth would actually have cooled slightly over the last 50 years. The clear message from fingerprint studies is that the observed warming over the last half-century can not be explained by natural factors, and is instead caused primarily by human factors.

The blue band shows how global average temperatures would have changed due to natural forces only, as simulated by climate models. The red band shows model projections of the effects of human and natural forces combined. The black line shows actual observed global average temperatures. As the blue band indicates, without human influences, temperature over the past century would actually have first warmed and then cooled slightly over recent decades(Fig. 6).

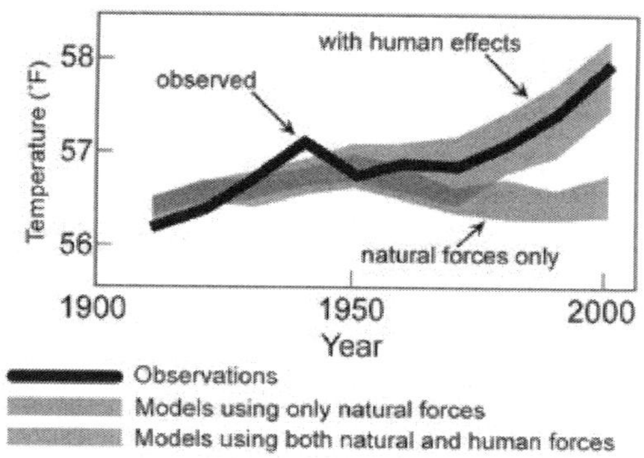

Fig. 6: Global average temperature

Another fingerprint of human effects on climate has been identified by looking at a slice through the layers of the atmosphere, and studying the pattern of temperature changes from the surface up through the stratosphere. In all climate models, increases in carbon di- oxide cause warming at the surface and in the troposphere, but lead to cooling of the stratosphere. For straightforward physical reasons, models also calculate that the human-caused depletion of stratospheric ozone has had a strong cooling effect in the stratosphere. There is a good match between the model fingerprint in response to combined carbon dioxide and ozone changes and the observed pattern of tropospheric warming and stratospheric cooling. In contrast, if most of the observed temperature change had been due to an increase in solar output rather than an increase in green- house gases, Earth's atmosphere would have warmed throughout its full vertical extent, including the stratosphere. The observed pattern of atmospheric temperature changes, with its pronounced cooling in the stratosphere, is therefore inconsistent with the hypothesis that changes in the Sun can explain the warming of recent decades. Moreover, direct satellite measurements of solar output show slight decreases during the recent period of warming.

The earliest fingerprint work focused on changes in surface and atmospheric temperature. Scientists then applied fingerprint methods to a whole range of climate variables, identifying human-caused climate signals in the heat content of the oceans, the height of the tropopause (the boundary be- tween the troposphere and stratosphere, which has shifted upward by hundreds of feet in recent decades), the geographical patterns of precipitation, drought, surface pressure, and the runoff from major river basins. Studies published after the appearance of the IPCC Fourth Assessment Report in 2007 have also found human fingerprints in the increased levels of atmospheric moisture (both close to the surface and over the full extent of the atmosphere), in the decline of Arctic Sea ice extent,61 and in the patterns of changes in Arctic and Antarctic surface temperatures(Fig. 7).

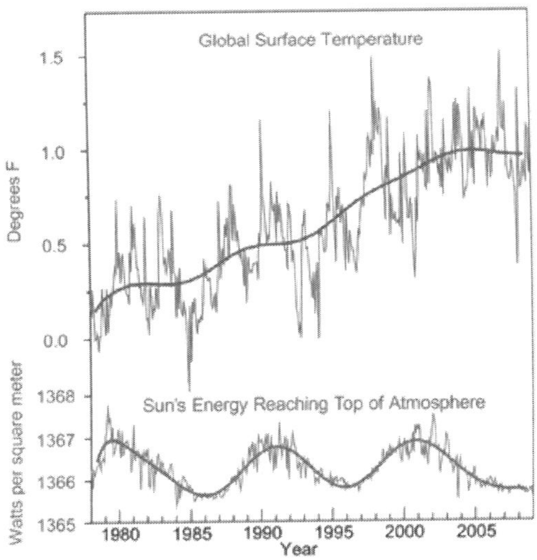

Fig. 7: The Sun's energy received at the top of Earth's atmosphere.

The Sun's energy received at the top of Earth's atmosphere has been measured by satellites since 1978. It has followed its natural 11-year cycle of small ups and downs, but with no net increase (bottom). Over the same period, global temperature has risen markedly (top). The message from this entire body of work is that the climate system is telling a consistent story of increasingly dominant human influence – the changes in temperature, ice extent, moisture, and circulation patterns fit together in a physically consistent way, like pieces in a complex puzzle.

Increasingly, this type of fingerprint work is shifting its emphasis. As noted, clear and compelling scientific evidence supports the case for a pronounced human influence on global climate. Much of the recent attention is now on climate changes at continental and regional scales, and on variables that can have large impacts on societies. For example, scientists have established causal links between human activities and the changes in snowpack, maximum and minimum temperature, and the seasonal timing of runoff over mountainous regions of the western United States. Human activity is likely to have made a substantial contribution to ocean surface temperature changes in hurricane formation regions. Researchers are also looking beyond the physical climate system, and are beginning to tie changes in the distribution and seasonal behavior of plant and animal species to human-caused changes in temperature and precipitation. For over a decade, one aspect of the climate change story seemed to show a significant difference between models and observations.

In the tropics, all models predicted that with a rise in greenhouse gases, the troposphere would be expected to warm more rapidly than the surface. Observations from weather balloons, satellites, and surface thermometers seemed to show the opposite behavior (more rapid warming of the surface than the troposphere). This issue was a stumbling block in our understanding of the causes of climate change. It is now largely resolved. Research showed that there were large uncertainties in the satellite and weather balloon data. When uncertainties in models and observations are properly accounted for, newer observational data sets (with better treatment of known problems) are in agreement with climate model results (Fig. 8).

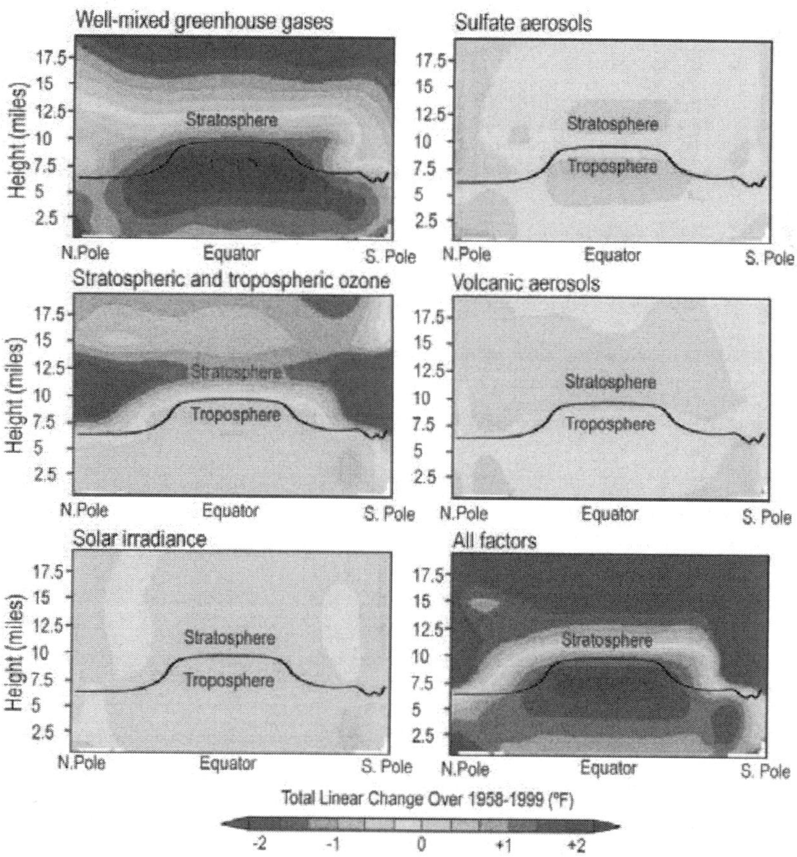

Fig. 8: Climate simulations of the vertical profile of temperature change due to various factors.

Climate simulations of the vertical profile of temperature change due to various factors, and the effect due to all factors taken together. The panels above represent a cross-section of the atmosphere from the north pole to the south pole, and from the surface up into the stratosphere. The black lines show the location of the tropopause, the boundary between the lower atmosphere (troposphere) and the stratosphere.

This does not mean, however, that all remaining differences between models and observations have been resolved. The observed changes in some climate variables, such as Arctic Sea ice, 61, 76 some aspects of precipitation, and patterns of surface pressure,56 appear to be proceeding much more rapidly than models have projected. The reasons for these differences are not well understood. Nevertheless, the bottom-line conclusion from climate fingerprinting is that most of the observed changes studied to date are consistent with each other, and are also consistent with our scientific understanding of how the climate system would be expected to respond to the increase in heat-trapping gases resulting from human activities.

Scientists are sometimes asked whether extreme weather events can be linked to human activities.24 Scientific research has concluded that human influences on climate are indeed changing the likelihood of certain types of extreme events. For example,an analysis of the European summer heat wave of 2003 found that the risk of such a heat wave is now roughly four times greater than it would have been in the absence of human-induced climate change.

Like fingerprint work, such analyses of human- caused changes in the risks of extreme events rely on information from climate models, and on our understanding of the physics of the climate system. All of the models used in this work have imperfections in their representation of the complexities of the "real world" climate system.79,80 These are due to both limits in our understanding of the climate system, and in our ability to represent its complex behavior with available computer resources. Despite this, models are extremely useful, for a number of reasons.

First, despite remaining imperfections, the current generation of climate models accurately portrays many important aspects of today's weather pat- terns and climate. Models are constantly being improved, and are routinely tested against many observations of Earth's climate system. Second, the fingerprint work shows that models capture not only our present-day climate, but also key features of the observed climate changes over the past century. Third, many of the large-scale observed climate changes (such as the warming of the surface and troposphere, and the increase in the amount of moisture in the atmosphere) are driven by very basic physics, which is well-represented in models. Fourth, climate models can be used to predict changes in climate that can be verified in the real world. Examples include the short-term global cooling subsequent to the eruption of Mount Pinatubo and the stratospheric cooling with increasing carbon dioxide. Finally, models are the only tools that exist for trying to understand the climate changes likely to be experienced over the course of this century. No period in Earth's geological history provides an exact analogue for the climate conditions that will unfold in the coming decades.

Global temperatures are projected to continue to rise over this century; by how much and for how long depends on a number of factors, including the amount of heat-trapping gas emissions and how sensitive the climate is to those emissions. Some continued warming of the planet is projected over the next few decades due to past emissions. Choices made now will influence the amount of future warming. Lower levels of heat-trapping emissions will yield less future warming, while higher levels will result in more warming, and more severe impacts on society and the natural world.

1.7 EMISSIONS SCENARIOS

The IPCC developed a set of scenarios in a Special Report on Emissions Scenarios (SRES).81 These have been extensively used to explore the potential for future climate change. None of these scenarios, not even the one called "lower", includes implementation of policies to limit climate change or to stabilize atmospheric concentrations of heat- trapping gases. Rather, differences among these scenarios are due to different assumptions about changes in population, rate of adoption of new technologies, economic growth, and other factors. The IPCC emission scenarios also do not encompass the full range of possible futures: emissions can change less than those scenarios imply, or they can change more. Recent carbon dioxide emissions are, in fact, above the highest emissions scenario developed by the IPCC82. Whether this will continue is uncertain. There are also lower possible emissions paths than those put forth by the IPCC. The Framework Convention on Climate Change, to which the United States and 191 other countries are signatories, calls for stabilizing concentrations of greenhouse gases in the atmosphere at a level that would avoid dangerous human interference with the climate system. What exactly constitutes such interference is subject to interpretation. A variety of research studies suggest that a further 2°F increase (relative to the 1980-1999 period) would lead to severe, widespread, and irreversible impacts. To have a good chance (but not a guarantee) of avoiding temperatures above those levels, it has been estimated that atmospheric concentration of carbon dioxide would need to stabilize in the long term at around today's levels.

Reducing emissions of carbon dioxide would re- duce warming over this century and beyond. Implementing sizable and sustained reductions in carbon dioxide emissions as soon as possible would significantly reduce the pace and the overall amount of climate change, and would be more effective than reductions of the same size initiated later. Reducing emissions of some shorter-lived greenhouse gases, such as methane, and some types of particles, such as soot, would begin to reduce the warming influence within weeks to decades. The graphs below show emissions scenarios and resulting carbon dioxide concentrations for three IPCC scenarios and one stabilization scenario.

The graphs show recent and projected global emissions of carbon dioxide in gigatons of carbon, on the left, and atmospheric concentrations on the right under five emissions scenarios. The top three in the key are IPCC scenarios that assume no explicit climate policies (these are used in model projections that appear throughout this report). The bottom line is a "stabilization scenario," designed to stabilize atmospheric carbon dioxide concentration at 450 parts per million. The inset expanded below these charts shows emissions for 1990-2010 under the three IPCC scenarios along with actual emissions to 2007 (in black) (Fig. 9).

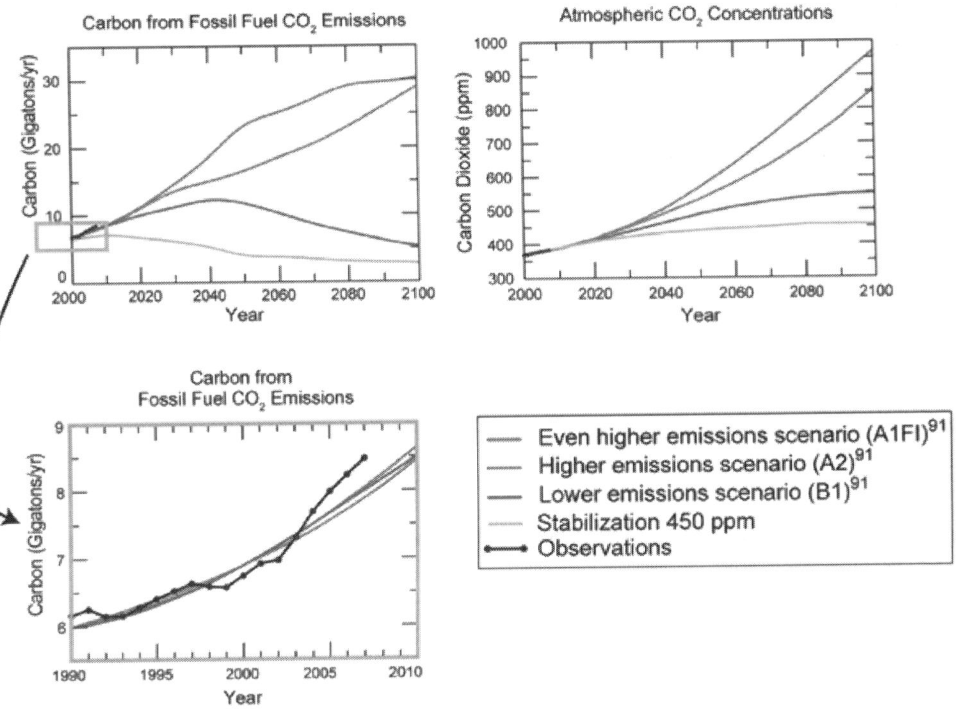

Fig. 9: The graphs show recent and projected global emissions of carbon dioxide in gigatons of carbon.

The stabilization scenario is aimed at stabilizing the atmospheric carbon dioxide concentration at roughly 450 parts per million (ppm); this is 70 ppm above the 2008 concentration of 385 ppm. Resulting temperature changes depend on atmospheric concentrations of greenhouse gases and particles and the climate's sensitivity to those concentrations. Of those shown on the previous page, only the 450 ppm stabilization target has the potential to keep the global temperature rise at or below about 3.5°F from preindustrial levels and 2°F above the current average temperature, a level beyond which many concerns have been raised about dangerous human interference with the climate system.88,89 Scenarios that stabilize carbon dioxide below 450 ppm (not shown in the figure) offer an increased chance of avoiding dangerous climate change.

Carbon dioxide is not the only greenhouse gas of concern. Concentrations of other heat-trapping gases like methane and nitrous oxide and particles like soot will also have to be stabilized at low enough levels to prevent global temperatures from rising higher than the level mentioned above. When these other gases are added, including the offsetting cooling effects of sulfate aerosol particles, analyses suggest that stabilizing concentrations around 400 parts per million of "equivalent carbon dioxide" would yield about an 80 percent chance of avoiding exceeding the 2°F above present temperature threshold. This would be true even if concentrations temporarily peaked as high as 475 parts per million and then stabilized at 400 parts per million roughly a century later. Reductions in sulfate aerosol particles would necessitate lower equivalent carbon dioxide targets.

1.8 RISING GLOBAL TEMPERATURE

All climate models project that human-caused emissions of heat-trapping gases will cause further warming in the future. Based on scenarios that +do not assume explicit climate policies to reduce greenhouse gas emissions, global average temperature is projected to rise by 2 to 11.5°F by the end of this century 90 (relative to the 1980-1999 time period). Whether the actual warming in 2100 will be closer to the low or the high end of this range depends primarily on two factors: first, the future level of emissions of heat-trapping gases, and second, how sensitive climate is to past and future emissions. The range of possible outcomes has been explored using a range of different emissions scenarios, and a variety of climate models that encompass the known range of climate sensitivity.

1.9 CHANGING PRECIPITATION PATTERNS

Projections of changes in precipitation largely follow recently observed patterns of change, with overall increases in the global average but substantial shifts in where and how precipitation falls.90 Generally, higher latitudes are projected to receive more precipitation, while the dry belt that lies just outside the tropics expands further poleward, and also receives less rain. Increases in tropical precipitation are projected during rainy seasons (such as monsoons), and especially over the tropical Pacific. Certain regions, including the U.S. West (especially the Southwest) and the Mediterranean, are expected to become drier. The widespread trend toward more heavy downpours is expected to continue, with precipitation becoming less frequent but more intense.90 More precipitation is expected to fall as rain rather than snow.

Currently rare extreme events are becoming more common

In a warmer future climate, models project there will be an increased risk of more intense, more frequent, and longer-lasting heat waves.90 The European heat wave of 2003 is an example of the type of extreme heat event that is likely to become much more common.90 If greenhouse gas emissions continue to increase, by the 2040s more than half of European summers will be hotter than the summer of 2003, and by the end of this century, a summer as hot as that of 2003 will be considered unusually cool. Increased extremes of summer dryness and winter wetness are projected for much of the globe, meaning a generally greater risk of droughts and floods. This has already been observed, and is projected to continue. In a warmer world, precipitation tends to be concentrated into heavier events, with longer dry periods in between. Models project a general tendency for more intense but fewer storms overall outside the tropics, with more extreme wind events and higher ocean waves in a number of regions in association with those storms. Models also project a shift of storm tracks toward the poles in both hemispheres. Changes in hurricanes are difficult to project be- cause there are countervailing forces. Higher ocean temperatures lead to stronger storms with higher wind speeds and more rainfall.98 But changes in wind speed and direction with height are also projected to increase in some regions, and this tends to work against storm formation and growth. It currently appears that stronger, more rain producing tropical storms and hurricanes are generally more likely, though more research is required on these issues.

Sea level will continue to rise

Projecting future sea-level rise presents special challenges. Scientists have a well-developed under- standing of the contributions of thermal expansion and melting glaciers to sea-level rise, so the models used to project sea-level rise include these process- es. However, the contributions to past and future sea-level rise from ice sheets are less well under- stood. Recent observations of the polar ice sheets show that a number of complex processes control the movement of ice to the sea, and thus affect the contributions of ice sheets to sea-level rise. Some of these processes are already producing substantial loss of ice mass. Because these processes are not well understood it is difficult to predict their future contributions to sea-level rise.

Because of this uncertainty, the 2007 assessment by the IPCC could not quantify the contributions to sea-level rise due to changes in ice sheet dynamics, and thus projected a rise of the world's oceans from 8 inches to 2 feet by the end of this century (Fig. 10).

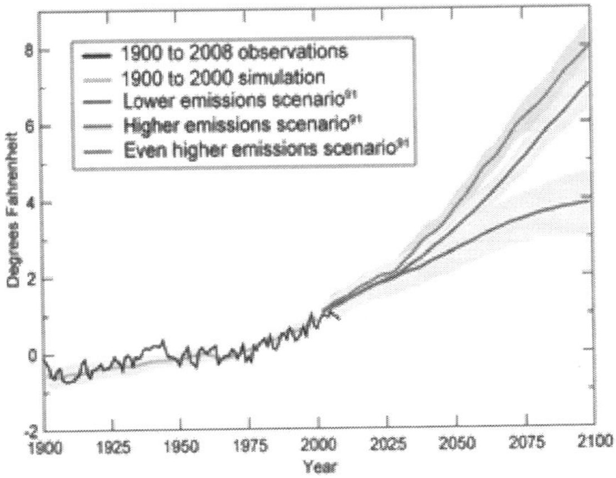

Fig. 10: Observed and projected changes in the global average temperature.

Observed and projected changes in the global average temperature under three IPCC no-policy emissions scenarios. The shaded areas show the likely ranges while the lines show the central projections from a set of climate models. A wider range of model types shows outcomes from 2 to 11.5°F. Changes are relative to the 1960-1979 average (Fig. 11).

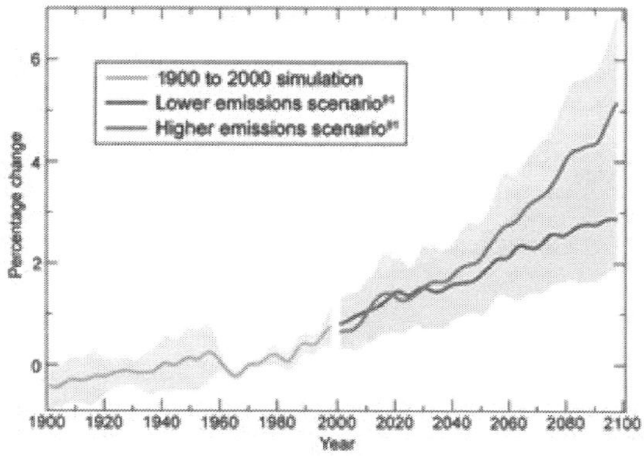

Fig. 11: Simulated and projected changes in the amount of precipitation falling.

Simulated and projected changes in the amount of precipitation falling in the heaviest 5 percent of daily events. The shaded areas show the likely ranges while the lines show the central projections from a set of climate models. Changes are relative to the 1960-1979 average.

More recent research has attempted to quantify the potential contribution to sea-level rise from the accelerated flow of ice sheets to the sea or to estimate future sea level based on its observed relationship to temperature.103 The resulting esti- mates exceed those of the IPCC, and the average estimates under higher emissions scenarios are for sea-level rise between 3 and 4 feet by the end of this century. An important question that is often asked is, what is the upper bound of sea-level rise expected over this century? Few analyses have focused on this question. There is some evidence to suggest that it would be virtually impossible to have a rise of sea level higher than about 6.5 feet by the end of this century.

The changes in sea level experienced at any particular location along the coast depend not only on the increase in the global average sea level, but also on changes in regional currents and winds, proximity to the mass of melting ice sheets, and on the vertical movements of the land due to geological forces. The consequences of sea-level rise at any particular location depend on the amount of sea- level rise relative to the adjoining land. Although some parts of the U.S. coast are undergoing uplift (rising), most shorelines are subsiding (sinking) to various degrees – from a few inches to over 2 feet per century.

1.10 ABRUPT CLIMATE CHANGE

There is also the possibility of even larger changes in climate than current scenarios and models project. Not all changes in the climate are gradual. The long record of climate found in ice cores, tree rings, and other natural records show that Earth's climate patterns have undergone rapid shifts from one stable state to another within as short a period as a decade. The occurrence of abrupt changes in climate becomes increasingly likely as the human disturbance of the climate system grows. Such changes can occur so rapidly that they would challenge the ability of human and natural systems to adapt.105 Examples of such changes are abrupt shifts in drought frequency and duration. Ancient climate records suggest that in the United States, the South- west may be at greatest risk for this kind of change, but that other regions including the Midwest and Great Plains have also had these kinds of abrupt shifts in the past and could experience them again in the future.

Rapid ice sheet collapse with related sea-level rise is another type of abrupt change that is not well understood or modeled and that poses a risk for the future. Recent observations show that melting on the surface of an ice sheet produces water that flows down through large cracks that create conduits through the ice to the base of the ice sheet where it lubricates ice previously frozen to the rock below.29 Further, the interaction with warm ocean water, where ice meets the sea, can lead to sudden losses in ice mass and accompanying rapid global sea-level rise. Observations indicate that ice loss has increased dramatically over the last decade, though scientists are not yet confident that they can project how the ice sheets will respond in the future.

There are also concerns regarding the potential for abrupt release of methane from thawing of frozen soils, from the sea floor, and from wetlands in the tropics and the Arctic. While analyses suggest that an abrupt release of methane is very unlikely to occur within 100 years, it is very likely that warming will accelerate the pace of chronic methane emissions from these sources, potentially increasing the rate of global temperature rise.

A third major area of concern regarding possible abrupt change involves the operation of the ocean currents that transport vast quantities of heat around the globe. One branch of the ocean circulation is in the North Atlantic. In this region, warm water flows northward from the tropics to the North Atlantic in the upper layer of the ocean, while cold water flows back from the North Atlantic to the tropics in the ocean's deep layers, creating a "conveyor belt" for heat. Changes in this circulation have profound impacts on the global climate system, from changes in African and Indian monsoon rainfall, to atmospheric circulation relevant to hurricanes, to changes in climate over North America and Western Europe. Findings indicate that it is very likely that the strength of this North Atlantic circulation will decrease over the course of this century in response to increasing greenhouse gases. This is expected because warming increases the melting of glaciers and ice sheets and the resulting runoff of fresh- water to the sea. This additional water is virtually salt-free, which makes it less dense than sea water. Increased precipitation also contributes fresh, less- dense water to the ocean. As a result, less surface water is dense enough to sink, thereby reducing the conveyor belt's transport of heat. The best estimate is that the strength of this circulation will decrease 25 to 30 percent in this century, leading to a reduction in heat transfer to the North Atlantic. It is considered very unlikely that this circulation would collapse entirely during the next 100 years or so, though it cannot be ruled out. While very unlikely, the potential consequences of such an abrupt event would be severe. Impacts would likely include sea-level rise around the North Atlantic of up to 2.5 feet (in addition to the rise expected from thermal expansion and melting glaciers and ice sheets), changes in atmospheric circulation conditions that influence hurricane activity, a southward shift of tropical rainfall belts with resulting agricultural impacts, and disruptions to marine ecosystems.

2

IMPACT OF CLIMATE CHANGE PROJECTIONS -2030 IN INDIA

2.1 INTRODUCTION

Global climate projections, given inherent uncertainties, indicate several changes in India's future climate:

- Global observations of melting glaciers suggest that climate change is well under way in the region, with glaciers receding at an average rate of 10–15 meters per year.
- If the rate of glacial melt increases, flooding is likely in the river valleys fed by these glaciers, followed by a diminished flow, resulting in a scarcity of water for drinking and agricultural irrigation.
- All models show a trend of general warming in mean annual temperature as well as decreased range of diurnal temperature and enhanced precipitation over the Indian subcontinent.
- A warming of 0.5oC is likely over all India by the year 2030 (approximately equal to the warming over the 20th century) and a warming of 2-4oC by the end of this century, with the maximum increase over northern India.
- Increased precipitation is likely to come in the form of fewer rainy days but more days of extreme rainfall events, with increasing amounts of rain in each event, leading to significant flooding. Fine precipitation (drizzle-type) that replenishes soil moisture is likely to decrease.
- Increased warming is likely to lead to higher levels of tropospheric ozone pollution and other air pollution in India's major cities.

- Most global models suggest that the Indian summer monsoons will intensify with a warming climate. The timing may also shift, causing a drying during the late summer growing season.
- Climate models also predict an earlier snow melt, which could have significant adverse effects on agricultural production, both irrigated and non-irrigated.
- Growing emissions of aerosols from energy production and other sources may suppress rainfall, leading to drier conditions with more dust and smoke from the burning of drier vegetation, affecting both regional and global hydrological cycles and agricultural production.

In both its greenhouse gas emissions and its vulnerability to climate change, India is one of the most significant countries in the world. With a large and growing population, India's emissions of greenhouse gases are increasing. Moreover, potential climate impacts in India are severe: sea level rise, changes in the monsoon, increased severe storms and flooding, more drought, and severe water stress. Recently, climate variability in the form of floods and cyclones has resulted in destruction of crops, property and infrastructure, as well as in negative impacts on human health and well-being. All of these impacts set back general socio-economic development.

Rural dwellers' continuing dependence upon agriculture for food and livelihood (17.5 percent of gross domestic product (GDP) and more than 60 percent of the labor force) makes the Indian people particularly vulnerable to climate variability and change. Nowhere is this more evident than in the linkage of the annual monsoon cycle and agricultural production, commonly referred to as, "Indian agriculture gambles with monsoon."

The diversity and extremes of India's climate and geography are characteristic of its society as well. Religious and cultural diversity is a major feature of Indian life. The strong Hindu traditions have been synthesized with and challenged by other religions, notably Islam, Christianity, and Sikhism. India has at least 300 known languages, 24 of which have at least one million speakers each. There are differences, sometimes amounting to estrangement, between the North, with its history of grand-scale invasions, and the relatively stable South. Religious divisions became geographical divides when Muslim Pakistan (1947), then Bangladesh (1971), were created, and ethnic and caste-related strife continues among groups. However, the connectedness of the extended family is a core feature of Indian life. Together with a sense of civil society's claims on individuals and families, the extended family knits the society together and emphasizes interdependence. Diversity and extremes are evident in India's patterns of economic development as well. In this sphere, two themes stand in contrast: modernist, democratic, and technical development, intensified by the economic liberalization beginning in the early 1990s; and persistent poverty, subsistence agriculture, and caste-based discrimination. Various models of development have been advanced. Das characterizes the country-level model thus: "Rather than adopting the classic Asian strategy—exporting labor-intensive, low-priced manufactured goods to the West—India has relied on its domestic market more than exports, consumption more than investment, services more than industry, and high-tech more than low-skilled manufacturing." Kerala's model of human development emphasizes education, health services, and equality; however, slow economic development has somewhat tarnished this model, as incomes remain low and the contributions of Keralans working abroad continue to be very much needed. The Karnataka model focuses on technology, centered in Bangalore, and historically participatory local governance by *Panchayat*. Yet Karnataka also has "enduring gender inequity and regional disparities, and a visibly increasing gap between urban and rural areas."

Despite substantial economic and general development progress, poverty, malnutrition, illiteracy, and inequality continue to plague India, as well as serious environmental issues. India has not only several Silicon Valleys but also several Nigerias. In addition, the ongoing dispute with Pakistan over Kashmir and ethnic strife (e.g., in Assam) claim national attention. Conflict with Pakistan has lessened by confidence-building measures since 2002, when nuclear war was hintedat, but such hints arose again after the recent terrorist attack in Mumbai, although the response sofar has been relatively temperate.

India's broad spectrum of highly articulated national policies includes inclusive growth goals in the areas of economic development, human development, and environmental protection.

National goals are, of course, differentially implemented in each of India's states, which exhibitwidely varying degrees of social and economic development. Limited growth has occurred in the areas of fiscal policy, privatization, small-scale industry, agriculture, and labor law.

At the national level, India's climate change policies are subsumed in its economic-industrial andhuman development policies, which come first. Local policies have had some success in limitingsignificant urban air pollution problems. Substantial improvements in local air quality in Delhi, for example, have resulted from recent government programs to improve the quality of petrol and diesel fuels, introduction of emissions standards for vehicles, and conversion of buses and three-wheelers to compressed natural gas (CNG) fuel—but Kolkata and other metropolitan centers experience worsening air quality, with increasing combustion of fossil fuels contributing to carbon dioxide emissions. Generally speaking, climate change policy has been reactive rather than proactive and focused largely on the energy sector. According to reports from the Organization for Economic Cooperation and Development (OECD) and the Pew Center, India, through normal policy developments, is "making significant progress in limiting greenhouse emissions" (i.e., from what emissions might have been) through energy efficiency improvements and environmentally friendly energy development. Also, India is participating in the Clean Development Mechanism of the United Nations Framework Convention on Climate Change (UNFCCC) and actively participating in the development of a proposed UNFCCC mechanism called Reducing Emissions from Deforestation and forest Degradation (REDD). The National Clean Development Mechanism Authority (NCDMA) is housed in the Ministry of Environment and Forests (MoEF); CDM India has operated since December 2003 as the Designated National Authority (DNA). More aggressive measures, India feels, should be financed by developed nations as they lead by reducing their own emissions and engaging in clean technology transfer, in accordance with the 1992 UNFCCC (ratified by 192 countries).

Internationally, India has played a key role in climate negotiations at several points. India brokethe impasse at the first Conference of the Parties by leading the development of a common statement that became the basis for the Berlin Mandate. More recently, India hosted the eighth Conference of the Parties to the Framework Convention on Climate Change in 2002. India, bolstered by nongovernmental organizations such as the Tata Energy Research Institute (TERI) and the Centre for Science and Environment (CSE), focuses on per capita emissions (low in India and high in most developed countries) and on cumulative emissions (also low in developing countries and high in developed countries), as the indicators that developed countries should undertake mitigation first. Government officials press developed nations to establish and conform to emissions reduction goals and to engage in technology transfer to developing countries.

CSE rebutted the second World Resources Report, making the distinction between "subsistence emissions" of the poor (mostly in developing countries) and the "luxury emissions"of the rich (mostly in developed countries). CSE has also characterized "green" policies dictated by the North (e.g., debt-for-nature swaps) as unwarranted interference in other nations, as exacerbating inequality among nations, and as likely to foster unsustainable management.

2.2 PROJECTED REGIONAL CLIMATE CHANGE

The current climate of India is highly diverse, ranging from the subfreezing Himalayan winters to the tropical climate of the south. The states of Assam and West Bengal experience extremelydamp, rainy, and humid conditions, while the regions of Rajasthan and Gujarat make up part of the arid Great Indian Desert. Based on precipitation and temperature, India can be divided into six climatic regions: the Himalayas, Assam and West Bengal, the Indo-Gangetic Plain, the Western Ghats and coast, the Deccan (the interior of the Peninsula south of the Narmada River),and the Eastern Ghats and coast.

The Indian Meteorological Service divides the year into four seasons, two of which are characterized by monsoon conditions. Winter occurs from December through February, whenconditions are generally relatively dry and cool. March through May is considered to be summer, as the conditions are usually hot and dry. During this period temperatures throughout non-Himalayan India reach the upper 30s°C and can reach as high as 48°C during the day in the pre-monsoon months.

The southwest monsoon season occurs from June through September, when the predominating southwest maritime winds bring rains to most of the country. One branch of the southwest monsoon, known as the Arabian Sea monsoon, generally breaks on the west coast early in the season and spreads across South Asia by early July. The other, known as the Bay of Bengal monsoon, spreads over Assam during June and travels along the Indo-Gangetic Plain toward New Delhi, merging with the Arabian branch to bring rains farther north. The southwest monsoon provides almost 80 percent of the annual rainfall to most of the country. It is criticallyimportant to agricultural production; predictions of its timing are used by agronomists and farmers to determine optimal dates for plantings.

The northeast monsoon occurs in October and November as the southwest monsoon retreats. The states of Tamil Nadu, Karnataka, and Kerala receive most of their rainfall from the northeast monsoon during November and December. Inter annual climate variability is linked to a global-scale, naturally occurring phenomenon known as the *El Niño*/Southern Oscillation (ENSO) cycle. ENSO can explain some of the inter-annual rainfall variability over the subcontinent of India and can affect the location and activity of tropical storms. Analysis of observational data shows a significant correlation between ENSOand tropical circulation and precipitation such that there is a tendency for less Indian summer monsoon rainfall in *El Niño* years and above-normal rainfall in *La Niña* years. However, the effect is not linear. The ability to accurately predict the timing and occurrence of the ENSO phenomenon is extremely important to agricultural production.

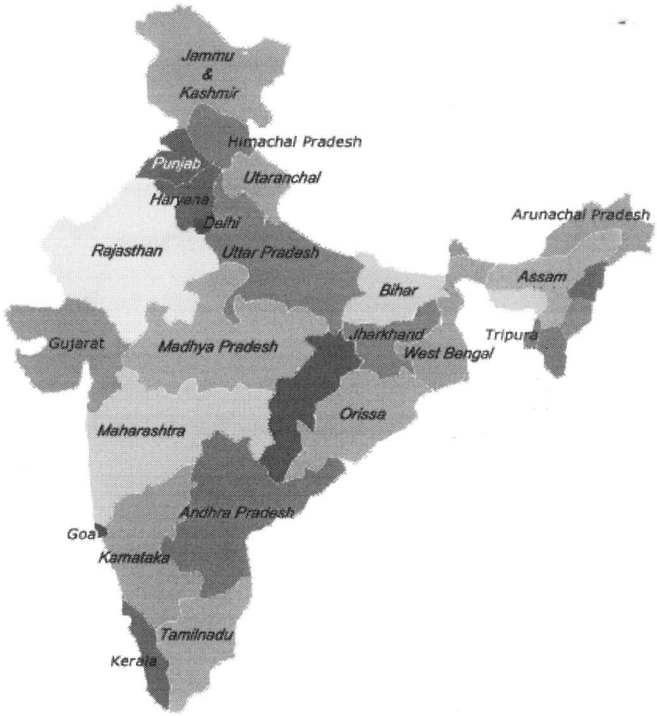

Fig.1: Indian States.

2.3 CLIMATE OBSERVATIONS

Global observations suggest that climate change is well under way. At continental, regional, and ocean basin scales, numerous long-term changes in climate have been observed, including widespread changes in precipitation amounts; ocean salinity; wind patterns; and aspects of extreme weather including droughts, heavy precipitation, heat waves, and the intensity of tropical cyclones. Studies of the tropical Indo-Pacific region show unusual warmth in the 20th century, and many isotope records show a trend toward warmer conditions in the tropical Indian Ocean. In most multi-centennial coral series, the late 20th century is shown to be warmer than any time in the past 100 to 300 years.

Some studies have suggested that the South Asian (Indian) monsoon, in the drier areas of its influence (northwest India), has recently reversed its millennia-long orbitally driven, low- frequency trend toward less rainfall. This recent reversal in monsoon rainfall also appears to coincide with a synchronous increase in inferred monsoon winds over the western Arabian Sea, achange that could be related to increased summer heating over and around the Tibetan Plateau.

Globally, estimates of the potential destructiveness of tropical storms and hurricanes show asubstantial upward trend since the mid-1970s, with a trend toward longer storm duration and greater storm intensity. Storm activity is generally correlated with tropical sea surface temperature.

The distributions of global minimum and maximum temperatures have shifted to higher values,consistent with overall warming. More warm extremes imply an increased frequency of heat waves. However, cold extremes have warmed more than the warm extremes over the past 50 years. Further indications include the observed trend toward fewer frost days associated with the average warming in most mid-latitude regions. A prominent indication of a change in extremes is the evidence of increases in heavy precipitation events over the mid-latitudes in the past 50 years, even in places where mean precipitation amounts are not increasing. For very heavy precipitation events, increases are reported as well, but results are available for only a few areas. Recent warming in sea surface temperatures (SSTs) is strongly evident at all latitudes, although there are inter-hemispheric differences. Much of the surface of the Indian Ocean has warmed since 1955, with the major exception of the 5°S to 20°S latitude belt. The Southern Ocean (south of 35°S) in the Atlantic, Indian, and Pacific sectors has generally warmed. The regions that exhibit cooling are mainly in the shallow equatorial areas and in some high-latitude regions. In the Indian Ocean, cooling occurs at subsurface depths centered on 12°S (South Equatorial Current) at 150 m depth and in the Pacific centered on the equator and 150 m depth. In the tropical and eastern subtropical Indian Ocean (north of 10°S), warming in the upper 100 m is consistent with the significant warming of the sea surface from 1900 to 1999. The surface warming trend during the 1900 to 1970 period was relatively weak but increased significantly in the 1970 to 1999 period, with some regions exceeding 0.2°C per decade. Models suggest that upper-ocean warming in the South Indian Ocean can be attributed to a reduction in the strength of the southeast trade winds and associated decrease in the southward transport of heat from the tropics to the subtropics.

Local and regional changes in the character of precipitation also depend a great deal on atmospheric circulation patterns determined by *El Niño,* the North Atlantic Oscillation (NAO),and other patterns of variability. India's rainfall features show strong variability but little in the way of a century-scale trend, even as the linear trends of rainfall decreases for 1900 to 2005 were 7.5 percent in western Africa and a similar decrease was observed when averaged over the broader southern Asia region as a whole. Over much of northwestern India, the 1901 to 2005period shows precipitation increases of more than 20 percent per century, but the same area shows a strong decrease in annual precipitation in the 1979 to 2005 period. Very dry land areas across the globe have more than doubled in area since the 1970s, an observation that has been associated with precipitation decreases related to ENSO and with subsequent increases primarily due to surface warming. The tendency of the warming to bemore pronounced in winter is a conspicuous feature of the observed temperature trends overIndia, one that is likely to continue.

2.4 CLIMATE PREDICTIONS (MODELING)

While Global Circulation (or Climate) Models (GCMs) can be used to infer climate changes inspecific regions, it is far preferable to develop models that have a high resolution sufficient to resolve local and regional scale changes. There are many challenges in reliably simulating andattributing observed temperature changes at regional and local scales. At these scales, natural climate variability can be relatively greater, making it harder to distinguish long-term changes expected due to external radiative forcings.

The procedure of estimating the response at local scales based on results predicted at larger scales is known as "downscaling." The two main methods for deriving information about the local climate are (1) dynamical downscaling (also referred to as "nested modeling" using "regional climate models" or "limited area models") and (2) statistical downscaling (also referred to as "empirical" or "statistical-empirical" downscaling). Chemical composition models include the emission of gases and particles as inputs and the simulation of their chemical interactions; global transport by the winds; and removal by rain, snow and deposition to the earth's surface.

Downscaled regional scale climate models rely on global models to provide boundary conditionsfor the region to be modeled. There are three primary approaches to numerical downscaling:

- Limited-area models,
- Stretched-grid models, and
- Uniformly high resolutionatmospheric GCMs (AGCMs).

Regional Climate Model (RCM) projections for climate change in India were compared for eightmodels using several different IPCC scenarios at time intervals of 20 years. All models show positive trends of widespread warming (Figure 2), with warming more pronounced during winter and post-monsoon months compared to the rest of the year (i.e., seasonal changes) consistent with recent observations.

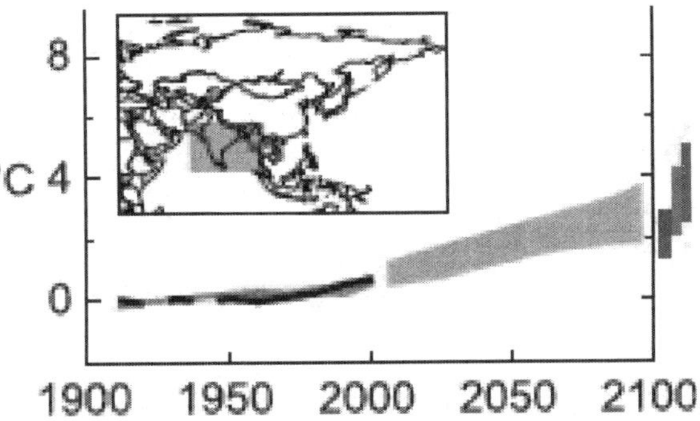

Fig. 2: Temperature anomalies with respect to 1901-1950 for six Asian land regions 1906-2005 (blackline) and as simulated (red envelope) by multi-model dataset (MMD) models.

Source: Intergovernmental Panel on Climate Change (IPCC), Climate Change 2007: the Physical Science Basis, eds. S. Solomon, D. Qin, M. Manning, M. Marquis, K. Averyt, M.M.B. Tignor, H.L. Miller Jr. and Z. Chen (Cambridge: Cambridge University Press 2007): 882.

Although some models show a slight decrease in precipitation over all India during the first few decades of the study, all models show increased precipitation during the monsoon season by the year 2100, especially over the northwestern parts of India, an area that is currently very dry. Onemodel projects that by 2050 there will be an overall decrease in the number of rainy days over allIndia but an increase in the number of one-day extreme rainfall events. The magnitudes and patterns of the projected rainfall changes differ significantly between models, probably due to their coarse resolution.

The Atlantic and Pacific Oceans are strongly influenced by natural variability occurring on decadal scales, but the Indian Ocean appears to be exhibiting a steady warming. Natural variability (from ENSO, for example) in ocean-atmosphere dynamics can lead to important differences in regional rates of surface-ocean warming that affect the atmospheric circulation andhence warming over land surfaces. New modeling efforts have shown improvements in climate forecasts in the southern hemisphere when the incidence of the Indian Ocean Dipole (IOD) is included. The IOD occurs periodically when there is a cold water upwelling in the eastern Indian Ocean, during which the eastern half of the ocean becomes much cooler than the western half.

The recent IPCC assessment report suggests that, in the future, circulation associated with the monsoon may slow down, but the moisture in the air may increase. However, the representationof the observed monsoon maximum rainfall along the west coast of India, northern parts of Bay of Bengal, and north India is poor in many models.

A comparison of four different GCM models applied to the Indian subcontinent found a significant spread in the summer monsoon precipitation anomalies. Soil moisture was determined to be an important parameter in model projections. Most models project increased precipitation during the monsoon season, particularly over the northwestern parts of India, with differences in the magnitudes of the increase from one model to the other. West central India shows maximum overall increase in rainfall. Extremes in both temperature and precipitation events are likely to increase out to 2030.

Including sulfate aerosols in the models reduces the regional climate sensitivity, but the greenhouse warming still dominates the changes at the larger scale and longer time scale. Models that include emissions of short-lived radiatively active gases and particles suggest that future climate changes could significantly increase maximum ozone levels in already polluted regions. Projected growth of emissions of radiatively active gases and particles in the models suggest that they may significantly influence the climate, even out to year 2100. Atmosphericbrown clouds, plumes of polluted air moving from the Asian continent out over the Pacific and Indian Ocean, may cause changes in the monsoon circulation and reduce summer monsoon precipitation in parts of South and East Asia.

Global models that are unable to include small-scale spatial geographical features and distortionsof albedo feedbacks are also unable to project precipitation over the Himalayas. It is expected that the net effect of a warming climate will be an increase in the rainfall associated with the monsoon. Several models suggest that throughout Asia there will be an overall decrease in the December-February precipitation, an increase in the remaining months, and more intense rainfallover larger areas in the future.

Stabilization emissions scenarios assume future emissions based on an internally consistent set ofassumptions about driving forces (such as population, socioeconomic development, and technological change) and their key relationships. These emissions are constrained so that the resulting atmospheric concentrations of the substance level off at a predetermined value in the future. For example, if one assumes the global CO_2 concentrations are stabilized at 450 parts permillion (ppm) (the current value is about 380 ppm), the climate models can simulate the climate consistent with the emissions of greenhouse gases and short-lived species of this scenario. The climate model predictions can be used to assess specific regional impacts at this stabilization level. A recent study of models from the Coupled Model Inter-comparison Project showed that the models' response to a +1 percent/yr increase in CO_2 includes a substantial warming of tropical storm basin SSTs, enhanced upper-tropospheric warming relative to the surface warming, and little change in lower-tropospheric relative humidity. The study included climate change scenarios from nine different global coupled climate models as inputs to an idealized hurricane model. The results demonstrate that there is a significant sensitivity in the models to the choiceof climate model and convective parameterization.

2.5 PROJECTIONS OF FUTURE TEMPERATURE AND PRECIPITATION

Current projections of climate change from the latest IPCC Assessment suggest increases intemperatures, precipitation rates, and the intensity of tropical storms over the Indian Ocean.

Most models either assume a doubling of CO_2 and their results can be compared using this assumption, or the models are run to the year 2100 and the results of the climate projections are assessed for that date. The IPCC has not included numerical model results specifically for the year 2030. Climate changes on the Indian subcontinent for intermediate scenarios at 2020 and 2050; however, only the results for the final scenario for 2080 have been published. An earlier study used eight global models to project temperature and precipitation changes over all India per decade for three different emission scenarios. Although the tables presented in this reference are useful, the uncertainties, especially in the prediction of precipitation changes, are great.

Most AGCMS predict general warming and enhanced rainfall over India, with these changes becoming particularly significant by 2040. The model Providing Regional Climates for Impacts Studies (PRECIS), developed by the Hadley Centre for Climate Predictionand Research, to yield projections for climate change during intermediate time scales (every 10 years until 2100); these projections can be averaged over model results to estimate climate change in 2030. The models project an average temperature increase over all India by the year 2030 of around 0.5°C, an increase comparable with that has occurred over the globe during the 20th century. The maximum temperature increase (i.e., annual mean temperature change) is expected to occur over northern India, with a secondary maximum over the eastern peninsula. By the year 2100, the temperature increase could be on the order of 2-4°C, with a maximum increase in the northern region of 4°C.

A regional model recently developed at the Indian Institute of Tropical Meteorology (IITM)projected that, averaged over the country, India could face a temperature increase of 4°C by 2100. However, the model predicts that, because of the longer time scales for system-wide changes in atmosphere-ocean interactions, a delay in the increase in monsoon rainfall will occur, resulting in drought in some areas.

2.6 SEASONAL WEATHER PATTERNS (MONSOONS)

Monsoon rains account for most of India's annual rainfall. Monsoons are generally defined as tropical and subtropical seasonal reversals in both the surface winds and associated precipitation.The strongest monsoons occur over the tropics of southern and eastern Asia and northern Australia as well as parts of western and central Africa. Rainfall is the most important monsoon variable because the associated latent heat release drives atmospheric circulations and because rainfall plays a critical role in the global hydrological cycle and is vital to socioeconomicimpacts.

Most global models suggest that the Indian monsoons will intensify with a warming climate. Since the continental-scale land-sea thermal contrast is expected to become larger in summer andsmaller in winter, one would expect that in the future the summer monsoon will be stronger and the winter monsoon will be weaker than they are presently. However, some models predict that a pronounced warming over the tropics will result in a weakening of the Asian summer monsoon circulations due to a reduction in the meridional thermal gradients between the Asian continent and adjacent oceans.

The global monsoon system embraces an overturning circulation that is intimately associated with the seasonal variation of monsoon precipitation over all major continents and adjacent oceans. The Asian monsoon can be divided into the East Asian and the South Asian or Indian monsoon systems. Although the Indian monsoons recur each year, their irregularity at a range oftime scales from weeks to years depends on feedback from the ocean in ways that are not fully understood. Intra-seasonal variability is associated with the Monsoon Intra-Seasonal Oscillation(MISO) and the Madden-Julian Oscillation (MJO), which are long-lasting weather patterns that evolve in a systematic way for periods of four to eight weeks. On an interannual and decadal scale, statistical methods have shown that, while there are periods of high correlation between ENSO and monsoon variation, there are decades in which little or no association is apparent.

The influence of ENSO on the position and strength of the subtropical high in the North Pacificinfluences both typhoons and other damaging heavy rainfall events and has been implicated in observed inter-decadal variations in typhoon tracks. This suggests that the spatial structure of warming in the Pacific will be relevant for changes.

For South Asia, the monsoon depressions and tropical cyclones generated over the Indian seas modulate the monsoon anomalies. For East Asia, the monsoonal circulations are strengthened by extratropical cyclones energized in the lee of the Tibetan Plateau and by the strong temperature gradient along the East Coast.

Although attention is often focused on the frequency or number of storms, the intensity, size, andduration are much more important. From an observational perspective, then, key issues are the tropical storm formation regions; the frequency, intensity, duration, and tracks of tropical storms;and associated precipitation. All of these can be influenced by climate teleconnections, especially those such as the Indian dipole oscillation (IOD) and the Southern Annular Mode (SAM) Index. Annular modes are hemispheric scale patterns of climate variability. SAM is linked to variations in temperatures over Antarctica, sea-surface temperatures throughout the Southern Ocean, and the distribution of sea-ice around the perimeter of Antarctica.

New evidence indicates that increased aerosol loading in the atmosphere may also have strong impacts on monsoon evolution through changes in local heating of the atmosphere and land surface. Polluted air can also have an effect on local circulation patterns. Heating of a lofted dust layer in the Tibetan Plateau could act as an elevated heat pump to strengthen the Asian summer monsoon circulation and cause a local increase in precipitation, despite the global reduction of evaporation that accompanies the aerosol-induced reduction in shortwave radiation at the surface. The dust-induced thermal contrast changes between the Eurasian continent and the surrounding oceans are postulated to trigger or modulate a rapidly varying or unstable Asian winter monsoon circulation, with a feedback to reduce the dust emission from its sources.

2.7 SEA LEVEL CHANGES

A significant fraction of sea level rise is due to thermal expansion of a warmed ocean (as much as 0.3 to 0.8 m over the past century, according to the 2007 IPCC report). Geographic patterns of sea level rise are due mainly to changes in the distribution of heat and salinity in the ocean, resulting in changes in ocean circulation. Precise satellite measurements since 1993 show that the largest sea level rise since 1992 has taken place in the western Pacific and eastern Indian Oceans, with the potential for significant impacts on the east coast of India. The 2004 Indian National Communication to the UNFCCC states that sea level rise is highest in the Gulf of Kutch(Gujarat) and on the coast of West Bengal. There is a large interannual variability in sea level rise associated with patterns of coupled ocean-atmosphere variability, including ENSO and the NAO.

Projected patterns of sea level rise display more similarity across models than in past assessments. Common features include a narrow band of pronounced sea level rise stretching across the southern Atlantic and Indian Oceans. Sea level is projected to rise between the present (1980-1999) and the end of this century (2090-2099) by 0.35 m (0.23 to 0.47 m). Due toocean density and circulation changes, the distribution will not be uniform.

2.8 AGRICULTURAL GROWING PERIODS

Many regions of India already face water scarcity. Productive agricultural regions in the Northdepend on the spring snow melt to replenish regional water supplies. Climate models predict anearlier snow melt, which could have a significant effect on agricultural production, especially if the levels of moisture in the soils are reduced during the growing season.

The retreat of snow and ice cover in and around the Himalayas is already having a drying effecton these regions. A recent study of the melting Naimona'ny glacier in the Himalayas, which provides water to the Indus, and Brahmaputra Rivers shows that the glacier has melted so muchthat the exposed surface of the glacier dated to 1944.

A study of glaciers in the region shows that they are now receding at an average rate of 10-15 meters per year. Himalayan glaciers collect water during the monsoon season and release itduring the dry season, providing irrigation water for crops. If the rate of glacial melt increases, flooding is likely to occur in the river valleys fed by the glaciers.

Later, as the river flows decrease to below previous rates, many people may be left without sufficient drinking water or water for irrigating crops.

Decreasing trends in evapotranspiration during recent decades are evident in records, even though such records are sparse. This is likely due to decreased sunshine duration related to increases in air pollution, atmospheric aerosols, and increases in cloud cover. An accelerating trend in sulfate deposition has been observed in Himalayan glaciers and is probably due to increased sulfur dioxide emissions from the increasing energy demand throughout Asia. The concentration of sulfate deposited in the glaciers in the past 50 years exceeded that for any prior 50-year period in the last millennium.

A model to predict changes in river flows due to future glacial melt from climate change was developed by the Centre for Ecology and Hydrology in the United Kingdom. Under different climate scenarios, the model predicts that in the upper Indus there will be an initial increase of between +14 percent and +90 percent in mean flows (compared to baseline) over the first fewdecades of the 100-year incremental scenario runs. By decade 10, river flows will decrease between –30 percent and –90 percent of baseline.

During the Indian Ocean Experiment that focused on emissions of human-generated aerosols from the Indian sub-continent, local forcing at the surface was observed to be significantly stronger than that at the top of the atmosphere. These results indicate that absorption of solar radiation by aerosols, primarily black carbon in the atmospheric column, is of great significance. It has been suggested that absorbing aerosols may have masked up to 50 percent of the surface warming in South Asia from the global increase in greenhouse gases. In cases where aerosols act to suppress rainfall (the second aerosol indirect effect), drier conditions tend to induce more dust and smoke due to the burning of drier vegetation, affecting both regional and global hydrological cycles and agricultural production. However, more research is needed to understand the combined effects of aerosols and dusts, which may influence the monsoon circulation and hydrological cycle in different ways.

2.9 CLIMATIC EVENTS

Extremes are the infrequent events at the high and low end of the range of values of a particular variable. The probability of occurrence of values in this range is called a probability distributionfunction (PDF) that for some variables is shaped similarly to a "Normal" or "Gaussian" curve (the familiar bell-shaped curve).

People affected by an extreme weather event (e.g., the heavy rainfall in Mumbai in July 2005) wonder whether climate changes due to human influences are responsible. It is difficult to attribute any individual event to a change in the climate. In most regions, instrumental records ofvariability typically extend only over about 150 years, so there is limited information to characterize how extreme rare climatic events could be. Further, several factors usually need to combine to produce an extreme event, so linking a particular extreme event to a single, specific cause is problematic. In some cases, it may be possible to estimate the anthropogenic contribution to such changes in the probability of occurrence of extremes.

As the climate changes and SSTs continue to increase, the conditions that cause tropical stormsto form are no longer the same. Higher SSTs are generally accompanied by increased water vapor in the lower troposphere; thus, the moist static energy that fuels convection and thunderstorms is also increased. Hurricanes and typhoons currently form from pre-existing disturbances only where SSTs exceed about 26°C; so, as SSTs have increased, the areas over which such storms can form are potentially expanded. However, many other environmental factors also influence the generation and tracks of disturbances.

The 2007 IPCC assessment concluded that there was a risk of increased temperature extremes inIndia, with more extreme heat episodes in a future climate. This result has been confirmed and expanded in more recent studies. Future increases in temperature extremes are projected to follow increases in mean temperature over most of the world except where surface properties (e.g., snow cover or soil moisture) change. There is still much debate over whether there is likely to be an increase in tropical cyclone intensity.

Changes in tropical storm and hurricane frequency and intensity are often masked by large natural variability. ENSO greatly affects the location and activity of tropical storms around the world. Globally, estimates of the potential destructiveness of hurricanes show a substantial upward trend since the mid-1970s, with a trend toward longer storm duration and greater storm intensity, and this activity is strongly correlated with tropical SSTs. One study found a large increase in numbers and proportion of hurricanes reaching categories four and five globally since1970, even as the total number of cyclones and cyclone days decreased slightly in most basins. The largest increase was in the North Pacific, Indian and Southwest Pacific Oceans.

Improved models, ones that prescribed convection constraints based on the relative humidity, were able to simulate the variability and extremes of rainfall quite well over most of India whencompared to satellite-derived rainfall but had a tendency to overestimate heavy rainfall events incentral India.

Several recent studies have addressed possible future changes in heat waves and found that in afuture climate, heat waves are expected to be more intense, longer-lasting and more frequent.

Based on an eight-member multi-model ensemble, heat waves are simulated to have increased for the latter part of the 20th century and are projected to increase globally and over most regions.

2.10 IMPACTS OF CLIMATE CHANGE ON NATURAL ECOSYSTEMS

2.10.1 Observed Changes

An analysis of seasonal and annual surface air temperatures for India, using data for 1881–2001 for 25 or more stations, shows a significant annual mean warming of 0.68°C per hundred years. Most of the warming occurs in the post-monsoon and winter seasons. The monsoon temperatures do not show a significant trend in most parts of the country except for a significantnegative trend over northwest India. Maximum daytime temperatures show more of a trend thanminimum nighttime temperatures, in contrast to general expectation.

Water supply is changing. Almost 67 percent of the glaciers in the Himalayan mountain ranges, the source of major rivers in India, have retreated in the past decade.

For India, the IPC reports increased frequency of hot days and multiple-day heat waves in the past century, with more deaths attributable to heat stress in recent years. Consecutive droughts in 1999 and 2000 led to a sharp decline in water tables in the northwest, and 2000-2002 droughts caused crop failures, leading to mass starvation and impacts on ~11 million people in Orissa.

Management decisions about natural hazards can cause conflicts. In 2002, under drought conditions in Andhra Pradesh, the state released dam water for electricity generation but not for irrigation. Poor farmers responded to such policies by smashing the pumps of their richer neighbors. Quarrels over water rights between states can be bitter, too. Tamil Nadu claimed thatits neighboring state, Karnataka, violated agreements about sharing water from the Cauvery River. Bangalore is facing acute water scarcity as it attempts to meet the drinking water needs of 7 million people in the city.

Public health is affected by currently experienced climate variability and change in the forms ofheat and flooding. Between 1980 and 1998, 18 heat waves were reported in India; one in 1988 affected ten states and caused 1,300 deaths. Heat waves in Orissa, in the 1998-2000 period caused an estimated 2,120 deaths, and heat waves in 2003 in Andhra Pradesh caused more than3,000 deaths. Flood-related increases in diarrheal disease have also been reported in India.

Observed trends in the mean sea level along the Indian coast indicate a rising trend of about 1 cm per decade, which is close to that recorded in other parts of the globe. Today, coastal regions in India and Bangladesh are subjected to stronger wind and flood damage than in the past because of stronger storm surges associated with tropical storms.

2.10.2 Projected Changes: The Example of Forests

Based on the IPCC's *Special Report on Emission Scenarios*, a study by Ravindranath et al. (2006) investigated the impacts of climate change on Indian forests into the year

2085 under twoemission scenarios: A2 (740 ppm CO_2) and B2 (575 ppm CO_2). Globally, A_2 is the more extreme scenario, representing a growing human population and slower and inequitable economic development whereby atmospheric CO_2 concentration is projected to double by 2050 and is likely to increase to 740 ppm by 2085. B_2 represents moderate population growth, intermediate levels of economic development, adoption of environmentally sound technologies, and greater social equity.

Using the B2 scenario projections and Forest Service of India's categories of forests, a pattern emerges: colder forests are subject to a larger increase of about 3°C, while the Western Ghat evergreen forests become warmer by only about 2.4°C on average—compared to the national average of 2.9°C. For the A2 scenario, the magnitudes of the impacts are larger. Most of the forests show an increase of about 4°C with the northern temperate forests undergoing a temperature increase of around 4.6°C. Western Ghats evergreen, semi-evergreen, and mangroveforest types show the least impacts under both A_2 and B_2 scenarios.

Even with a conservative temperature increase of 1–2°C, most ecosystems and landscapes will be impacted through changes in species composition, productivity and biodiversity. Impacts to nearly 200,000 forest villages—naturally heavily dependant on forest resources—will be innumerable. Impacts to the country as a whole are also projected by way of economically important forest types, such as *Tectona grandis*, *Shorea robusta*, bamboo, upland hardwoods andpine. A clear possibility of a large-scale shift in forest types in India is projected for the period 2070 to 2100, with adverse implications for biodiversity and a nearly 70 percent increase in net primary productivity of forest types, with implications for biomass production and timber markets.

India's forests are already changing because of socioeconomic pressures; virgin forest areas are less dense and monocultures and plantations are preferred to native species. These conditions will be greatly exacerbated by climate change. Specifically, biodiversity is likely to be reduced under the projected climate scenarios representing changes or shifts in forest or vegetation types,forest dieback during the transient phase, and different response changes of species to climate changes even when there is no change in forest type.

2.11 IMPACTS OF CLIMATE CHANGE ON HUMAN SYSTEMS

2.11.1 Energy System

Primary energy demand in 2005 was roughly equivalent to that of Japan—but of course with many more people, India's per capita demand remains at about one-tenth of the OECD average.However, Indian demand is growing at a fast pace, 3.2 percent per year (2000-2005). The International Energy Agency (IEA) projects India's primary energy demand to more than double by 2030 in a reference scenario (i.e., no policies to slow demand); this projection is attributable largely to a projected annual growth rate for GDP of 6.3 percent. The primary energy demand growth areas will be industrial (mostly steel) and transport (although two- and three-wheelers will still be over half of vehicle stock). In this scenario, India will become the world's third-largest carbon dioxide emitter by 2030, although per-capita emissions will still becomparatively low.

Inequality is such an issue in the energy sector that the IEA devoted a chapter to it, "Focus on Energy Poverty." Approximately 412 million people are still without access to electricity in India. The use of fuelwood and dung for cooking and heating causes more than 400,000 premature deaths annually, mostly of women and children. Income disparities account for most of the energy access disparities, but other barriers include "unreliable energy service delivery, ineffective and regressive subsidies, gender discrimination in policy planning, inadequate information about the health impacts of current fuels and technologies, and administrative hurdles in getting connections." "Energy poverty" is one aspect of a concern in India that its booming economic conditions have benefited the "haves" but not the "have-nots."

India's current electrical system runs mostly on domestic coal: 82.7 percent fossil fuel, 14.5 percent hydropower, and 3.4 percent nuclear. The transportation sector runs mostly on imported oil; domestic production is 785,000 bbl/day against a demand of 2.45 million bbl/day (2004 estimate). OECD/IEA characterized the overall energy system as fueled "largely by coal and combined renewables and waste, with much smaller but growing shares of oil, gas, nuclear, and hydro." Future increases in energy production will likely include the following:

Domestic coal—although deposits are located at some distance from population centers and the coal is of low quality. (Higher quality coal is imported for steel making; India is the seventh-largest steel producer in the world.) Coal gasification could also be used to make diesel fuel. However, the use of coal contributes to anthropogenic climate change and bad air quality, so pressures to reduce greenhouse gas and other emissions could limit both options for coal use in India, unless carbon dioxide capture and storage technologies could be implemented.

Imported oil and gas—with associated issues of investing in oil fields, exploitation rights, and refineries in Myanmar, Sudan, Iraq, Russia, Vietnam, Venezuela, and Libya, as well asraising the problematic prospect of a gas pipeline running through either Iran and Pakistan orTurkmenistan, Afghanistan, and Pakistan. Nuclear power—in part as a fruit of the 2005 "strategic partnership" with the United States,raising concerns about nuclear nonproliferation. Renewable sources—such as biomass fuels (waste and purpose-grown crops) and hydropower. Residential use of biomass (fuelwood, dung, and agricultural waste) is projected to change from a current 54 percent share to 12 percent in urban households, but asmaller 92-to-79 percent decline in rural households still means that the absolute amount of biomass used will increase. Biofuel crops, like other crops, will be affected by climate change; hydropower, like other water uses, will be subject to changes in precipitation under climate change. Domestic gas reserves—with 2002 and subsequent discoveries in the Krishna-Godavari basin(in Andhra Pradesh on the east coast) and more discoveries expected.

In all of these areas, advanced technologies will be essential to increasing production and meeting environmental regulations, including those related to climate change mitigation and adaptation. The Indian Government has introduced clean coal technologies like coal washingand the use of cleaner and less carbon-intensive fuels, *e.g.*, liquefied petroleum gas (LPG) forautomobiles and motor spirit-ethanol blending projects in selected states.

2.11.2 Economic Growth and Development

Since 1990, India as a country has moved aggressively from a centrally planned economy to private ownership of businesses and trade liberalization. It has "developed a diversified industrial base and sophisticated financial sector. Its software subsector—one of the most dynamic in the world—has experienced a sustained and rapid growth." Over the past 25 years, the annual growth rate has been 6 percent, accelerating to a five-year annual average of percent, investment at 30 percent of GDP, and booming foreign direct investment. Indiahas made substantial strides in fostering human capital by reducing infant mortality, increasinglife expectancy, and improving literacy. The central, democratic government provides stabilityand some curbs to unbridled free market-ism (or barriers to growth in an alternative characterization), and hordes of entrepreneurs provide the impetus for growth.

On the positive side, India's democracy results in equity slightly higher than the global average.The dependency ratio (the percentage of the population dependent on the percentage of the population in the work force) is relatively high, indicating that many people are available for thework force, supporting relatively few people other than themselves.

However, the poor condition of people engaged in agriculture and/or born into lower castes reduces the robustness of the overall economy. Climate change, adding to existing problems of the agricultural system, may worsen conditions for the large poor segment of the population enough to severely tax the economic and industrial resources of the central and state governments.

Thus, the impacts of climate change are likely to be felt first and foremost in the agricultural sector and associated water availability, with many people affected by lower food productivity (*e.g.*, hunger, malnutrition, and its consequences for education and productive economic life) andburdens on the central and state governments in dealing with smallholders and landless workers who will require assistance. Educational and employment inequalities will exacerbate these conditions. Some (or many) of these workers will migrate to urban areas, placing stress on cities. The need to add to or replace infrastructure affected by climate change (*e.g.*, in the energyand transportation sectors, as well as irrigation systems) will present additional economic costs. Finally, migrants, particularly from Bangladesh, will affect India's economy by providing competition for low cost labor.

2.11.3 Food Production and Drinking Water Supply

Agriculture and water are inseparable issues, as almost all the water use in India is for irrigation to support high productivity in agriculture. Thus, both the monsoon onset and active/break phasemean a great deal to the agriculture sector. Research indicates that the monsoon active/break phase is related to a tropical phenomenon called the Madden Julian Oscillation (MJO), which hasa periodicity of 30-70 days, with a predictability of 20-30 days. Predicting the MJO has important implications to predicting subseasonal[5] rainfall variability. If farmers sow too earlyand the crop needs water when the monsoon break phase occurs, the whole crop is lost.

At the national level, India has abundant water resources for agriculture and drinking, as well as for cooking and sanitation uses. India withdraws 34 percent of its available water annually. However, at state and local levels, there are wide-ranging differences; overexploitation is alreadyproblematic and will likely become worse under climate change conditions.

Increasing temperatures and increased seasonal variability will likely cause Himalayan glaciers to melt more and more quickly, also leading to increasing danger from floods as glacial lakes burst out of their natural bounds. Changes over decades will include less flow in rivers fed today by snow and ice, with impacts on hydropower, urban water supply, and agriculture. Althoughthere is clear evidence of de-glaciation across the Himalayas, the effect on river flows is likely tobe substantially different in different areas. The extent of the changes is very uncertain because, as Sengupta puts it, "River flow data is so scant and recent that it is impossible for scientists to predict how the current rates of glacial retreat will affect river volume."

There are major regions, including many of the most highly productive agricultural and industrialregions of India, where water scarcity is already felt in day-to-day life. The retreats of snow and ice cover are important for several local climates, especially those near the Himalayas. The drying effect of an earlier spring snow melt and, more generally, the earlier reduction in soil moisture is a continuing concern. Already 15 percent of aquifers are in critical condition, a figure that is projected to increase to a "frightening" 60 percent by the year 2030.

Especially in important agricultural areas that have benefited from the Green Revolution, irrigation has led to exploitation of groundwater resources—e.g., 94 percent in Punjab, 84 percent in Haryana, 60 percent in Tamil Nadu, and 51 percent in Rajasthan. Farmers who borrow to dig bore wells for crop irrigation may find that when groundwater tables decline, they cannot repay their loans; some probably will commit suicide. Groundwater depletion has been accelerated by the provision of free or subsidized electricity for agriculture: subsidized rates on irrigation pumps, some state-level free power policies, and supports for agricultural consumers. In many places (northwest India as well as Pakistan) irrigation has resulted in notonly overexploitation of water resources, but also increased soil salinity, water logging, and siltation of river basins. These problems have been intensified by declining soil fertility, increasing pest damage, and decreasing genetic diversity in crops.

Environmental and economic leaders have called for a second Green Revolution as rates of growth in agricultural production have slowed. This would place more pressure on water resources for irrigation. To partially compensate for increased water demand, the efficiency ofirrigation systems can be improved. And in recent years, Water Users Associations (WUAs) have formed to conduct participatory irrigation management; these may promote conservation and water use efficiency.

Rainfed agriculture (65 percent of cropped land) will experience climate change impacts, too.Without irrigation, smallholder farmers and landless agricultural workers will likely face household food shortages and loss of livelihoods, requiring government assistance or even international aid, as well as pressures on cities to accommodate migrants from rural areas.

The Indo-Gangetic Plain, the "bread basket" of the region, comprises both highly modernized and traditional agriculture. The western region (the Punjab in Pakistan and India) is characterized by high productivity, high investment in infrastructure, and widespread use of fertilizers and groundwater irrigation. The eastern region (northeastern India and Bangladesh)exhibits low productivity, with poor infrastructure and low inputs of fertilizer and water, high risk of flooding, and chronic poverty. Rising population and climate change will have a combined impact in each of these regions. Changes and uncertainties in water supply (especially monsoonal changes) will affect irrigated and rainfed crops alike.

During the "wheat panic" of 2007, India purchased nearly 800,000 tonnes of wheat to buffer its reserves of this essential grain. India sees keeping a buffer stock as a necessity in a country that, although theoretically food sufficient, has historically experienced shortages. Climate is one of the key components influencing agricultural production in small Indian farming systems,accounting for two-thirds of the variation in production. ENSO can explain some of the inter annual rainfall variability that affects overall production. ENSO-based climate forecasts may be used to aid vulnerable smallholder agriculture production systems in India.

However, unless projections of future climate change are able to account for how ENSO maychange in a warming climate, these forecasts may not be able to prevent the significant agricultural losses that are possible.

Over the longer term, most studies project decreased yields in non-irrigated wheat and in rice and a loss in farm-level net revenue between 9 and 25 percent for a temperature increase of 2.0– 3.5°C. Considering a range of equilibrium climate change scenarios which project a temperature rise of 2.5°C to 4.9°C for India, Kumar and Parikh (2001) estimated that the impacts of climate change on Indian agriculture would be significant across the scenarios. They estimated that, with a temperature change of +2°C and an accompanying precipitation change of +7 percent, farm-level total net revenue would fall by 9 percent, whereas with a temperature increase of +3.5°C and precipitation change of +15 percent, the fall in farm-level total net revenue would be nearly 25 percent. Aggarwal et al. explored agricultural adaptations to global environmental change in the Indo-Gangetic Plan, concluding that "new information and tools are needed to analyze the trade-offs between the joint socioeconomic and environmental goals, and possible adaptation strategies."

Livestock, too, will be affected by climate change. India has the largest livestock population in the world, with animals used as milk producers (especially cattle and buffalo), draft animals, nutrient recycling (manure) and seeding, and as household capital, particularly in landless households. Heat stress lowers production and reproduction, reduces feed and fodder, and increases conditions favorable to disease. For example, outbreaks of foot and mouth disease in cattle are explained 52 percent (in Andhra) and 84 percent (in Maharashtra) by temperature, humidity, and rainfall; mastitis increases in dairy animals during hot and humid weather, which also is associated with increases in flies and cattle ticks.

Water availability will be a crucial factor in the continued viability of both crops and livestock, but management is diffuse. Under the Indian constitution, water is the responsibility of the states, not the central government. Thus, water management institutions at the state level have the biggest say in planning and allocation. At the national level, at least five ministries (Water Resources, Environment and Forests, Rural Development, Power, and Urban Development and Poverty Alleviation) are concerned with water, but no organizational mechanism exists to coordinate water management among these ministries. Moreover, efforts toadapt to more variability in water flows by constructing dams and catchments may increase tensions with Bangladesh and Pakistan.

While freshwater resources will be affected, the marine environment is and will be changing too.Fisheries stocks may collapse or move in the already-contested waters of the Indian Ocean, potentially affecting livelihoods and food supply for millions of people. The international ocean management regimes will require renegotiation.

2.11.4 Human Health

Currently, India's public health care system produces relatively poor health outcomes. ForIndia, Healthy Life Expectancy (HALE), which includes adjustment for time spent in poor health, is 53 years for children born in 2003. Despite perceived strengths of the national-levelpublic health system (good expertise, written guidelines and standards, and network of researchand training institutions), implementation and monitoring of services are weak. Funding for control of communicable diseases has been deemphasized since the 1980s; several infectious diseases, such a tuberculosis and malaria, have reemerged as public health care concerns. InIndia, unplanned urbanization has contributed to the spread of *Plasmodium vivax* malaria and dengue.

Climate change impacts on health include an expected increase in communicable diseases, suchas malaria. Malaria is projected to move to higher latitudes and altitudes in India. An assessment in India projected shifts in the geographical range and duration of the transmission window for *Plasmodium falciparum* and *Pvivax malaria.*

2.11.5 Coping Capabilities in Facing Natural Disasters

India is highly vulnerable to natural disasters, and people live on marginal lands or in coastal and delataic cities where they are at greater risk. Floods, regional droughts, cyclones, and earthquakes affect millions of Indians.

In the area of disaster mitigation, much has been done to document conditions leading to vulnerability. For example, a *Flood Atlas of India* and a *Vulnerability Atlas of India* have been produced; the latter assesses the vulnerability of housing and infrastructure to earthquakes, cyclones and floods to improve zoning and construction.

In India, as in the United States, the primary responsibility for responding to disasters lies at the state level. The Disaster Management Act of 2005 set up the Natural Disaster Management Division in the Ministry of Home Affairs and proclaimed a new emphasis on disaster prevention.The United Nations Development Program has developed a plan for disaster preparedness and response that attempts to integrate government and other activities.

Many Indian states have limited resources and lack their own disaster management plans. Because of these factors, India's disaster response record has been mixed, with delayed response,lack of early warning systems and resources to undertake measures like mass evacuation, inadequate coordination among various government departments, failure to keep essential stores (e.g., sandbags, medicines, and life-saving equipment) on hand, and inadequate coordination with the Army and other service organizations, as well as donors.

As an example of disaster response, in 2007 the *New York Times* reported a United Nationsofficial as saying that about 2,800 people in India, Bangladesh, Nepal, and Pakistan had died inmonsoonal floods ("the worst in living memory"), from drowning, waterborne illnesses, snakebites or hunger. For survivors, the land damage meant there would be no near-term agricultural work for millions of landless laborers, leaving them to rely on the sporadic supportof aid agencies and government relief organizations. The condition of the 31 million people affected in India was covered internationally but was not featured either in New Delhi newspapers or on national news channels.

Infrastructure is affected by climate hazards as well. For instance, 14 percent of the annual repair and maintenance budget of the newly built 760 km Konkan Railway in India is spent repairing damage to track, bridges, and cuttings due to extreme weather events such as rain- induced landslides. In spite of preventive targeting of vulnerable stretches of the line, operations must be suspended for an average of seven days each rainy season because of such damage. Systematic disaster preparedness at the community level has helped reduce death tolls; for instance, new warning systems and evacuation procedures in Andhra Pradesh after 1977 reduceddeaths from coastal tropical cyclones to 10 percent of the 1977 total by 1997.

2.11.6 Climate Change in the Neighborhood

The most obvious regional climate change impacts will likely concern water. The Indo-Gangetic Plain stretches from Pakistan's coast across northern India to almost the whole of Bangladesh.

The Indus River drains in Pakistan, the Ganges in India, and Bangladesh. Rivers fed by Himalayan snow and ice provide irrigation water for this important agricultural area. India haswater agreements with both neighboring countries, and alleged violations of these agreements have caused international disputes.

Bangladesh, in particular, depends upon India to allow sufficient fresh water to flow into Bangladesh for every water use and for preventing salt water intrusion. Sea level rise and management decisions have reduced freshwater flows and increased salt-water intrusion in the Indus delta and Bangladesh. This adversely affects the Sundarbans, the largest mangrove forestin the world, which is shared by India and Bangladesh. In addition, flooding has displaced thousands to millions of Bangladeshis, who are subject to outbreaks of xenophobic violence if they resettle in India. Also potentially important are changes in Arabian Sea (*e.g.*, loss of fresh water input due to channeling of Indus river water to agriculture in Pakistan, leading to biogeochemical changes in the Western Arabian Sea); the impacts of aerosols (the "Atmospheric Brown Cloud"), which may diminish winter precipitation in western India and in Pakistan); and changes in the carbon cycle in the Bay of Bengal.

2.12 ADAPTIVE CAPACITY

The impacts of climate change will be felt differentially, depending upon how well a society can cope with or adapt to climate change, that is, its adaptive capacity. Adaptive capacity is defined by the IPCC as, "The ability of a system to adjust to climate change (including climate variability and extremes) to moderate potential damages, to take advantage of opportunities, or tocope with the consequences." Thus, adaptive capacity is distinguished from both climate change impacts and the degree to which those impacts affect the systems that are in place.

Although the specific determinants (or "drivers") of adaptive capacity are a matter of debate among researchers, there is broad agreement that economic, human, and environmental resources are essential elements. Some components of this adaptive capacity are near term, such as the ability to deliver aid swiftly to those affected by, *e.g.*, flooding or droughts. Other components include a high enough level of education so that people can change livelihoods, a quantity of unmanaged land that can be brought into food production, and institutions that provide knowledge and assistance in times of change. For instance, Yohe and Tol identified eight qualitative "determinants of adaptive capacity," many of which are societal in character, although the scientists draw on an economic vocabulary and framing:

- The range of available technological options for adaptation.
- The availability of resources and their distribution across the population.
- The structure of critical institutions, the derivative allocation of decision-making. authority, and the decision criteria that would be employed.
- The stock of human capital, including education and personal security.
- The stock of social capital, including the definition of property rights.
- The system's access to risk-spreading processes.

The ability of decision makers to manage information, the processes by which these decision makers determine which information is credible, and the credibility of the decision makers themselves. The public's perceived attribution of the source of stress and the significance of exposureto its local manifestations. Researchers have only recently taken on the challenge of assessing adaptive capacity in a comparative, quantitative framework. A global comparative study of resilience to climatechange (including adaptive capacity) was conducted using the Vulnerability-Resilience Indicators Model (VRIM—see box below).

Adaptive capacity, as assessed in this study, consists of seven variables (in three sectors), chosento represent societal characteristics important to a country's ability to cope with and adapt to climate change:

- Human and Civic Resources
    - *i.* ***Dependency ratio:*** proxy for social and economic resources available for adaptation aftermeeting basic needs.
    - *ii.* ***Literacy:*** proxy for human capital generally, especially the ability to adapt by changingemployment.
- Economic Capacity
    - *iii.* ***GDP (market) per capita:*** proxy for economic well-being in general, especially access tomarkets, technology, and other resources useful for adaptation.
    - *iv.* ***Income equity:*** proxy for the potential of all people in a country or state to participate inthe economic benefits available.

- Environmental Capacity
  - *v. Percent of land that is unmanaged:* proxy for potential for economic use or increased crop productivity and for ecosystem health (e.g., ability of plants and animals to migrateunder climate change).
  - *vi. Sulfur dioxide per unit land area:* proxy for air quality and, through sulfur deposition,other stresses on ecosystems.
  - ***vii. Population density:*** proxy for population pressures on ecosystems (e.g., adequate foodproduction for a given population).

Adaptive capacity for a sample of 11 countries from the 160-country study is shown in Figure 3(base year of 2005). There is a wide range of adaptive capacity represented by these countries; India ranks low, both in the sample and overall:

- Russia ranks 32nd and Libya 34th (in the highest quartile).
- Indonesia ranks 45th, Belize 48th, Mexico 59th, and China 75th (in the second quartile).
- The Philippines ranks 91st and India 119th (in the third quartile).
- Morocco ranks 136th and Haiti 156th (in the lowest quartile).

Any country-level analysis must take into account the comparative ranking of the country.

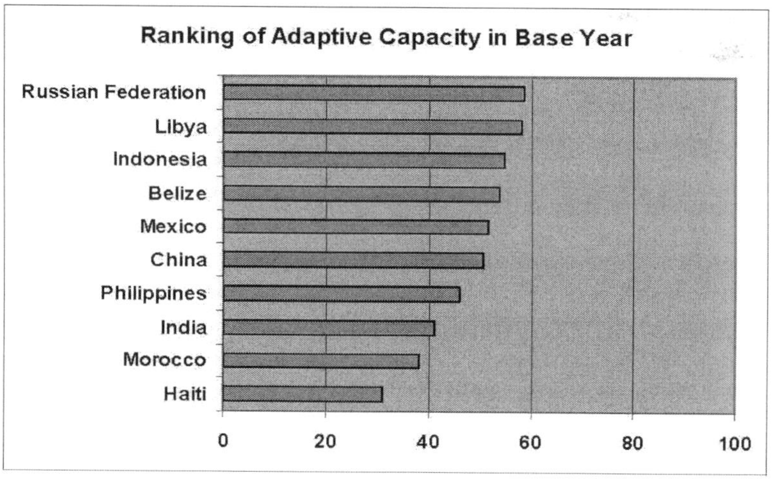

Figure 3. Sample of 11 countries' rankings of adaptive capacity (2005).

Figure 3 shows the contribution of each variable to the overall ranking (slight differences occurring because of the methodology (see box on page 25). India ranks low in comparison withRussia and China because of lower human resources (dependency ratio and literacy levels) than China, and both lower human resources and environmental capacity (non-managed land, emissions per total land area, and population density) than Russia.

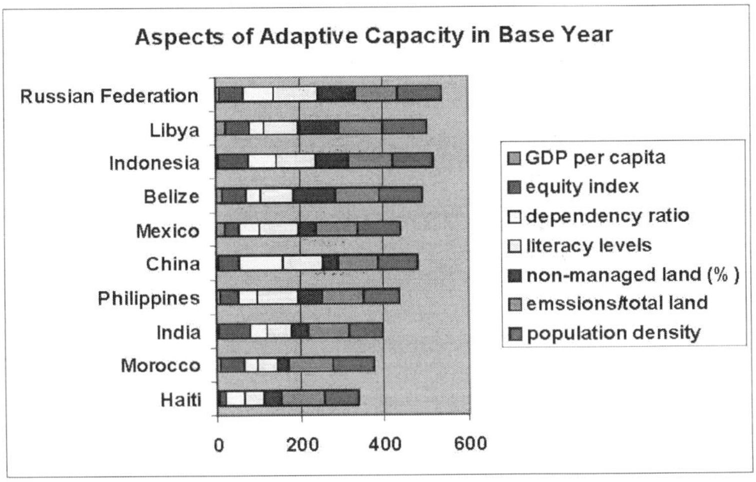

Fig. 4: Variables' contributions to adaptive capacity rankings.

Figure 4 shows projected adaptive capacity growth over time for the 11-country sample. Projections are made for two scenarios; rates of growth are based on the IPCC's A1 scenario in its *Special Report on Emissions Scenarios.* Both scenarios feature moderate population growthand a tendency toward convergence in affluence (with market-based solutions, rapid technological progress, and improving human welfare). The scenarios used in this study differ inthe rate of economic growth, one modeling high-and-fast economic growth, the other delayed growth. In the high-growth scenario, China overtakes Russia in adaptive capacity, but India retains its relatively low position throughout the entire period in both scenarios.

2.13 STRENGTHS/WEAKNESSES IN ADAPTIVE CAPACITY ASSESSMENTS

Even comparative measures of adaptive capacity only allow analysts to ask improved, more focused questions about area or local conditions that contribute to or reduce resilience. It is likely, for instance, that for particular places (e.g., states in India) important variables or domainsare not included. For agricultural regions, this might include the extent of irrigation; for urban areas, better measures of education could be important. The measure of unmanaged land does not account for the potential usefulness of that land. However, comparative measures such as these can be an important first step toward determining where to direct resources—for further analysis or additional factors.

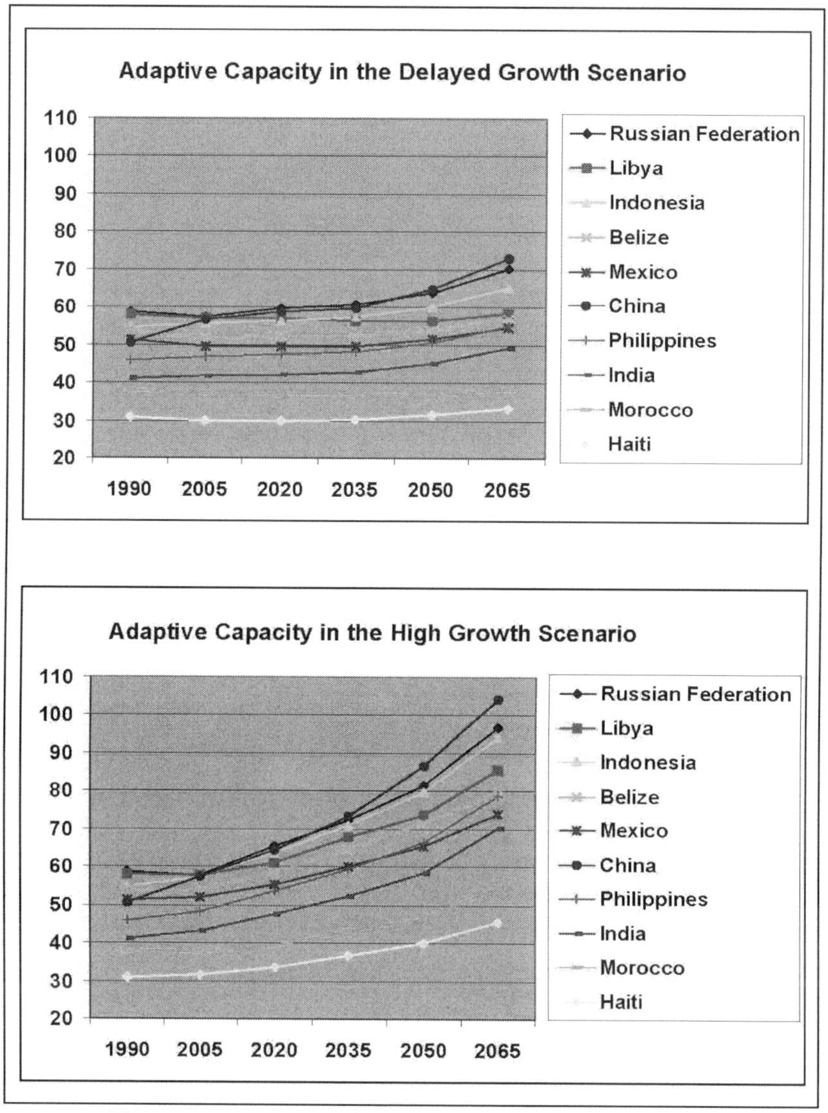

Fig. 5: Projections of adaptive capacity for 11 countries.

2.14 HIGH RISK IMPACTS

2.14.1 Agriculture

There are impacts to high-input, high-output agriculture as a result of various changes in the Indian climate, which will affect productivities negatively even as demand for food and other agricultural products rises because of both rising population and standard of living. On the other hand, millions of subsistence and smallholders farmers will experience immediate hardship and hunger because they will be less able to predict climate well enough to make adequate decisionsabout when to sow, what to grow, and how to time inputs.

2.14.2 Water:

Glacier melt may yield more runoff in the short term but less in the medium and long terms. More severe storms (especially cyclones) will cause more damage to infrastructure and livelihoods and exacerbate saltwater intrusion[6] in storm surges. Changes in the timing and amount of monsoon rain will make the production of food and other agricultural products moreuncertain, so that, even in good-weather years, farmers will be more likely to make decisions resulting in lower-productivity outcomes.

2.14.3 Energy:

As India searches for additional sources of energy to meet rising demand, climate change mitigation efforts may constrain its use of indigenous and imported coal, oil, and gas, while development of nuclear energy will be slow at best and likely to encounter opposition. Other non-emitting technologies will require technology transfer and capacity-building.

2.14.4 Exacerbation of Inequality:

In a country so prone to natural hazards, the welfare of those who are affected and who have limited means to recover will loom large under climate change, as a force that can change governments, strain public budgets, and foster unrest. The proportions are significant: about a quarter of the population lives on less than $1 a day; at least one-third face poverty, discrimination (often scheduled castes or tribes and Muslims), and lack of educational opportunities; and 60 percent are dependent upon agriculture for their livelihoods. Those who lack financial resources and adequate education and who depend on agriculture for sustenance and livelihood will be disadvantaged under climate change because they have few choices but continued dependence on shrinking or uncertain resources. India is attempting to reduce inequality by using quotas in government jobs and educational institutions; the National Rural Employment Guarantee Act (2005) also guarantees that at least one member of a rural householdwill have 200 days of work per year; and India's 11th five-year plan adopts the policy of sectoral,spatial, and socioeconomic growth that includes all people.

2.14.5 Migration:

India receives immigrants from a number of countries. For instance, since 1960 the government has hosted approximately 110,000 de facto refugees from Tibet. The government considers Tibetans and Sri Lankans to be refugees and provides assistance to them. However, the government regards most other groups, especially Bangladeshis, as economic migrants and doesnot provide them with aid.

3

CLIMATE CHANGE: A 'GLOCAL' PROBLEM REQUIRING 'GLOCAL' ACTION

3.1 INTRODUCTION: AN OVERVIEW

It is generally assumed that global problems like climate change call for global solutions, which are then implemented at different administrative levels. This paper attempts to deconstruct climate change in physical terms to assess its global through to local characteristics, presents existing policy frameworks at different levels, and then discusses some theoretical and practical aspects of policymaking. It investigates the policy space at different administrative levels to see if it is possible to complement the slow global action with local action that goes beyond that strictly speaking needed for implementation. The paper argues that climate change is a problem that is both global and local in nature, that policy responses are possible at multiple administrative levels, and that there is considerable politics in the way responsibilities are assigned to different levels.

With substantial attention paid to climate change at the 2007 G-7 Summit, the publication of the 2007 Fourth Assessment Report of the Intergovernmental Panel on Climate Change (IPCC), the recent Stern Report (2006) and the Al Gore book (2006), climate change is firmly on the international agenda. There is clear consensus on the need to accelerate action considerably over the next few years if we are to reduce emissions consistent with a maximum of a 28C warming by the end of the century (IPCC 2007).

3.2. IMPACTS, EMISSIONS AND GOVERNANCE AT DIFFERENT LEVELS

Over the course of its history, the earth's climate has varied between extreme cold and extreme warm conditions (Herman & Spicer 1996), mainly forced by plate tectonics and orbital variations. Over the last two hundred years, land- use changes and anthropogenic emissions of GHGs became new factors influencing the global climate system (IPCC

2007). This section looks at climate change impacts and emissions on different levels and how governance is presently organized at these levels.

The global level

The key climate impact at the global scale is the increasing mean atmospheric temperature. This temperature rise leads to increasing sea surface temperatures and melting of glaciers and ice sheets, causing the water bodies to expand, and thus to rising sea levels worldwide. Increased atmospheric CO_2 also acidifies the oceans, influencing marine ecosystems. The projected magnitude and pace of this acidification is alarming and the effects largely unknown.

The global emission problem results from the cumulative increase in global concentrations of GHGs and its systemic impacts. While international aviation, shipping and rail transport have a distinct global dimension, emissions are emitted primarily locally in energy, agricultural, industrial and waste management systems. However, the policy processes that impact on these systems may extend from local to global level.

The early political discussions saw climate change as essentially a global level issue, since it was irrelevant where the GHG emissions came from; once they accumulated in the atmosphere they could potentially lead to global warming. The problem could not be addressed if some countries stayed out of the system and, hence, to prevent the free-rider problem, all countries would have to take measures to reduce (the rate of growth of) their emissions. The initial political declarations on climate change – the Toronto Declaration (1988), the Noordwijk Declaration (1989), the Second World Climate Conference Declaration (1990) and the UN General Assembly Declaration of 1990 – emphasized this. The first IPCC report reiterated the global nature of the problem and the need for global policymaking.

This led to the adoption of the United Nations Framework Convention on Climate Change in 1992 and the Kyoto Protocol in 1997 (KP 1997). These agreements bind 189 and 168 countries, respectively, and list a number of policies and measures that all countries should consider taking and should report on and include quantitative targets for the developed countries.

Although many countries are involved and there is a continuous learning process taking place, the results in terms of actual impact are quite low. In 2003, global CO2 emissions were at 25,938.3 $MtCO_2$ per year; while 1990 levels were at 21,427.9 MtCO2 per year. After 13 years of policy, the environmental impact was marginal (IPCC 2007).

The slowdown in global activity can be partly attributed to the US decision not to ratify the Kyoto Protocol, which reflects the US withdrawal from multilateralism in many fields. This forced the European Union to take on the leadership role; while nevertheless reducing the motivation of other developed countries to take action, since unilateral action is perceived to negatively impact on their competitiveness at the international level. The climate negotiations are going far too slowly to have any significant impact on the problem, given that the Stern report (2006) demonstrates that the costs of taking action now are only 1/20th the costs of the potential impacts worldwide and that global adaptation methods including geo-engineering (e.g. injecting sulphate aerosols to block incoming sunlight, are hypothetical and extremely expensive.

The subcontinental level

In this paper, we use the term subcontinental, instead of regional, to indicate the level between global and national to avoid confusion with the subnational meaning of regional used in the other papers in this issue. At this level, the relatively high rise in temperature at high latitudes causes permafrost to degrade and sea-ice to melt leading, inter alia, to seaways opening up in the Arctic. The degradation of permafrost has major consequences for the stability of constructions and infrastructure. At more moderate latitudes, climate change affects rainfall patterns and, hence, land use, water and food security. At low latitudes, arid and semiarid conditions seem to be expanding northward as well, making agriculture more difficult. There are few emission sources with a distinct subcontinental character. International transport can be considered one and sometimes energy supply is arranged at a subcontinental level. At this subcontinental level, there are some agreements promoted by the USA, including the International Partnership on the Hydrogen Economy and the Asia Pacific Partnership on Clean Development (APPCD). The EU has a climate change programme and several directives on emissions trading, the promotion of biofuels and combined heat and power, as well as the voluntary agreement with the European car industry are being implemented. A follow-up programme (ECCP II) develops policies on impacts. These policies have to be transposed into national policies of EU member states. Although few other regions of the world have climate policies, regional banks like the Asian Development Bank do have relevant measures.

The national level

Climate change has severe potential implications at the national level. Changing precipitation patterns can affect agricultural potential and thus food and water security. In addition, changing weather patterns and increased extreme event frequency could cause more large scale flooding and possibly droughts and storms with national consequences. Rising sea levels threaten low-lying coastal zones, which are the most densely inhabited and developed areas.

GHG emissions at national level are mostly from national industry and energy production facilities, waste disposal systems, transport and land use change. The top emitters of GHGs at national level are the United States, China, the Russian Federation, India, Japan, Germany, Brazil, Canada, United Kingdom and Italy. At a per capita level, however, the figures differ significantly.

Most developed countries have adopted policies in the energy, transport, agricultural and waste management sectors. Although the USA and Australia did not ratify the Kyoto Protocol, they too have some policies in place. For example, the USA aims to reduce its GHG intensity by 18% in 2012 and promotes technologies on hydrogen, zero emission coal plants and fusion energy.

Between 1990 and 2000, total aggregated GHG emissions of the developed countries decreased by 3% UNFCCC 2003). However, most of the decrease can be attributed to decreases in 'economies in transition' (EIT) countries. Although there are many policies in place, the existing and planned portfolio of measures may be inadequate to achieve the Kyoto targets, especially because of a large projected increase in emissions in the transport sector. Hence, additional measures are needed.

Although they have no binding targets, the majority of the developing countries have adopted, or are considering adopting, policies on halting deforestation, energy demand side management, energy supply technologies and possibly fuel switching. Individual countries have specific programmes, such as Brazil's national alcohol programme for transport, China's green light programme and India's renewable energy programme (UNFCCC 2002). Nevertheless, their emissions are still rising.

The local level

It is, however, at the local level that the effects of climate change are mostly experienced. Changes in climate cause weather patterns to change affecting local agriculture, household practices and vector borne diseases. Global warming also causes extreme events like floods, droughts and tropical storms to become more frequent. These impacts are already devastating in developed countries, as the impact of hurricane Katrina in the USA has demonstrated. On top of that, in developing countries, such impacts tend to have multiplier effects, adding to their long list of existing problems.

At the same time, all emissions come from the local level, though not always governed at that scale as many sources are often linked to sectors working on a national or higher scale. However, in some countries these have been decentralized and should thus be considered local sources.

As mentioned, governments often delegate policies to lower levels or individual sectors. The question—how are these policy processes further regulated—becomes increasingly important as national governments are often politically unable to adopt far-reaching and long- term measures on global change issues because of economic, political and practical reasons. At the domestic level, the distribution of powers is often a black box in relation to climate change. Different countries have different domestic systems and power is shared accordingly.

Yet, initiatives to reduce emissions can come from the local level. Besides generating the social support for national policy, the cumulative effects of such initiatives can be substantial. The motivation of such local action may include the need to implement national policy, but may also be driven by local concerns with respect to climate change. Very often though, co-benefits like cost savings or local environmental benefits are important drivers. These initiatives provide insight into the potential policy space for climate action. The International Coalition of Local Environ- mental Initiatives (ICLEI), for instance, has over 500 cities worldwide participating in its programme and developing local policies on sustainable development and climate change.

Some lower governments are adopting far-reaching policies. In the USA, New York State aims to reduce its GHG emissions by 5% by 2010 and 10% by 2020 with respect to 1990 levels. In 2001, the New England Governors and Eastern Canadian Premiers adopted a climate change plan for Connecticut, Maine, Massachusetts, New Hampshire, Rhode Island, Vermont, Newfoundland, New Brunswick, Nova Scotia, Prince Edward Island and Quebec to reduce their emissions by 10% by 2020 with respect to 1990 levels. California aims to reduce its GHG emissions to 80% below 1990 levels by 2050.1 There are many other examples in the developed world.

Besides mitigation, there are many local adaptation practices, for instance to cope with extreme heat (e.g. air-conditioning in sanatoriums), droughts (e.g. water harvesting), extreme rainfall (e.g. storing water) and flood risks. Thirty such practices have been developed and evaluated in the European urban environment within the AMICA project. Many cities and lower governments are taking policies because of the structure of national policymaking in these countries. This special issue discusses policy initiatives in Italy, France, the Netherlands and China.

Inferences

We can conclude that in the area of impacts, climate change has different, although interrelated effects that manifest themselves at all levels of administration. GHG emissions are essentially local in nature, but it is their cumulative concentration level that is a critical factor in causing climate change. While emissions are locally emitted, they are often the result of local through to global processes and policies.

As illustrated in this section, policies are being taken from global through to the local level. These policies are not necessarily the result of top-down translation of measures to local level, but are also often the result of external influences from epistemic communities or networks, through which local communities are galvanized into developing possible policy options. Where national governments are reluctant to take action, provincial and local governments have stepped into the vacuum to develop policies.

3.3. THE POLITICS OF SCALING

Introduction

This section analyses what is the optimal level to manage policy, if at all, and why social actors are motivated to scale up/down the problem and, hence, the response strategies. In addition, possible frameworks for policy responses with respect to climate change are discussed.

Climate—global or local problem; global or local governance?

Is climate change a global or local problem? "Climate change is a global phenomenon, but global or even national 'thinking' averages together too many distinctive local trajectories of greenhouse gas emissions and making local actions less specific. But local 'thinking' is also insufficient for action because, for the most part, decisions about major emission reducing actions are made far from the local community". In 1985, Clark advised, with respect to the geo-bio-physical aspects, that "the challenge is not to establish the pre-eminence of any particular scale, but rather to match scales of explanations, processes, and patterns in a realistic and effective way". While many have claimed that the appropriate level of governance for climate is the global arena, argues that "the appropriate governance scale is at the level of the resource user and their management of the climate- impacted natural resource or livelihood source, rather than a global commons".

The literature reveals that problems are essentially politically framed and defined; and the political framing process defines problems to suit the political interests of national entities or of the actors engaged in the framing process. For example, it was argue that scaling up is necessary, since vulnerable coastal cities worldwide will never be able to adapt to the problem all by themselves. Other reasons for scaling up are to enhance understanding of the problem, to improve governance, to promote domestic interests, and the desire to promote extraterritorial interests.

Reasons for scaling down problems include the need to enhance understanding of local causes and impacts, nested sets of explanations the local cultural aspects and also sometimes to avoid liability.

We conclude that climate change occurs at multiple levels, the driving forces are to be found at multiple levels, the authority to deal with the problem occurs at multiple levels and, hence, the governance debate should not be about which is the most appropriate level, but about how can policies be developed and initiatives be taken simultaneously at different levels effectively.

Models of policy interaction

Since climate change is a glocal phenomenon combining global with local characteristics, we argue that the only appropriate response is a multilevel governance response in which concurrent policy processes at all levels identify policy space and foster initiatives as well as put pressure on the other governance levels.

Three ideal typical models of policy interaction can be envisaged. In the first, prescriptive model, policy making is highly top down, with global threshold values determining global targets, which are then divided into subcontinental and national targets, which are then divided into sectoral and/or provincial and local targets. In the second bottom-up model, local policy potential determines what national governments are willing to do, and global policy- making is hostage to national willingness to take action. In the third exploratory model, policy processes at each level seek their own space and equilibrium. Global policies are a function of global politics and small political windows of opportunity can shape the nature of the policies. National politics are shaped by national level politics and bargaining, while still trying to balance local with global pressures. Local policies may be determined by national ones but may also evolve according to specific local dynamics that occur in a particular context influenced by many transnational influences.

With respect to the top-down model, although there are 'global', sub continental and national targets, targets for local government are more or less nonexistent. The bottom-up model is also limited in that in many cases, countries have not developed national policy on the basis of what is locally feasible nor has international policy always resulted from what is nationally feasible. The exploratory model is perhaps more relevant in the 21st century, since at each level, different dynamics are visible and mechanisms often develop to compensate for shortcomings at other levels. Thus, while the USA and Australia have not ratified the Kyoto Protocol, lower governments are exploring the potential for further action (see also papers in this Special Issue). Furthermore, many organisations are compensating their own emissions by planting trees.

Table 1. Reasons for scaling up and down.

	Reasons
Scaling up	To enhance understanding of the cumulative and systemic nature of a problem accounting for all
	Externalities, and to understand global threshold levels and driving forces including ideologies
	To improve governance for the global good and achieve sustainable development
	To promote domestic interests at the international level by, e.g. postponing decision at national
	Level till international consensus is reached
	To promote extraterritorial interests by, e.g. gaining control over policies over foreign resources
Scaling down	To enhance understanding of the contextual details of the problem
	To improve the effectiveness of implementation
	To meet strategic interests such as avoiding liability for extraterritorial impacts

Source: Based on Gupta 2007 (forthcoming).

Table 2. EXamples of different impacts and policy options for different scales.

Level	Impacts	Emissions	Authorities andnetworks	Reasons	Policies
Global	Temperature rise; sea level rise; ocean acidification	Global Concentrations of GHGs; aviation and shipping	UN agencies andtreaties; G7 and G77; Bi and plurilateral agreements; Int. NGOs	Prevent free-riding; share information/ experiences; transfer technologies and resources	International targets, policies andmeasures; flexibility mechanisms
Subcontinental	Sea ice melting; opening of searoutes; permafrost melting; species migration	International energy and transport systems	UN regional agreements; European Union; subcontinental NGOs	To create a level playing field	Subcontinental policies and measures; subcontinental policies, regulations and directives; flexibility mechanisms
National	Loss in coastalzone and offshore islands; impacts on food, water and, human security	Energy Transport; agriculture; animal husbandry; land-use andwaste management	National governments; National NGOs	State is the negotiating unit in international relations; regulatory functions restwith the state	Sectoral policies on mitigation and adaptation; nuclear power; regulatory, economic and suasive instruments; spatial planning
Local	Impacts on localaccess to waterand food; impacts on local income; vector borne diseases	Consumption patterns; local spatial planning; local infrastructure and housing design	Local Governments; local communities; local and transnational NGOs (ICLEI)	Laboratories of policymaking; capable of own initiative; building constructions, reducing the emissions of municipal services, promoting renewables	Spatial design infrastructure, housing and transport policy; management of local government services; municipal taxes; influence of consumption patterns

At different levels, different types of policies are possible (Table II). Global and national level policies are enumerated in the UNFCCC (1992) Kyoto Protocol (1997) and in the IPCC reports, but the potential at the local level is less clear. The literature reveals that it is precisely at the local level that consumption patterns at homes can be influenced, that small-scale energy production systems can be influenced through local incentives and spatial rules, that participatory planning can help design new sustainable localities, and that the ability of local authority to provide a forum for change is often a function of statutory obligation and power. Action can far better be taken at the local level, simply because it may be easier to mobilize people more effectively when issues are designed closer to their own interests.

3.4. A 'GLOCAL' PROBLEM REQUIRING 'GLOCAL' ACTION: AT A GLANCE

This chapter has argued first that climate change is a problem that is caused locally, but cumulates into a global problem; a problem with global impacts that are experienced locally. Hence, it is a glocal problem calling for a multilevel governance solution. Second, for a comprehensive understanding of climate policy, it is vital to understand the policy processes at all levels and to understand how these feed into each other. To see policies as occurring within 'nested territorial containers' would be missing the point. The political dynamics and influences at each level are very different and the windows of opportunity may open up at different moments of time. Third, while national and international policies often have a symbiotic relationship, local policies have a different driving force and often take different dimensions. They are not merely aiming to implement national policy, they may go beyond; they may complement or contradict national policy! Local and provincial policies are 'laboratories of democracy' and can incrementally influence national policy through their own exploration of what is possible. Fourth, where countries are reluctant to participate in the process, whether developed or developing, exploring possible measures at local and provincial level may be very useful. Fifth, the policy instruments that are available at the different levels are different; and the policy mandate is also accordingly different. This overview paper has thus shown that much is possible at different levels of policy to address a global problem such as climate change.

4

MODELLING THE CLIMATE

4.1 MODELLING THE WEATHER

An English mathematician, Lewis Fry Richardson, set up the first numerical model of the weather. During his spare moments while working for the Friends' Ambulance Unit (he was a Quaker) in France during the First World War he carried out the first numerical weather forecast. With much painstaking calculation with his slide-rule, he solved the appropriate equations and produced a six-hour forecast. It took him six months – and then it was not a very good result. But his basic methods, described in a book published in 1922,2 were correct. To apply his methods to real forecasts, Richardson imagined the possibility of a very large concert-hall filled with people, each person carrying out part of the calculation, so that the integration of the numerical model could keep up with the weather. But he was many years before his time! It was not until some forty years later that, essentially using Richardson's methods, the first operational weather forecast was produced on an electronic computer. Computers more than one million times faster than the one used for that first forecast (Figure 1) now-run the numerical models that are the basis of all weather forecasts.

Numerical models of the weather and the climate are based on the fundamental mathematical equations that describe the physics and dynamics of the movements and processes taking place in the atmosphere, the ocean, the ice and on the land. Although they include empirical information, they are not based to any large degree on empirical relationships– unlike numerical models of many other systems, for instance in the social sciences.

Setting up a model of the atmosphere for a weather forecast (see Figure 2) requires a mathematical description of the way in which energy from the Sun enters the atmosphere from above, some being reflected by the surface or by clouds and some being absorbed at the surface or

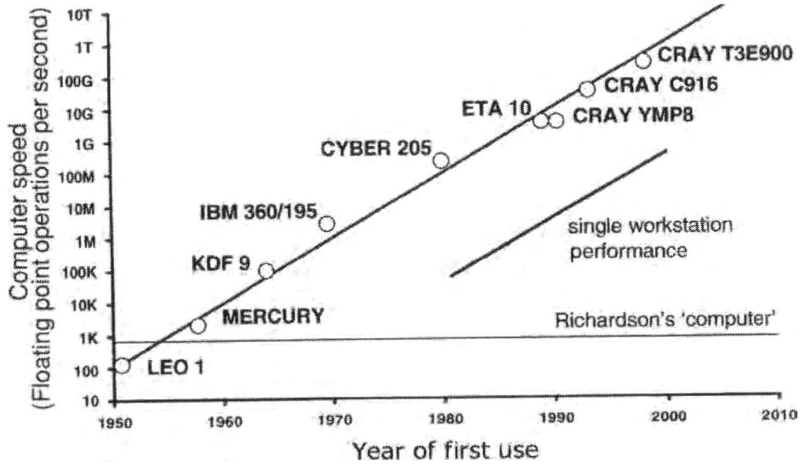

Fig. 1: Illustrating the growth of computer power available at major forecasting centres. The computers are those used by the UK Meteorological Office for numerical weather prediction research, since 1965 for operational weather forecasting and most recently for research into climate prediction.

in the atmosphere. The exchange of energy and water vapour between the atmosphere and the surface must also be described.

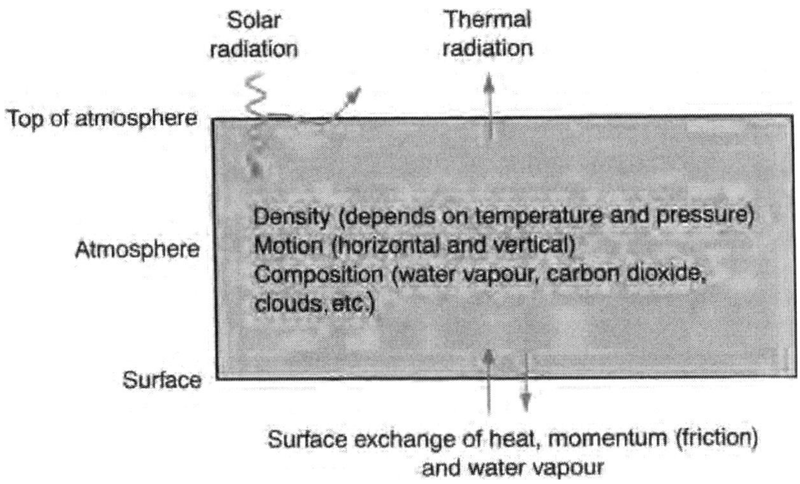

Fig. 2: Schematic illustrating the parameters and physical processes involved in atmospheric models.

Water vapour is important because of its associated latent heat (in other words, it gives out heat when it condenses) and also because the condensation of water vapour results in cloud formation, which modifies substantially the interaction of the atmosphere with the incoming energy from the Sun. Variations in both these energy inputs modify the atmospheric temperature structure, causing changes in atmospheric density (since warmed gases expand and are therefore less dense). It is these density changes that drive atmospheric motions such as winds and air currents, which in their turn alter and feed back on atmospheric density and composition. More details of the model formulation are given in the box below.

To forecast the weather for several days ahead a model covering the whole globe is required; for example, the southern hemisphere circulation today will affect northern hemisphere weather within a few days and vice versa. In a global forecasting model, the parameters (i.e. pressure, temperature, humidity, wind velocity and so on) that are needed to describe the dynamics and physics (listed in the box below) are specified at a grid of points (Figure 3) covering the globe. A typical spacing between points in the horizontal would be 100 km and about 1 km in the vertical; typically there would be twenty levels or so in the model in the vertical. The fineness of the spacing is limited by the power of the computers currently available.

Having set up the model, to generate a forecast from the present, it is started off from the atmosphere's current state and then the equations are integrated forward in time (see box below) to provide new descriptions of the atmospheric circulation and structure up to six or more days ahead.

For a description of the atmosphere's current state, data from a wide variety of sources (see box below) have to be brought together and fed into the model.

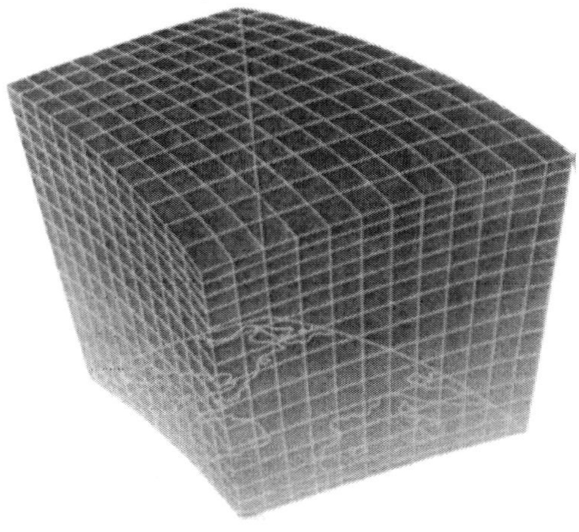

Fig. 3: Illustration of a model grid. The levels in the vertical are not equally spaced; the top level is typically at about 30 km in altitude.

Since computer models for weather forecasting were first introduced, their forecast skill has improved to an extent beyond any envisaged by those involved in the development of the early models. As improvements have been made in the model formulation, in the accuracy or coverage of the data used for initialisation or in the resolution of the model (the distance between grid points), the resulting forecast skill has increased. For instance, for the British Isles, three-day forecasts of surface pressure today are as skilful on average as two-day forecasts of ten years ago, as can be seen from Figure 5.

When looking at the continued improvement in weather forecasts, the question obviously arises as to whether the improvement will continue or whether there is a limit to the predictability we can expect. Because the atmosphere is a partially chaotic system, even if perfect observations of the atmospheric state and circulation could be provided, there would be a limit to our ability to forecast the detailed state of the atmosphere at some time in the future. In Figure 6 current forecast skill is compared with the best estimate of the limit of the forecast skill for the British Isles (similar results would be obtained with any other midlatitude situation) with a perfect model and near perfect data. According to that estimate, the limit of significant future skill is about 20 days ahead.

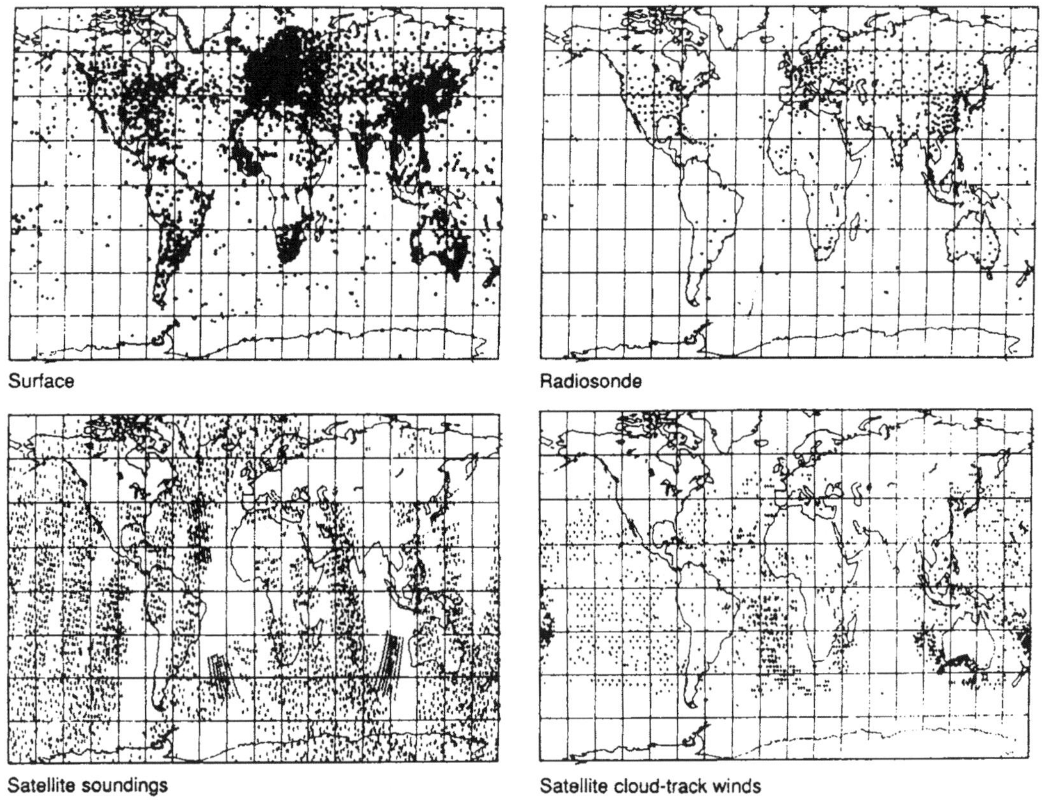

Fig. 4: Illustrating some of the sources of data for input into the UK Meteorological Office global

Weather forecasting model on a typical day. Surface observations are from land observing stations(manned and unmanned), from ships and from buoys. Radiosonde balloons make observations at up to 30 km altitude from land and from ship-borne stations. Satellite soundings are of temperature and humidity at different atmospheric levels deduced from observations of infrared or microwave radiation.

Satellite cloud-track winds are derived from observing the motion of clouds in images from geostationary satellites. Forecast skill varies considerably between different weather situations or patterns. In other words some situations are more 'chaotic' than others.

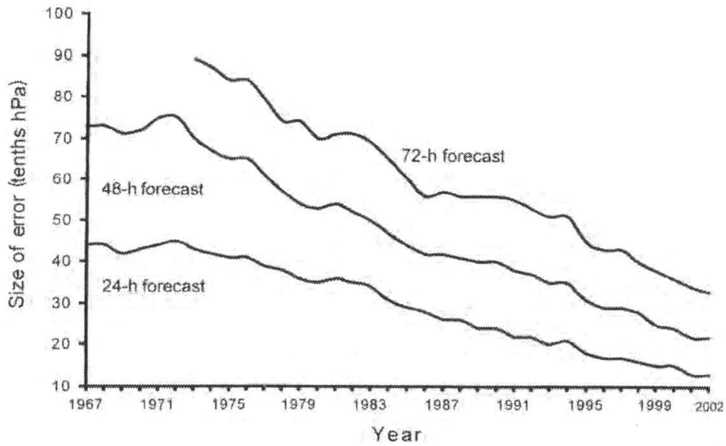

Fig. 5: Errors (root mean square differences of forecasts of surface pressure compared with analyses) of UK Meteorological Office forecasting models for the north Atlantic and Western Europe since 1966 for 24-h, 48-h and 72-h forecasts. Note that 1 hPa = 1 mbar.

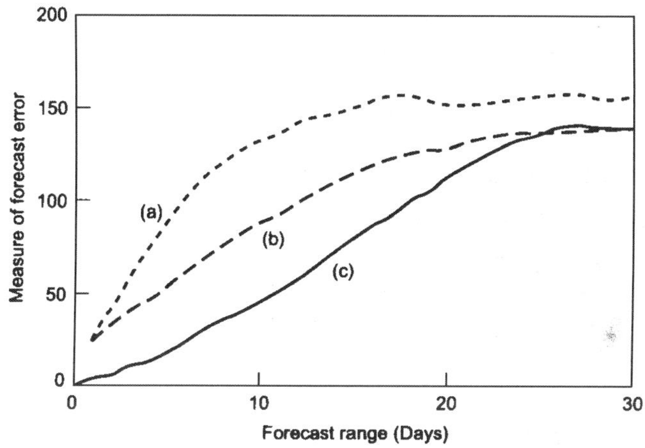

Fig. 6: Illustrating potential improvements in forecast skill if there were better data or a better model. The ordinate (vertical axis) is a measure of the error of model forecasts (it is the root mean square differences of forecasts of the 500 hPa height field compared with analyses). Curve (a) is the error of 1990 UK Meteorological Office forecasts as a function of forecast range. Curve (b) is an estimate showing how, with the same initial data, the error would be reduced if a perfect model could be used. Curve (c) is an estimate showing the further improvement which might be expected if near-perfect data could be provided for the initial state. After a sufficiently long period, all the curves approach a saturation value of the average root mean square difference between any forecasts chosen at random.

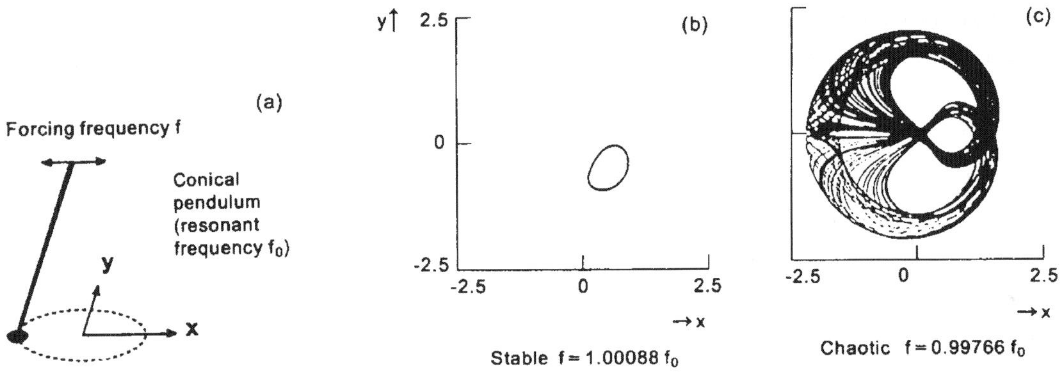

Fig. 7: (a) Illustrating a simple pendulum consisting of a bob at the end of a string of length 10 cm attached to a point of suspension which is moved with a linear oscillatory forcing motion at frequencies near the pendulum's resonance frequency *f*0. (b) and (c) show plots of the bob's motion on a horizontal plane, the scale being in centimetres. (b) For a forcing frequency just above *f*0 the motion of the bob settles down to a simple, regular pattern. (c) For a forcing frequency just below *f*0 the bob shows 'chaotic' motion (although contained within a given region) which varies randomly and discontinuously as a function of the initial conditions.

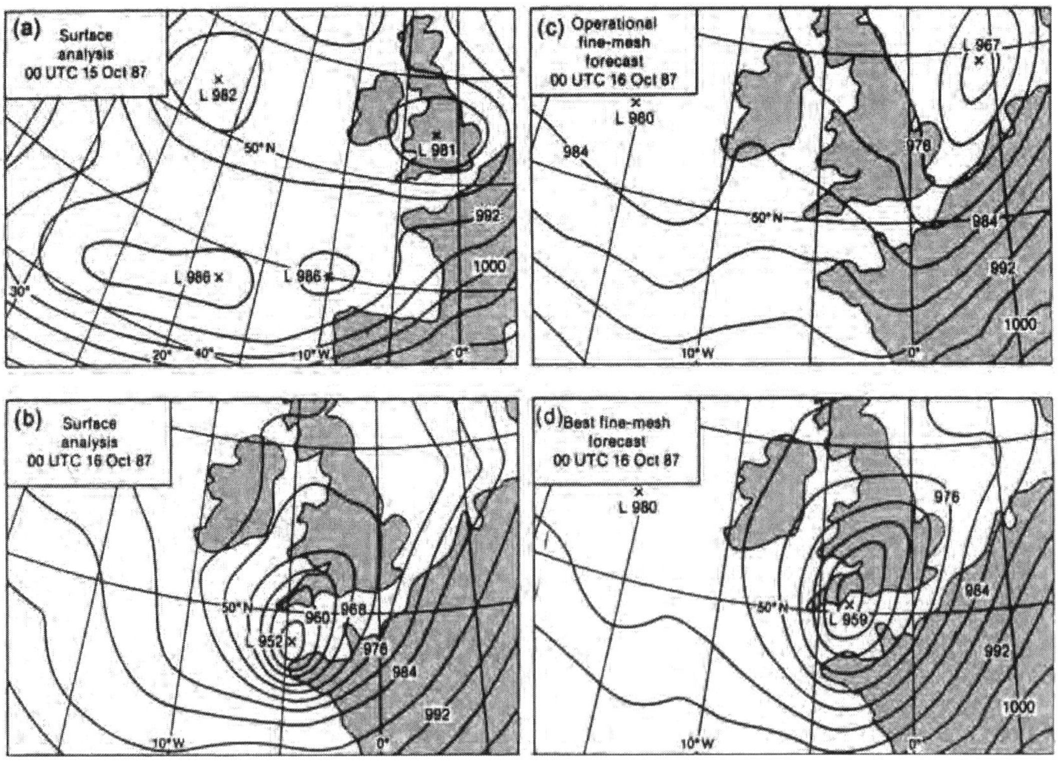

Figure 8 Analyses and forecasts of the surface pressure in millibars (1 mbar = 100 Pa = 1hPa) for the storm which passed over southern England on 15/16 October 1987. The fully developed storm is shown in (b) about 4 hours before it reached the London area; (a) is the situation twenty-four hours earlier from which the storm developed. (c)

and (d) are twenty-four-hour forecasts with the Meteorological Office operational fine-mesh model, (c) being the one available at the time and (d) being one produced after the event using more complete data and better assimilation procedures.

Given situation is to employ *ensemble* forecasting in which an ensemble of forecasts is run from a cluster of initial states that are generated by adding to an initial state small perturbations that are within the range of observational or analysis errors. The forecasts provided from the means of such ensembles show significant improvement compared with individual forecasts. Further, ensemble forecasts where the spread amongst the ensemble is low possess more skill that those where the spread in the ensemble is comparatively high.

4.2 SEASONAL FORECASTING

So far short-term forecasts of detailed weather have been considered. After twenty days or so they run out of skill. What about further into the future? Although we cannot expect to forecast the weather in detail, is there any possibility of predicting the average weather, say, a few months ahead? As this section shows, it is possible for some parts of the world, because of the influence of the distribution of ocean surface temperatures on the atmosphere's behaviour. For seasonal forecasting it is no longer the initial state of the atmosphere about which detailed knowledge is required. Rather, we need to know the conditions at the surface and how they might be changing.

In the tropics, the atmosphere is particularly sensitive to sea surface temperature. This is not surprising because the largest contribution to the heat input to the atmosphere is due to evaporation of water vapour from the ocean surface and its subsequent condensation in the atmosphere, releasing its latent heat. Because the saturation water vapour pressure increases rapidly with temperature – at higher temperatures more water can be evaporated before the atmosphere is saturated – evaporation from the surface and hence the heat input to the atmosphere is particularly large in the tropics.

It is during El Ni˜no events in the east tropical Pacific that the largest variations are found in ocean temperature. Anomalies in the circulation and rainfall in all tropical regions and to a lesser extent at mid latitudes are associated with these El Ni˜no events (fig. 4). A good test of the atmospheric models described above is to run them with an El Ni˜no sequence of sea surface temperatures and see whether they are able to simulate these climate anomalies. This has now been done with a number of different atmospheric models; they have shown considerable skill in the simulation of many of the observed anomalies, especially those in the tropics and sub-tropics.

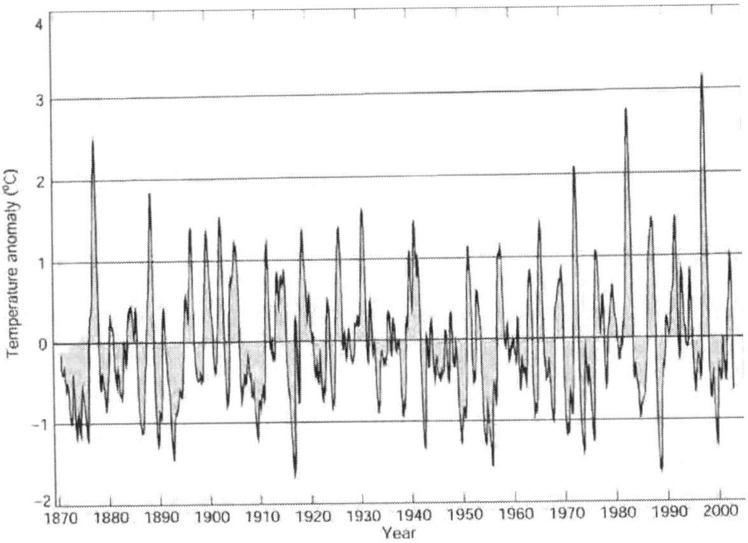

Fig. 9: Changes in sea surace temperature 1871–2002, relative to the 1961–90 average, for the eastern tropical Pacific off Peru.

Because of the large heat capacity of the oceans, anomalies of ocean surface temperature tend to persist for some months. The possibility therefore exists, for regions where there is a strong correlation between weather and patterns of ocean surface temperature, of making forecasts of climate (or average weather) some weeks or months in advance. Such seasonal forecasts have been attempted especially for regions with low rainfall; for instance, for north east Brazil and for the Sahel region of sub-Saharan Africa, a region where human survival is very dependent on the marginal rainfall.

The ability to make accurate seasonal forecasts significantly in advance depends critically on being able to forecast changes in ocean surface temperature. To do that requires understanding of, and the ability to model, the ocean circulation and the way it is coupled to the atmospheric circulation. Because the largest changes in ocean surface temperature occur in the tropics and because there are reasons to suppose that the ocean may be more predictable in the tropics than elsewhere, most emphasis on the prediction of ocean surface temperature has been placed in tropical regions, in particular on the prediction of the *El Nino* events themselves.

Later on in this chapter the coupling of atmospheric models and ocean models is described. For the moment it will suffice to say that, using coupled models together with detailed observations of both atmosphere and ocean in the Pacific region, significant skill in the prediction of El Ni˜no events up to a year in advance has been achieved.

4.3 THE CLIMATE SYSTEM

So far, the forecasting of detailed weather over a few days and of average weather for a month or so, up to perhaps a season ahead, has been described in order to introduce the science and technology of modelling, and also because some of the scientific confidence in the more elaborate climate models arises from their ability to describe and forecast the processes involved in day-to-day weather. Climate is concerned with substantially longer periods of time, from a few years to perhaps a decade or longer. A description of the climate Since we live in the atmosphere the variables commonly used to describe climate are mainly concerned with the atmosphere. But climate cannot be described in terms of atmosphere alone. Atmospheric processes are strongly coupled to the oceans (see above); they are also coupled to the land surface. There is also strong coupling to those parts of the Earth covered with ice (the cryosphere) and to the vegetation and other living systems on the land and in the ocean (the biosphere). These five components – atmosphere, ocean, land, ice and biosphere – together make up the climate system (Figure 13).

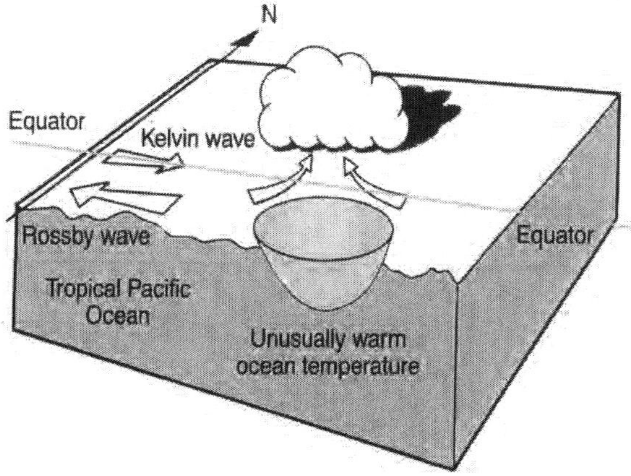

Fig. 12: Schematic to illustrate El Ni˜no oscillation over a period involves the averages of appropriate components of the weather (for example, temperature and rainfall) over that period together with the statistical variations of those components. In considering the effect of human activities such as the burning of fossil fuels, changes in climate over periods of decades up to a century or two ahead must be predicted.

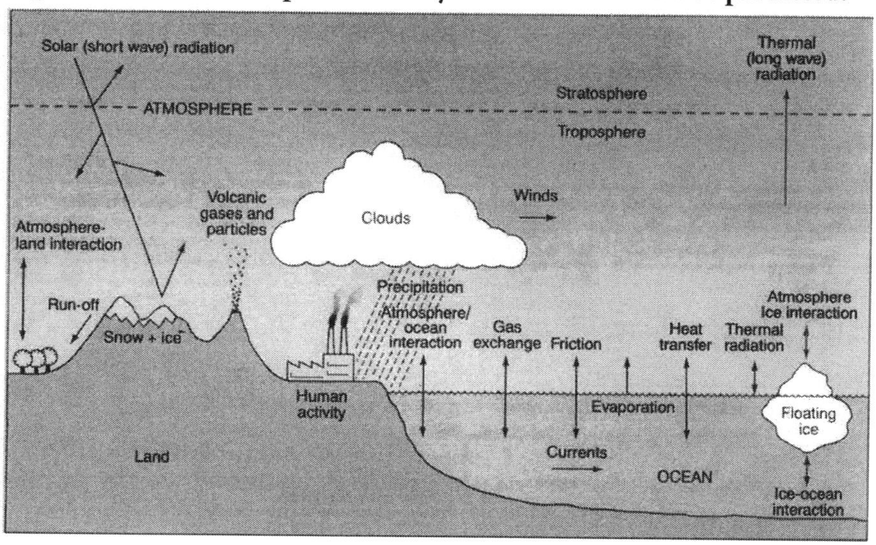

Fig. 13: Schematic of the climate system.

4.4 FEEDBACKS IN THE CLIMATE SYSTEM

As it considered the rise in global average temperature which would result from the doubling of the concentration of atmospheric carbon dioxide assuming that no other changes occurred apart from the increased temperature at the surface and in the lower atmosphere. The rise in temperature was found to be 1.2 C. However, it was also established that, because of feedbacks (which may be positive or negative) associated with the temperature increase, the actual rise in global average temperature was likely to be approximately doubled to about 2.5 °C.

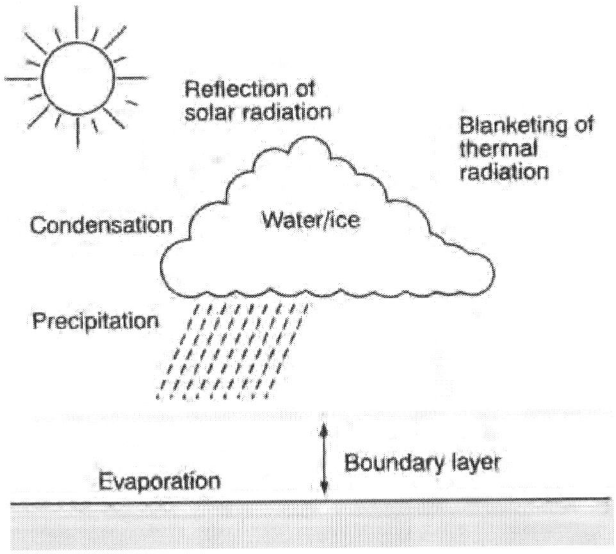

Figure.14 Schematic of the physical processes associated with clouds

4.4.1 Water vapour feedback

With a warmer atmosphere more evaporation occurs from the ocean and from wet land surfaces. On average, therefore, a warmer atmosphere will be a wetter one; it will possess a higher water vapour content. Since water vapour is a powerful greenhouse gas, on average a positive feedback results of a magnitude that models estimate to approximately double the increase in the global average temperature that would arise with fixed water vapour. Cloud-radiation feedback This is more complicated as several processes are involved. Clouds interfere with the transfer of radiation in the atmosphere in two ways (Figure 14). Firstly, they reflect a certain proportion of solar radiation back to space, so reducing the total energy available to the system. Secondly, they act as blankets to thermal radiation from the Earth's surface in a similar way to greenhouse gases. By absorbing thermal radiation emitted by the Earth's surface below, and by themselves emitting thermal radiation, they act to reduce the heat loss to space by the surface. The effect that dominates for any particular cloud depends on the cloud temperature (and hence on the cloud height) and on its detailed optical properties (those properties which determine its reflectivity to solar radiation and its interaction with thermal radiation). The latter depend on whether the cloud is of water or ice, on its liquid or solid understood. It is through comparisons of this kind that further elucidation of cloud radiation feedback will be achieved.

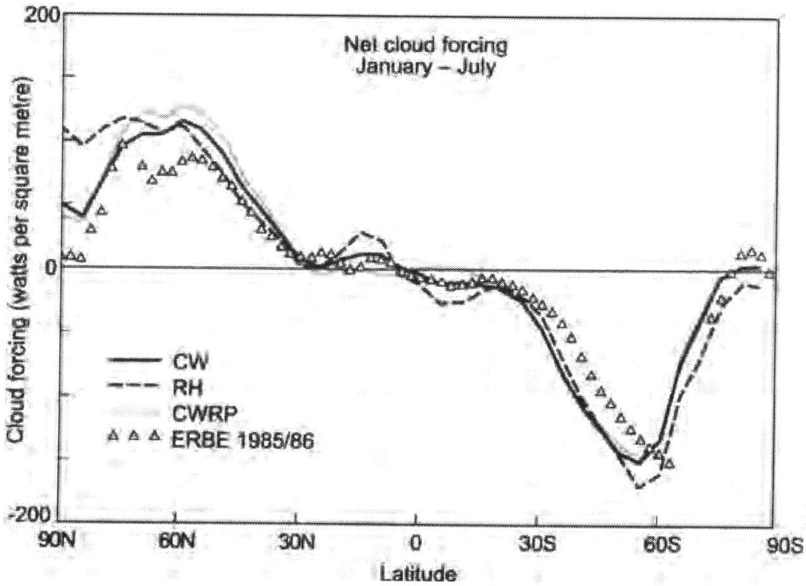

Fig. 15: The cloud radiative forcing is made up of a solar radiation and a thermal radiation component, which generally act in opposite senses, each typically of magnitude between 50 and 100 W.m−2. The average net forcing is shown here for the period January to July as a function of latitude as observed from satellites in two years (1985–6) of the Earth Radiative Budget Experiment (ERBE) and also as simulated in a climate model with different schemes of cloud formulation (a simple threshold relative humidity scheme (RH); a scheme that includes cloud water as a separate variable (CW); as CW but also with cloud radiative properties dependent on the cloud water content – thick clouds with a large water content are more reflective than thin clouds (CWRP)). There is encouraging agreement between the models' results and observations, but also differences that need to be water content (how thick or thin it is) and on the average size of the cloud particles. In general for low clouds the reflectivity effect wins so they tend to cool the Earth-atmosphere system; for high clouds, by contrast, the blanketing effect is dominant and they tend to warm the system. The overall feedback effect of clouds, therefore, can be either positive or negative.

Table 1 Estimates of global average temperature changes under different assumptions about changes in greenhou se gases and clouds

Source	Best estimate	Uncertainty
Natural		
Wetlands	150	(90–240)
Termites	20	(10–50)
Ocean	15	(5–50)
Other (including hydrates)	15	(10–40)
Human-generated		
Coal mining, natural gas, petroleum industry	100	(75–110)
Rice paddies	60	(30–90)
Enteric fermentation	90	(70–115)
Waste treatment	25	(15–70)
Landfills	40	(30–70)
Biomass burning	40	(20–60)
Sinks	545	
Atmospheric removal		(450–550)
Removal by soils	30	(15–45)
Atmospheric increase	22	(35–40)

Climate is very sensitive to possible changes in cloud amount or structure, as can be seen from the results of models discussed in later chapters. To illustrate this, Table 5.1 shows that the hypothetical effect on the climate of a small percentage change in cloud cover is comparable with the expected changes due to a doubling of the carbon dioxide concentration.

4.4.2 Ocean-circulation feedback

The oceans play a large part in determining the existing climate of the Earth; they are likely therefore to have an important influence on climate change due to human activities.

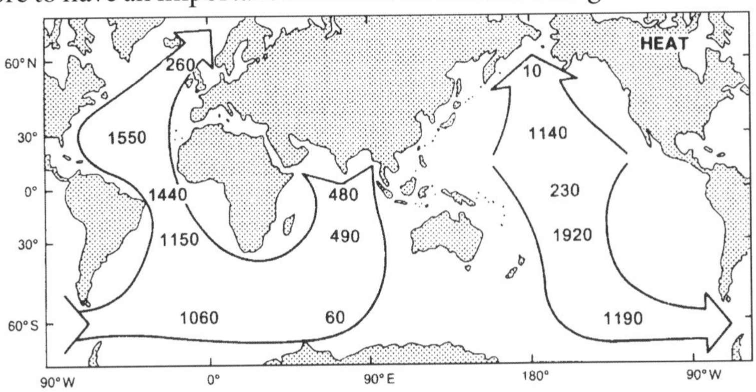

Fig. 16: Estimates of transport of heat by the oceans. Units are terawatts (1012 W or million watts). Note the linkages between the oceans and that some of the heat transported by the north Atlantic originates in the Pacific.

The oceans act on the climate in four important ways. Firstly, there are close interactions between the ocean and the atmosphere; they behave as a strongly coupled system. As we have already noted, evaporation from the oceans provides the main source of atmospheric water vapour which, through its latent heat of condensation in clouds, provides the largest single heat source for the atmosphere. The atmosphere in its turn acts through wind stress on the ocean surface as the main driver of the ocean circulation.

Secondly, they possess a large heat capacity compared with the atmosphere, in other words a large quantity of heat is needed to raise the temperature of the oceans only slightly. In comparison, the entire heat capacity of the atmosphere is equivalent to less than three metres depth of water. That means that in a world that is warming, the oceans warm much more slowly than the atmosphere. We experience this effect of the oceans as they tend to reduce the extremes of atmospheric temperature.

For instance, the range of temperature change both during the day and seasonally is much less at places near the coast than at places far inland.

The oceans therefore exert a dominant control on the rate at which atmospheric changes occur. Thirdly, through their internal circulation the oceans redistribute heat throughout the climate system. The total amount of heat transported from the equator to the polar regions by the oceans is similar to that transported by the atmosphere. However, the regional distribution of that transport is very different (Figure 5.16). Even small changes in the regional heat transport by the oceans could have large implications for climate change. For instance, the amount of heat transported by the north Atlantic Ocean is over 1000 terawatts (1 terawatt = 1 million million watts = 1012 watts). To give an idea of how large this is, we can note that a large power station puts out about 1000 million watts and the total amount of commercial energy produced globally is about twelve terawatts. To put it further in context, considering the region of the north Atlantic Ocean between north west Europe and Iceland, the heat input (Figure 16) carried by the ocean circulation is of similar magnitude to that reaching the ocean surface there from the incident solar radiation.

Any accurate simulation of likely climate change, therefore, especially of its regional variations, must include a description of ocean structure and dynamics.

4.4.3 Ice-albedo feedback

An ice or snow surface is a powerful reflector of solar radiation (the albedo is a measure of its reflectivity). As some ice melts, therefore, at the warmer surface, solar radiation which had previously been reflected back to space by the ice or snow is absorbed, leading to further increased warming. This is another positive feedback which on its own would increase the global average temperature rise due to doubled carbon dioxide by about twenty per cent. Four feedbacks have been identified, all of which play a large part in the determination of climate, especially its regional distribution. It is therefore necessary to introduce them into climate models. Because the global models allow for regional variation and also include the important non-linear processes in their formulation, they are able in principle to provide a full description of the effect of these feedbacks. They are, in fact, the only tools available with this potential capability. It is to a description of climate prediction models that we now turn.

4.5 MODELS FOR CLIMATE PREDICTION

For models to be successful they need to include an adequate description of the feedbacks we have listed. The water vapour feedback and its regional distribution depend on the detailed processes of evaporation, condensation and advection (the transfer of heat by horizontal air flow) of water vapour, and on the way in which convection processes (responsible for showers and thunderstorm) are affected by higher surface temperatures.

All these processes are already well included in weather forecasting models, andwater vapour feedback has been very thoroughly studied. The most important of the others are cloud radiation feedback and ocean-circulation feedback. How are these incorporated into the models?

For modelling purposes, clouds divide into two types – *layer* clouds present on scales larger than the grid size and *convective* clouds generally on smaller scales than a grid box. For the introduction of layer clouds, early weather-forecasting and climate models employed comparatively simple schemes. A typical scheme would generate cloud at specified levels whenever the relative humidity exceeded a critical value, chosen for broad agreement between model-generated cloud cover and that observed from climatological records. More recent models parameterise the processes of condensation, freezing, precipitation and cloud formation much more completely. They also take into account detailed cloud properties (e.g. water droplets or ice crystals and droplet number and size) that enable their radiative properties (e.g. their reflectivity and transmissivity) to be specified sufficiently well for the influence of clouds on the atmosphere's overall energy budget to be properly described. The most sophisticated models also include allowance for the effect of aerosols on cloud properties – denoted the indirect aerosol effect in Figure.8. The effects of convective clouds are incorporated as part of the model's scheme for the parametrisation of convection. The amount and sign (positive or negative) of the average cloudradiation feedback in a particular climate model is dependent on many aspects of the model's formulation as well as on the particular scheme used for the description of cloud formation. Different climate models, therefore, can show average cloud-radiation feedback that can be either positive or negative; further, the feedback can show substantial regional variation. For instance, models differ in their treatment of low cloud such that in some models the amount of low cloud increases with increased greenhouse gases, with other models it decreases. Uncertainty regarding cloud-radiation feedback is the main reason for the wide uncertainty range in what is called *climate sensitivity*) or the change in global average surface temperature due a doubling of carbon dioxide concentration.

The second important feedback is that due to the effects of the ocean circulation. Compared with a global atmospheric model for weather forecasting, the most important elaboration of a climate model is the inclusion of the effects of the ocean. Early climate models only included the ocean very crudely; they represented it by a simple slab some fifty or one hundred metres deep, the approximate depth of the 'mixed layer' of ocean which responds to the seasonal heating and cooling at the Earth's surface. In such models, adjustments had to be made to allow for the transport of heat by ocean currents. When running the model with a perturbation such as increased carbon dioxide, it was not possible to make allowance for any changes in that transport which might occur. Such models therefore possess severe limitations.

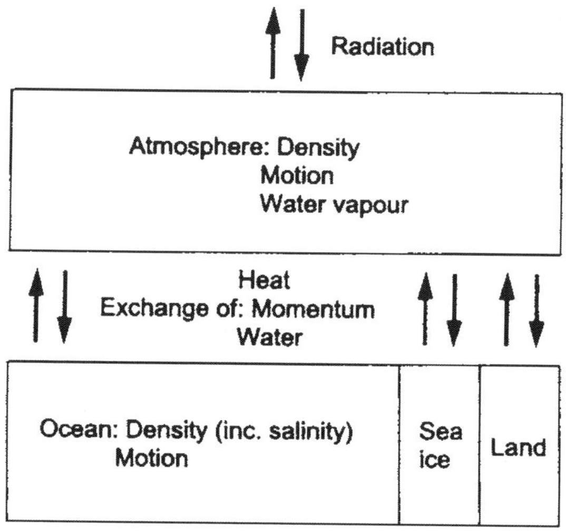

Fig. 17: Component elements and parameters of a coupled atmosphere–ocean model including the exchanges at the atmosphere–ocean interface.

For an adequate description of the influence of the ocean it is necessary to model the ocean circulation and its coupling to the atmospheric circulation. Figure 5.17 shows the ingredients of such a model. For the atmospheric part of the model, in order to accommodate long runs on available computers, the size of the grid has to be substantially larger, typically 300 km in the horizontal. Otherwise it is essentially the same as the global model for weather forecasting described earlier. The formulation of the dynamics and physics of the ocean part of the model is similar to that of its atmospheric counterpart. The effects of water vapour are of course peculiar to the atmosphere, but the salinity (the salt content) of the oceans has to be included as a parameter together with its considerable effects on the water density. Because dynamical systems, e.g. large scale eddies in the oceans, are of smaller scale than their atmospheric counterparts, the grid size of the ocean component is typically about half that of the atmospheric component. On the other hand, because changes in the ocean are slower, the time step for model integration can be greater for the ocean component.

At the ocean–atmosphere interface there is exchange of heat, water and momentum (exchange of momentum leads to friction) between the two fluids. The importance of water in the atmosphere and its influence on the atmospheric circulation have already been shown. The distribution of fresh water precipitated from the atmosphere as rain or snow also has a large influence on the ocean's circulation through its effect on the distribution of salt in the ocean, which in turn affects the ocean density.

It is not surprising, therefore, to find that the 'climate' described by the model is quite sensitive to the size and the distribution of water exchanges at the interface. Before the model can be used for prediction it has to be run for a considerable time until it reaches a steady 'climate'. The 'climate' of the model, when it is run unperturbed by increasing greenhouse gases, should be as close as possible to the current actual climate. If the exchanges are not correctly described, this will not be the case. Getting the exchanges right in the model has proved to be difficult. Until recently, many coupled models introduced artificial adjustments to the fluxes (known as flux adjustments) so as to ensure that the model's 'climate' is as identical as possible to the current climate. However, the ocean component of the model has been improved especially through introducing higher resolution (150 km or less), so the need for such adjustments has diminished and many models are now able to provide an adequate description of the climate with no such adjustments. Before leaving the oceans, there is a particular feedback which should be mentioned between the hydrological cycle and the deep ocean circulation. Changes in rainfall, by altering the ocean salinity, can interact with the ocean circulation. This could affect the climate, particularly of the North Atlantic region; it may also have been responsible for some dramatic climate changes in the past.

The most important feedbacks belong to the atmospheric and the ocean components of the model. They are the largest components, and, because they are both fluids and have to be dynamically coupled together, their incorporation into the model is highly demanding. However, another feedback to be modelled is the ice-albedo feedback, which arises from the variations of sea-ice and of snow.

Sea-ice covers a large part of the Polar Regions in the winter. It is moved about by both the surface wind and the ocean circulation. So that the ice-albedo feedback can be properly described, the growth, decay and dynamics of sea-ice have to be included in the model. Land ice is also included, essentially as a boundary condition – a fixed quantity – because its coverage changes little from year to year. However, the model needs to show whether there are likely to be changes in ice volume, even though these are small, in order to find out their effect on sea level. Interactions with the land surface must also be adequately described.

The most important properties for the model are land surface wetness or, more precisely, soil moisture content (which will determine the amount of evaporation) and albedo (reflectivity to solar radiation). The models keep track of the changes in soil moisture through evaporation and precipitation. The albedo depends on soil type, vegetation, snow cover and surface wetness.

4.5.1 The ocean's deep circulation

For climate change over periods up to a decade, only the upper layers of the ocean have any substantial interaction with the atmosphere. For longer periods, however, links with the deep ocean circulation become important. The effects of changes in the deep circulation are of particular importance Experiments using chemical tracers, for instance those illustrated in Figure 5.20 (see next box), have been helpful in indicating the regions where strong coupling to the deep ocean occurs. To sink to the deep ocean, water needs to be particularly dense, in other words both cold and salty. There are two main regions where such dense water sinks down to the deep ocean, namely in the north Atlantic Ocean (in the Greenland Sea between Scandinavia and Greenland and the Labrador Sea west of Greenland) and in the region of Antarctica. Salt-laden deep water formed in this way contributes to a deep ocean circulation that involves all the oceans (Figure 5.18) and is known as the *thermohaline circulation* (THC).

Increases in the ice melt can lead to the ocean surface water becoming less salty and therefore less dense. It will not sink so readily; the deep-water formation will be inhibited and the THC is weakened. In Chapter 6, the link between the THC and the hydrological (water) cycle in the atmosphere is mentioned. Increased precipitation in the north Atlantic region, for instance, can lead to a weakening of the THC.

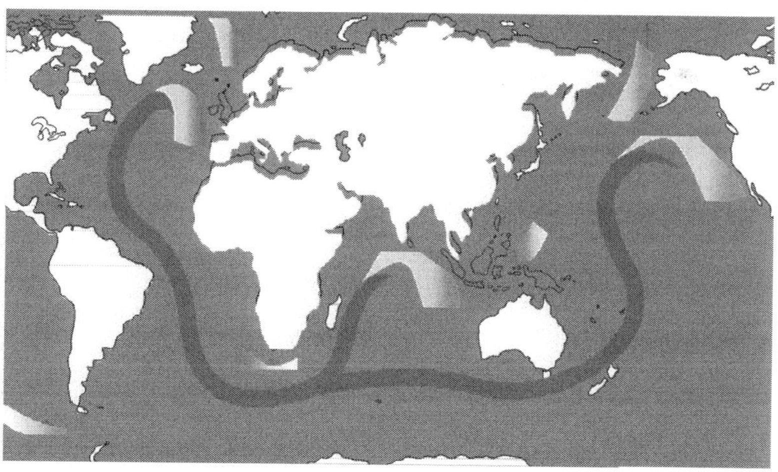

Fig. 18: Deep water formation and circulation. The deep salty current originates in the north Atlantic. Northward flowing water near the surface that is unusually salty becomes cooler and even saltier through evaporation, so increasing its density and causing it to sink.

4.5.2 Validation of the model

In discussing various aspects of modelling we have already indicated how some validation of the components of climate models may be carried out. The successful predictions of weather- orecasting models provide validation of important aspects of the atmospheric component, as do the simulations mentioned earlier in the chapter of the connections between sea surface temperature anomalies and precipitation patterns in some parts of the world. Various tests have also been carried out of the ocean component of climate models; for instance, through comparisons between the simulation and observation of the movement of chemical tracers (see box below).

Once a comprehensive climate model has been formulated it can be tested in three main ways. Firstly, it can be run for a number of years of simulated time and the climate generated by the model compared in detail to the current climate. For the model to be seen as a valid one, the average distribution and the seasonal variations of appropriate parameters such as surface pressure, temperature and rainfall have to compare well with observation. In the same way, the variability of the model's climate must be similar to the observed variability. Climate models that are currently employed for climate prediction stand up well to such comparisons.

Secondly, models can be compared against simulations of past climates when the distribution of key variables was substantially different than at present; for example, the period around 9000 years ago when the configuration of the Earth's orbit around the Sun was different (see Figure 19). The perihelion (minimum Earth–Sun distance) was in July rather than in January as it is now; also the tilt of the Earth's axis was slightly different from its current value (24 rather than 23.5). Resulting from these orbital differences (see Chapter 4), there were significant differences in the distribution of solar radiation throughout the year. The incoming solar energy when averaged over the northern hemisphere was about seven per cent greater in July and correspondingly less in January.

When these altered parameters are incorporated into a model, a different climate results. For instance, northern continents are warmer in summer and colder in winter. In summer a significantly expanded low

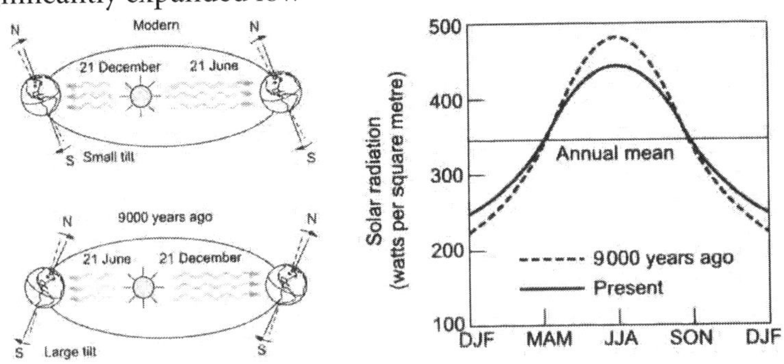

Fig. 19: Changes in the Earth's elliptical orbit from the present configuration to 9000 years ago and (right hand side) changes in the average solar radiation during the year

over the northern hemisphere. these regions are strengthened and there is increased rainfall. These simulated changes are in qualitative agreement with pale climate data; for example, these data provide evidence for that period (around 9000 years ago) of lakes and vegetation in the southern Sahara about 1000 km north of the present limits of vegetation. The accuracy and the coverage of data available for these past periods are limited. However, the model simulations for 9000 years ago, described above, and those for other periods in the past have demonstrated the value of such studies in the validation of climate models.

A third way in which models can be validated is to use them to predict the effect of large perturbations on the climate. Good progress is being achieved with the prediction of El Ni˜no events and the associated climate anomalies up to a year ahead (see earlier in the chapter). Other short-term perturbations are due to volcanic eruptions, the effects of which were mentioned in Chapter 1. Several climate models have been run in which the amount of incoming solar radiation has been modified to allow for the effect of the volcanic dust from Mount Pinatubo, which erupted in 1991 (Figure 21). Successful simulation of some of the regional anomalies of climate which followed that eruption, for instance the unusually cold winters in the Middle East and the mild winters in western Europe, has also been achieved by the models.

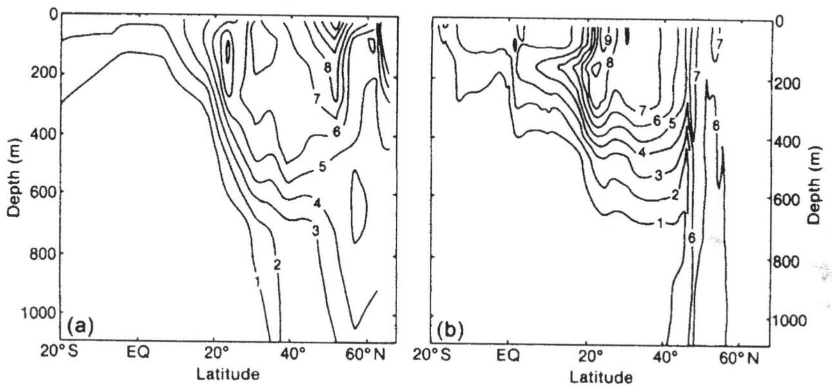

Fig. 20: The tritium distribution in a section of the western north Atlantic Ocean approximately one decade after the major atomic bomb tests, as observed in the GEOSECS programme (a) and as modelled (b).

In these three ways, which cover a range of timescales, confidence has been built in the ability of models to predict climate change due to human activities.

4.5.3 Comparison with observations

More than fifteen centres in the world located in ten countries are currently running climate models of the kind we have described in which the circulations of the atmosphere and the ocean are fully coupled together. Some of these models have been employed to simulate the climate of the last 150 years allowing for variations in aspects of natural forcing (e.g., solar variations and volcanoes) and the increases in the concentrations of greenhouse gases and aerosols.

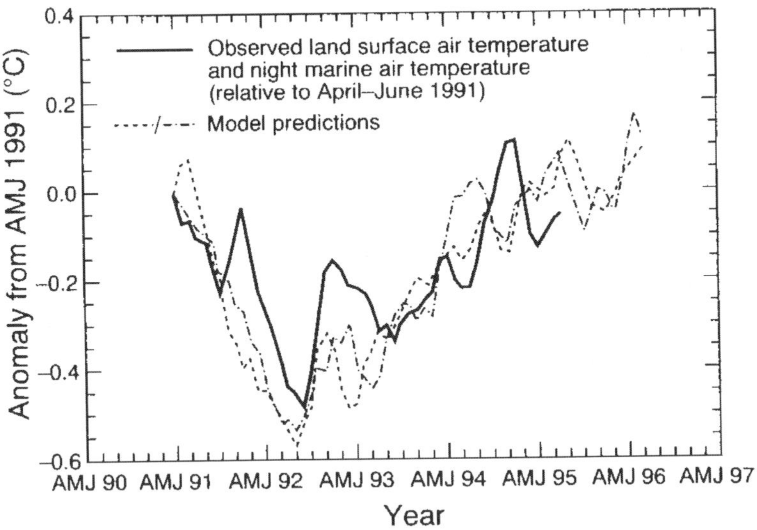

Fig. 21: The predicted and observed changes in global land and ocean surface air temperature after the eruption of Mount Pinatubo, in terms of three-month running averages from April to June1991 to March to May 1995.

An example of such simulations is shown in Figure 22, where the observed record of global average surface air temperature is compared with model simulations taking into account in turn natural forcings, anthropogenic forcings (i.e. the increase in greenhouse gases and aerosols) and the combination of natural and anthropogenic forcings. Although the simulations in Figure 22 are based on one model, similar results have been obtained from many models.

Three interesting features of Figure 22 can be noted. Firstly, that the inclusion of anthropogenic forcings provides a plausible explanation for a substantial part of the observed temperature changes over the last century (especially for the latter part of the century), but that the best match with observations occurs when both natural and anthropogenic factors are included. In particular it is likely that changes in solar output and the comparative absence of volcanic activity were important variations in natural forcing factors during the first part of the twentieth century.

Secondly, the model simulations show variability up to a tenth of a degree Celsius or more over periods of a few years up to decades. This variability is due to internal exchanges in the model between different parts of the climate system, and is not dissimilar to that which appears in the observed record. Thirdly, due to the slowing effect of the oceans on climate change, the warming observed or modelled so far is less than would be expected if the climate system were in equilibrium under the amount of radiative forcing due to the current increase in greenhouse gases and aerosols.

There remains, however, a large amount of natural variability in both the observations and the simulations and much debate has taken place over the last decade or more about the strength of the evidence that global warming due to the increase in greenhouse gases has actually been observed in the climate record. In other words has the 'signal' that can be attributed to global warming risen sufficiently above the 'noise' of natural variability? The Intergovernmental Panel on Climate Change (IPCC) has been much involved in this debate. The IPCC's first Report published in 199016 made a carefully worded statement to the effect that, although the size of the observed warming is broadly consistent with the predictions of climate models, it is also of similar magnitude to natural climate variability. An unequivocal statement that anthropogenic climate change had been detected could not therefore be made at that time. By 1995 more evidence was available and the IPCC 1995 Report17 therefore reached the cautious conclusion as follows.

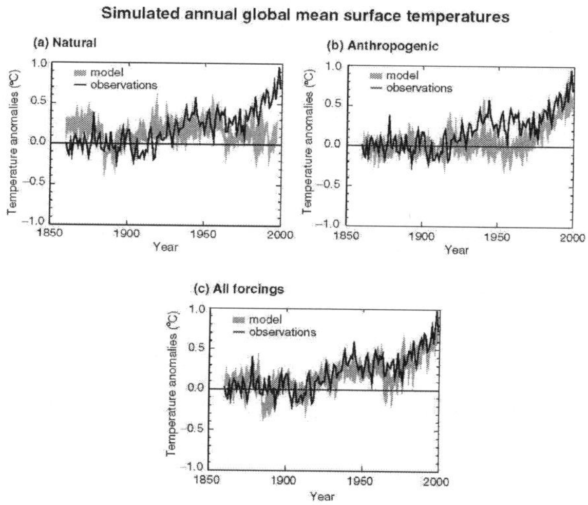

Fig. 22: Annual global mean surface temperatures simulated by a climate model compared with observations for the period 1860–2000. The simulations in (a) were done with only natural forcings – solar variation and volcanic activity; in (b) with only anthropogenic forcings – greenhouse gases and

Sulphate aerosols; and in (c) with both natural and anthropogenic forcings combined. The simulations are shown in a band that covers the results from four runs with the same model and therefore illustrates the range of natural variability within the model.

Our ability to quantify the human influence on global climate is currently limited because the expected signal is still emerging from the noise of natural climate variability, and because there are uncertainties in key factors. These include the magnitude and patterns of long-term natural variability and the time-evolving pattern of forcing by, and response to, changes in the concentrations of greenhouse gases and aerosols, and land surface changes. Nevertheless, the balance of evidence suggests a discernible influence on global climate.

Since 1995 a large number of studies have addressed the problems of *detection* and *attribution* of climate change. Better estimates of natural variability have been made, especially using models, and the conclusion reached that the warming over the last one hundred years is very unlikely to be due to natural variability alone. In addition to studies using globally averaged parameters, there have been detailed statistical studies using pattern correlations based on optimum detection techniques applied to both model results and observations. For example, Figure 23 shows a comparison between simulated and observed estimates of zonal mean temperature change as a function of altitude. Taking these studies into account, the conclusion reached in the IPCC 2001 Report20 is as follows.

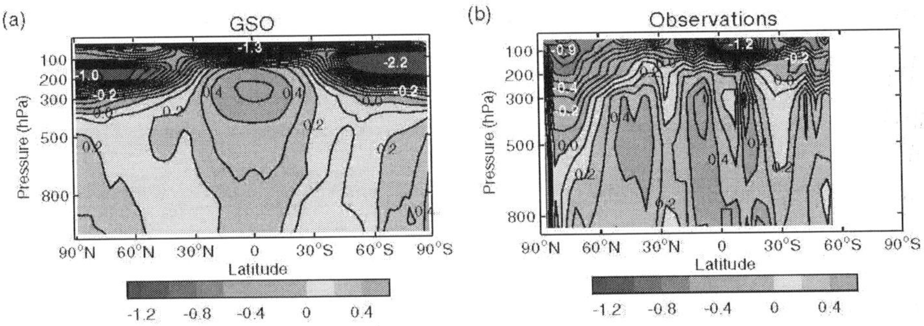

Fig. 23: Simulated and observed zonal mean temperature change as a function of latitude and height. The changes plotted are the differences between the 1986–1995 decadal average and the twenty-year average from 1961 to 1980. The contour interval is 0.1 C. (a) simulated changes taking into account increases in carbon dioxide, sulphate aerosols and the effect of observed changes in stratospheric ozone; (b) observed changes. A common pattern of stratospheric cooling and tropospheric warming is evident in the observations and both model experiments; the stratospheric cooling is partially due to the increase of carbon dioxide and the reduction in ozone.

In the light of new evidence and taking into account the remaining uncertainties, most of the observed warming over the last 50 years is likely to have been due to the increase in greenhouse gas concentrations. Confidence having been established in climate models in the ways we have outlined in the last two sections, these models can now be used to generate projections of the likely climate change in the future due to human activities.

Before leaving comparison with observations, I should mention some recent work relating to the warming of the ocean that adds further confirmation to the picture that has been presented. The effect of an increase of greenhouse gases was expressed in terms of radiative forcing or, in other words, a net input of heat energy into the earth-atmosphere system. Most of this extra energy is stored in the ocean. The amount of this extra energy has been estimated from measurements of the temperature increase in the ocean at different locations and depths down to 3 km. From 1957 to 1994 the estimate is $(19 \pm 9) \times$ 1022 joules, the uncertainty relating mostly to inadequate sampling of some large ocean areas. This amount is a little over 0.5% of the solar energy received by the Earth over this period. Within the limits of uncertainty, it agrees well with the estimates of radiative forcing presented earlier and also with model estimates of ocean heat uptake.

Throughout this chapter the implicit assumption has been made that climate change is predictable and that models can be used to provide predictions of climate change due to human activities. The capability of the models themselves has been demonstrated so far as weather forecasting is concerned. They also possess some skill in seasonal forecasting. They can provide a good description of the current climate and its seasonal variations. Further, they provide predictions which on the whole are reproducible and which are reasonably consistent between different models. But it might be argued, this consistency could be a property of the models rather than of the climate. Is there any other evidence to support the view that the climate is predictable? Correlation between the Milankovitch cycles in the Earth's orbital parameters and the cycles of climate change over the past half million years provides strong evidence to substantiate the Earth's orbital variations as the main factor responsible for the triggering of climate change. The nature of the feedbacks which control the very different amplitudes of response to the three orbital variations still need to be understood.

Some 60 ± 10% of the variance in the record of global average temperature from paleontological sources over the past million years occurs close to frequencies identified in the Milankovitch theory. The existence of this surprising amount of regularity suggests that the climate system is not strongly chaotic so far as these large changes are concerned, but responds in a largely predictable way to the Milankovitch forcing.

This Milankovitch forcing arises from changes in the distribution of solar radiation over the Earth because of variations in the Earth's orbit. Changes in climate as a result of the increase of greenhouse gases are also driven by changes in the radiative regime at the top of the atmosphere. These changes are not dissimilar in kind (although different in distribution) from the changes that provide the Milankovitch forcing. It can be argued therefore that the increases in greenhouse gases will also result in a largely predictable response.

4.6 REGIONAL CLIMATE MODELLING

The simulations we have so far described in this chapter are with global circulation models (GCM) that typically possess a horizontal resolution (grid size) of around 300 km – the size being limited primarily by the availability of computer power. Weather and climate on scales large compared with the grid size are described reasonably well. However, at scales comparable with the grid size, described as the regional scale, the results from global models possess serious limitations. The effects of forcings and circulations that exist on the regional scale need to be properly represented. For instance, patterns of precipitation depend critically on the major variations in orography and surface characteristics that occur on this scale. Patterns generated by a global model therefore will be a poor representation of what actually occurs on the regional scale.

To overcome these limitations regional modelling techniques have been developed. That most readily applicable to climate simulation and prediction is the *Regional Climate Model* (RCM). A model covering an appropriate region at a horizontal resolution of say 25 or 50 km can be 'nested' in a global model. The global model provides information about the response of the global circulation to large-scale forcings and the evolution of boundary information for the RCM. Within the region, physical information, for instance concerning forcings, is entered on the scale of the regional grid and the evolution of the detailed circulation is developed within the RCM. The RCM is able to account for forcings on smaller scales than are included in the GCM (e.g. due to topography or land cover inhomogeneity, see Figure 24) and can also simulate atmospheric circulations and climate variables on these smaller scales.

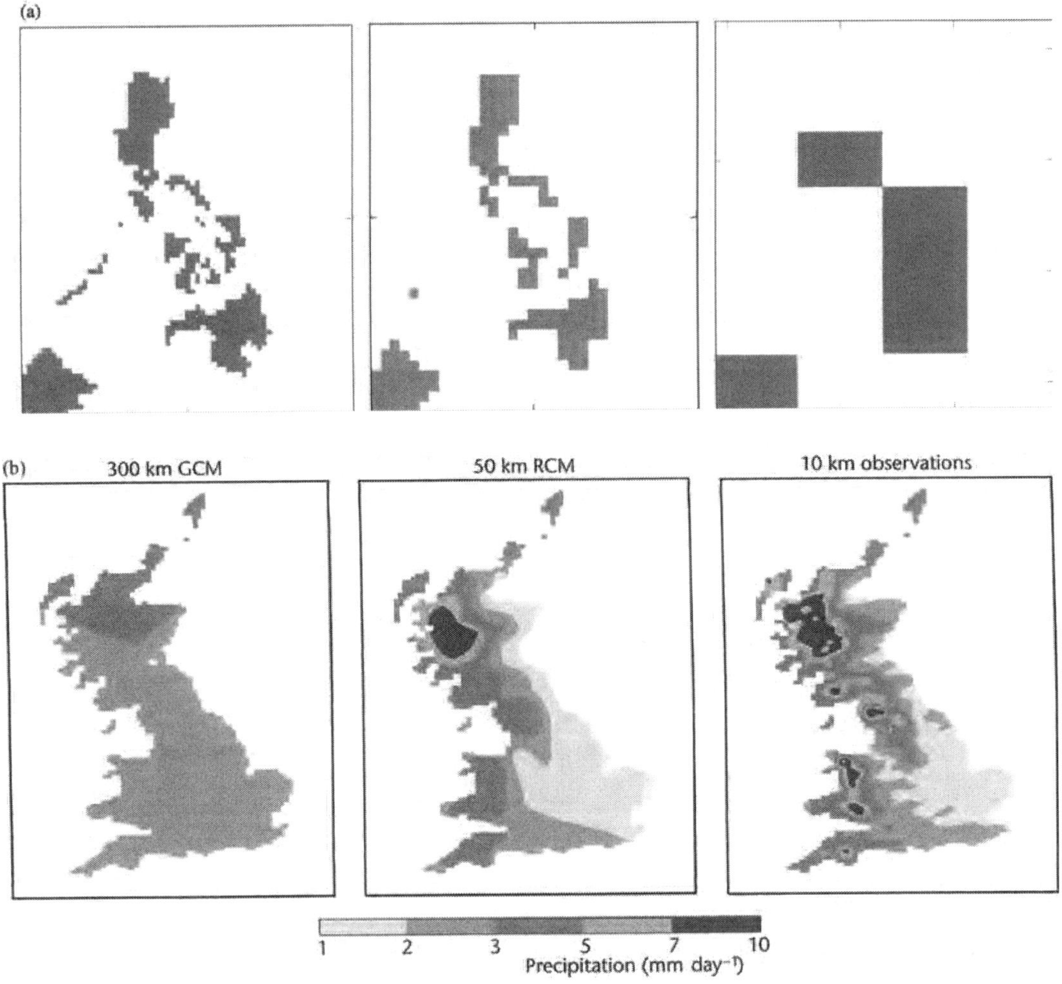

Fig. 24: (a) Representation of the Philippines in RCMs with resolutions of 25 km and 50 km and in a GCM with 400 km resolution; (b) patterns of present-day winter precipitation over Great Britain, (i) as simulated with a 300 km resolution global model, (ii) with 50 km resolution regional model, (iii) as observed with 10 km resolution.

A limitation of the regional modelling technique we have described is that, although the global model provides the boundary inputs for the RCM, the RCM provides no interaction back on to the global model. As larger computers become available it will be possible to run global models at substantially increased resolution so that this limitation becomes less serious; at the same time RCMs will acquire an ability to deal with detail on even smaller scales.

Another technique is that of *Statistical Downscaling* that has been widely employed in weather forecasting. This uses statistical methods to relate large scale climate variables (or 'predictors') to regional or local variables. The predictors from a global circulation climate model can be fed into the statistical model to estimate the corresponding regional climate characteristics. The advantage of this technique is that it can easily be applied. Its disadvantage from the point of view of simulating climate change is that it is not possible to be sure how far the statistical relations apply to a climate-changed situation.

4.7 THE FUTURE OF CLIMATE MODELLING

Very little has been said in this chapter about the biosphere. The large three dimensional global circulation climate models described in this chapter contain a lot of dynamics and physics but no interactive chemistry or biology. As the power of computers increases, global dynamical and physical circulation models that couple in the biological and chemical processes that make up the caron cycle and the chemistry of other gases are now being developed. Before very long we can expect that models will be available that are fully interactive and comprehensive in their inclusion of dynamical, physical, chemical and biological processes in the atmosphere, the ocean and on the land. Climate modelling continues to be a rapidly growing science. Although useful attempts at simple climate models were made with early computers it is only during the last ten years or so that computers have been powerful enough for coupled atmosphere–ocean models to be employed for climate prediction and that their results have been sufficiently comprehensive and credible for them to be taken seriously by policy makers. The climate models which have been developed are probablythe most elaborate and sophisticated of computer models developed in any area of natural science. Further, climate models that describe the natural science of climate are now being coupled with socio-economic information in integrated assessment models.

As the power of computers increases it becomes more possible to investigate the sensitivity of models by running a variety of ensembles that include different initial conditions, model parameterisations and formulations. A particularly interesting project involves thousands of computer users around the world in running state-of-the art climate prediction models on their home, school or work computers. By collating data from thousands of models it will generate the world's largest climate modelling prediction experiment. A great deal remains to be done to narrow the uncertainty of model predictions. The first priorities must be to improve the modelling of clouds and the description in the models of the ocean–atmosphere interaction. Larger and faster computers are required to tackle this problem, especially to enable the resolution of the model grid to be increased, as well as more sophisticated model physics and dynamics. Much more thorough observations of all components of the climate system are also necessary, so that more accurate validation of the model formulations can be achieved. Further, regional climate modelling techniques will develop rapidly as they are applied to a wide variety of situations. Very substantial national and international programmes are underway to address all these issues.

5

THE IMPACT OF CLIMATE CHANGE ON THE AGRICULTURAL SECTOR

5.1. INTRODUCTION

Climate change refers to changes beyond the average atmospheric condition that are caused both by natural factors such as the orbit of earth's revolution, volcanic activities and crustal movements and by artificial factors such as the increase in the concentration of greenhouse gases and aerosol. Climate change by global warming, which refers to the average increase in global temperature, has become a megatrend that will lead to significant global changes in the future. Concerning its impacts, the UN Intergovernmental Panel on Climate Change (IPCC) presented considerable scientific evidences in its fourth report on climate change (2007) and they have become clearly recognized worldwide. In addition, people have become more aware of the fact that global warming cannot be avoided due to the continued increase in greenhouse gas emissions and the changes in the climate system. The Club of Rome Report 1972 officially raised global warming as an international issue and, in 1985, World Meteorological Organization (WMO) and United Nations Environment Programme (UNEP) officially declared carbon dioxide as the principal cause of global warming. In order to effectively cope with the global warming issue, Intergovernmental Panel on Climate Change (IPCC) was organized in 1988 and has carried out systematic research and in-depth studies on climate change.

According to the fourth report of UN IPCC (2007) on climate change, it is indisputable that global warming has serious impacts on the earth and it is very likely that the increase in greenhouse gas emission by anthropogenic activities has caused global warming since the mid-20th century. Especially, this report warns us that, if mankind continues its present level of consumption of fossil fuels (e.g., oil and coal), the average temperature of the earth will rise by up to 6.4°C by the end of the 21st century (2001~2100) and the sea level will rise by 59㎝. In fact, the average temperature of the earth has risen 0.74°C over the past 100 years (1906~2005) (Korea Meteorological Agency, 2008).

Global warming not only causes a change in average temperature and precipitation but also increases the frequency of floods, droughts, heat waves, and the intensity of typhoons and hurricanes following the change in temperature and precipitation patterns. The impacts of climate change are also shown in various other forms throughout the world, including the rise of sea level, decrease in glaciers, northward movement of plant habitats, changes in animal habitats, rise of ocean temperature, shortened winter and early arrival of spring.

As the acceleration of global warming affects not only ecological systems but also human life, it has become an important issue both nationally and internationally. Approaches to deal with the issue of global warming are divided largely into mitigation measures, focusing on reduction and absorption of greenhouse gases, the causative factors, and adaptation measures to minimize the damages by climate change. So far, the global warming issue has focused on the mitigation of greenhouse gases based on international environmental conventions such as IPCC and Kyoto Protocol. For agriculture, however, the focus has been shifted to adaptation and adaptability based on the assessment of the impacts of climate change and vulnerability to it. IPCC emphasizes that it is very important for the agricultural sector to adapt to climate change. This is because even if greenhouse gas emissions decrease, global warming will still continue for the next several decades due to its previously emitted greenhouse gases. It takes at least 5 to 10 years to assess the impacts of climate change and the vulnerability to it and prepare proper countermeasures against it. Especially, as agriculture is climate-dependent and thus susceptible to climate change, it is very urgent to prepare adaptation measures against climate change. Proper countermeasures drawn based on scientific diagnosis and assessment of the impacts of climate change on East Asian countries' agriculture are essential in establishing the vision and administrative policies of future agriculture. This will also provide valuable information for local governments in establishing mid to long-term agricultural development plans and for farming households to prepare their production plans.

5.2 DIAGNOSIS AND FORECAST OF GLOBAL CLIMATE CHANGE

5.2.1 Current Conditions of Global Warming

Climate refers to a long-term variation in the atmospheric condition of a specific region or regions, and climate change means a gradual change in the climate system both by natural and artificial causes. Climate change is caused by the change in each component of the climate system such as atmosphere, hydrosphere, biosphere, cryosphere and lithosphere or by complicated interactions among those components. The causes of climate change are largely divided into natural causes and artificial causes. Natural causes include the change in solar activity, volcanic eruption, sea water temperature, ice cap distribution, westerly waves and atmospheric waves. On the other hand, artificial causes include carbon dioxide emission from industry and agricultural production activities, deforestation, acid rain and the destruction of the ozone layer by Freon gas, with global warming by the increase of greenhouse gases as the representative. Global warming refers to the average increase of the Earth's temperature due to the greenhouse effect caused by carbon dioxide (CO_2), methane (CH_4), nitrous oxide (N_2O), hydro fluorocarbon (HFCs), perfluorocarbon (PFCs) and sulfur hexafluoride (SF_6). 1 Global warming, meaning a continuous increase of the Earth's temperature due to the greenhouse effect, started from the time of the Industrial Revolution which was accompanied by a rapid increase of fossil fuel consumption. This issue has attracted international interests as the scientific knowledge of climate has accumulated since the 1970s and it has been widely accepted by scientists that the anthropogenic greenhouse gas emissions are the cause of global warming. The global greenhouse gas concentration based on carbon dioxide is estimated to have increased from 280ppm before the Industrial Revolution (1750) to 379ppm in 2005. According to an analysis of the average temperatures of the Earth (Climate Research Unit, 2009), the increase of the Earth's average temperature so far, since the Industrial Revolution, appears to be much higher than the increase before the Industrial Revolution. Specifically, global warming has significantly accelerated since 1980 and the average temperature of 1998 was shown to be 0.58°C higher than the average temperatures of 1960~1990. As shown in the Figure 1, 11 out of 12 hottest years since 1850 were recorded to be in the last 12 years.

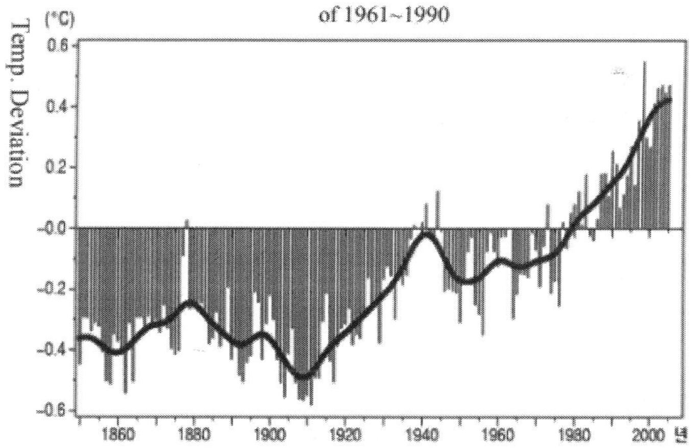

Source: Climate Research Unit (2006).

Fig. 1: Temperature deviation from the average temperature

https://www.unescap.org/sites/default/files/5.%20The-Impact-of-Climate-Change-on-the-Agricultural-Sector.pdf(LINK FOR DTP Fig.)

To make a systematic and reliable diagnosis of global warming, scientific analysis of climate change has been periodically made by IPCC since 1990. So far, IPCC has published its First (1990), Second (1995) and Third (2001) Assessment Report and its Fourth Assessment Report was being prepared as its Working Group I (Physical Science of Climate Change), Working Group II (Impacts, Adaptation, Vulnerability) and Working Group III (Mitigation of Climate Change) announced their reports in April 2007 (IPCC, 2007).[2]

IPPC WGI Report, which was prepared based on physical science, suggests that the atmosphere's carbon dioxide concentration has increased by about 1.4 times (379ppm in 2005) over the past 100 years, in comparison to the pre-industrialization concentration (280ppm). Accordingly, it is estimated that the average global temperature has risen 0.74°C (0.56~0.92°C) over the past 100 years (1906~2005) . Especially, the average temperature of the Northern hemisphere in the late 20th century, which appears to be the highest temperatures recorded since 1850 and the temperature rise during the last 20 years is shown to be more than twice that of the past 100 years. This report states that there is no doubt that global warming is occurring in the climate system and affirms that greenhouse gases are an artificial cause of global warming.[3]

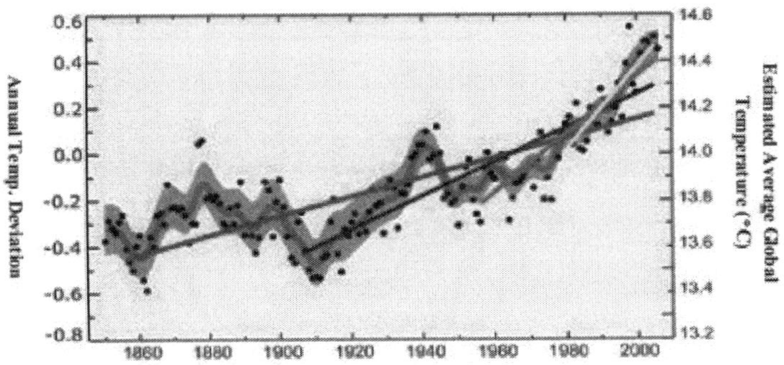

Source: IPCC(2007)

Fig. 2: Average global temperature and change trend

As global warming continues, the temperature of the North Pole and the South Pole have risen, accelerating the rate of ice cap melting, shortening the ice-breaking period in the polar lakes and thus causing a significant rise in sea level. Furthermore, global warming causes extreme climatic phenomena such as flood, drought and heat waves, increasing the occurrence of natural disasters worldwide.

5.2.2 Current Conditions and Forecasts of Global Climate Change

As global climate change is affected by various factors such as regional characteristics, socioeconomic variables and meteorological variables, it is forecasted by making several feasible scenarios. The IPCC Assessment Report provides greenhouse gas emission scenarios by forecasting the change in greenhouse gas concentration according to demographics and socioeconomic development. The Special Report on Emission Scenario

(SRES) presents four main scenarios (A1, A2, B1, and B2) and three other scenarios (A1F, A1T, A1B) modified according to their technological emphasis in the A1 scenario .

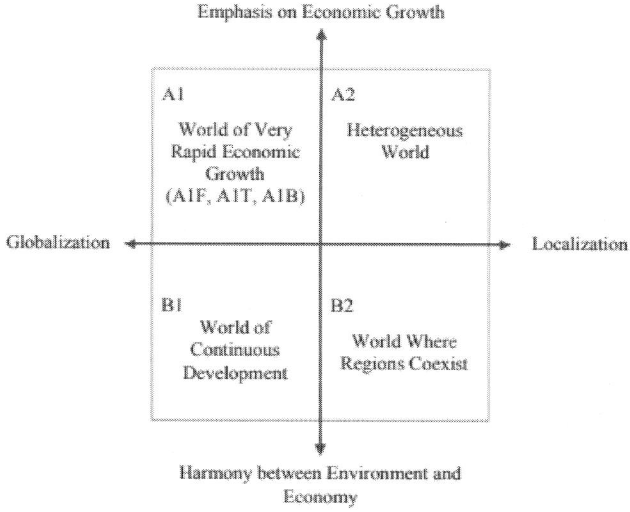

Fig. 3: Conceptual diagram of scenarios for estimating climate change

The A1 scenario assumes a very-rapid economic growth, in which the rapid growth of the global economy and population peaks in 2050 and declines thereafter, in which then new efficient technologies are introduced. It is divided into three groups according to the alternative development of energy technology. The three scenarios are the fossil intensive scenario (A1F1), non-fossil energy scenario (A1T), and balanced-energy source scenario (A1B).

A2 is the scenario for a heterogeneous world with a high population growth rate, a low economic growth rate, and the most diversified but slowly developing technologies. The B1 scenario assumes the same population growth rate as that of the A1 scenario but at a lower economic growth rate. In this scenario, the economic structure changes toward a service and information economy and the sustainable development is pursued with an emphasis on clean and resource-efficient technologies.

Table 1. Estimated temperature rise in 2100 under each scenario

Scenario	Temperature Change (C)		Sea Level Rise
	Optimal Estimation	Expected Range	(cm)
Very Rapid economic growth (AIFI)	4.0	2.4~6.4	26~59
Non-fossil intensive enegy(AIT)	2.4	1.4~3.8	20~45
Heterogeneous world (A2)	2.8	1.7~4.4	21~48
Continuous development (B1)	3.4	2.0~5.4	23~51
	1.8	1.1~2.9	18~38
	2.4	1.4~3.8	20~43

Source: IPCC (2007), p.8

Estimates of future climate change vary greatly from scenario to scenario. In the continuous development scenario (B1), in which environmental conservation and economic development are compatible with each other, temperature change is estimated to be about 1.8°C (1.1~2.9°C), while the rise of about 4.0°C (2.4~6.4°C) is expected under the very rapid economic growth scenario (A1) based on fossil-intensive energy sources. By 2030, however, it is estimated in all scenarios that the temperature will rise at the rate of 0.2°C for every 10 years.

Figure 4. Estimated trend of temperature rise under each scenario

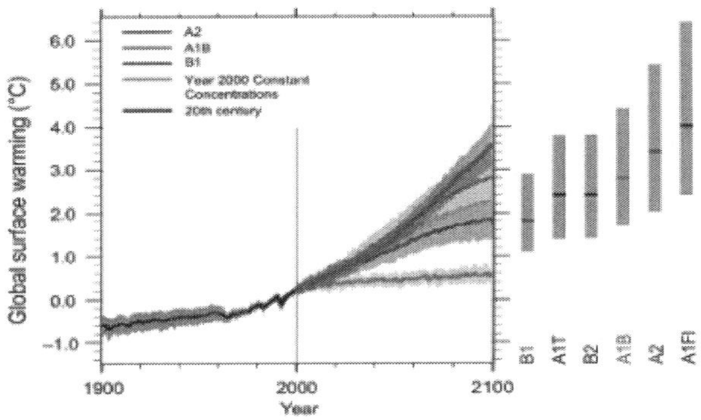

Source: IPCC (2007).

Fig. 4:Estimated trend of temperature rise under each scenario

According to the fourth IPCC Assessment Report, it is anticipated that the impacts of climate change will vary greatly with the degree of temperature rise and the latitudinal locations. When the temperature rise is less than 1°C, damages by natural disasters such as water shortage and floods are foreseen in some areas. However, the report warns that if the temperature rises by 2~3°C, most areas will be subject to damages by natural disasters and about 20~30% of animals and plants will be subject to endangerment. Furthermore, if the temperature rises by more than 3°C, substantial economic and environmental damages are expected, including chronic water shortages, ecosystem destruction, reduced food production, and increased occurrences of diseases.

5.3. IMPACTS OF CLIMATE CHANGE ON THE AGRICULTURAL SECTOR

5.3.1 Conceptual Approach

Agricultural production is carried out through the selection of crops suitable for the climate of a specific region and application of proper farming methods. Therefore, agriculture is a climate dependent bio-industry with notable regional characteristics. Regional characteristics refer to the ecosystem characteristics determined by the climate of the region. Climate change disturbs the agricultural ecosystem, resulting in the change in agricultural climatic elements such as temperature, precipitation, and sunlight, while further influencing the arable, livestock, and hydrology sectors. The flow of the impacts of climate change on the agricultural sector can be illustrated as shown in Fig. 5 . First of all, the impacts of climate change on the arable and livestock sector are made known by biological changes including the change of flowering and harvesting seasons, quality change, and shift of areas suitable for cultivation.5 Climate change affects the agricultural ecosystem, giving rise to blights and pests and causing population movement and change in biodiversity. In the livestock sector, climate change brings about biological changes in areas such as fertilization and breeding and also affects the growing pattern of pastures.

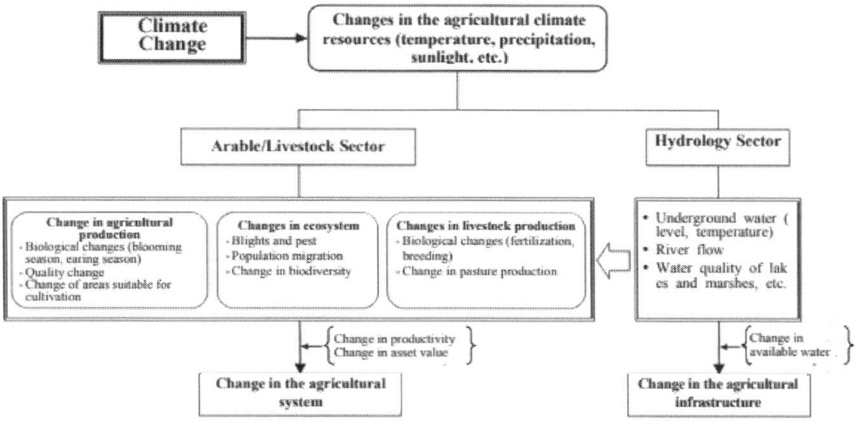

Source: Kim, Chang-Gil and et al. (2009), p.36.

Fig.5: Flow of the climate change impact on the agricultural sector

Climate change affects the hydrology including underground water level, water temperature, river flow, and water quality of lakes and marshes, by impacting precipitation, evaporation, and soil moisture content. In particular, the increase of precipitation by climate change leads to an increase of outflow while the temperature rise increases evaporation, resulting in the reduction of outflow. In order to understand the quantitative impacts of climate change on water resources, a deterministic hydrology model, based on the general circulation model, is used.

As illustrated above, climate change has a wide range of impacts on the rural economy including agricultural productivity, revenues of the farm household and asset values, and it also affects the agricultural infrastructure through the change in water sources available for agriculture.

So far, the quantitative analyses on the impacts of climate change on the agricultural sector have been experimental centering on cross-sectional analysis. The experimental analyses are carried out on the basis of agro-economic simulation models. They are similar to the controlled experiments in which related variables are regulated, in that variables related to greenhouse gases such as temperature levels and carbon dioxide emissions levels are regulated. In these experiments, the impact of climate change on agricultural production can also be estimated.

Agro-ecological zone analysis is carried out by using the crop simulation model (called the crop model for short) that tracks the changes in agricultural production and agro-ecological zones that have resulted from climate change. Crop growth is determined by the combined action of three elements, those being the genetic characteristics of crop, cultivation technology, and environment (climate, soil, etc.). The crop model refers to a computer program that can estimate the crop growth and its quantity when these three elements are entered. Using the crop model, it is possible to estimate and analyze agricultural production under climate change. The Crop estimation through the Resource and Environment Synthesis (CERES) model developed in the USA by integrating the crop model and the resource environment can assume a certain situation that is likely to happen and forecast its possible results.

To analyze how and to what extent the change in temperature and precipitation following global warming affects the agricultural sector various experiments, simulations and research methods are carried out both in laboratories and in the field. As the impacts of climate change on the agricultural sector vary with the related variables, it is difficult to generalize certain analytical results. Therefore, what is attempted here is to classify the impacts of climate change into positive and negative ones based on the results that researches have gathered thus far in related fields.

The positive impacts of global warming include the increase in crop productivity due to fertilization effect caused by the increase in carbon dioxide concentration in the atmosphere, expansion of the areas available for production of tropical and/or subtropical crops, expansion of two-crop farming due to the increased cultivation period, reduction of damages of winter crops by low temperature, and reduction of heating cost for agricultural crops grown in the protected cultivation facilities.

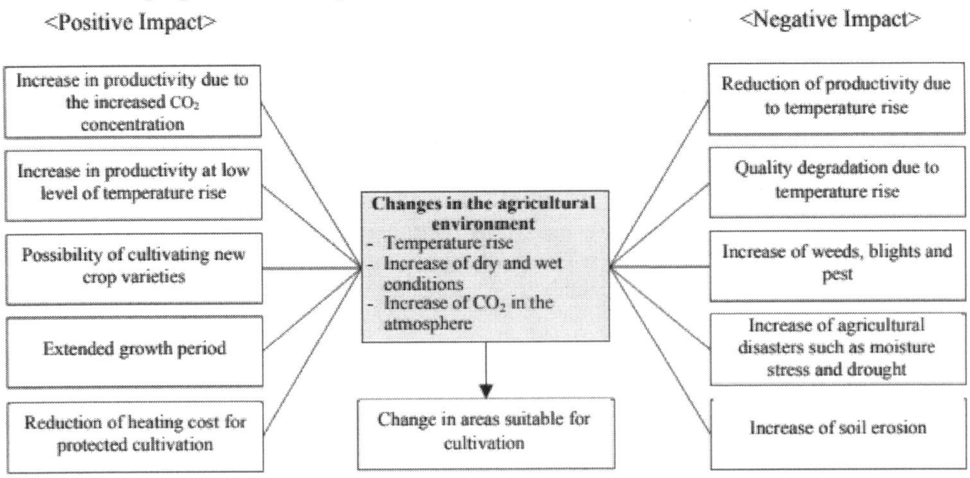

Source: Kim, Chang-Gil and et al. (2009), p.38.

Fig. 6: Potential impacts of global warming on the agricultural sector

Negative impacts of global warming include reduced crop quantity and quality due to the reduced growth period following high levels of temperature rise; reduced sugar content, bad coloration, and reduced storage stability in fruits; increase of weeds, blights, and harmful insects in agricultural crops; reduced land fertility due to the accelerated decomposition of organic substances; and increased soil erosion due the increased rainfall.

In addition, each crop requires different climate and environmental conditions to grow. So, if climate change like temperature rise occurs, the boundary and suitable areas for cultivation move north and thus the main areas of production also change. The change in the main areas of production might be as a crisis for certain areas but might be an opportunity for other areas, so it cannot be classified either as a positive or as a negative impact. In sum, the impacts of climate change on the agricultural sector have ambivalent characteristics of positive impacts creating opportunities and of negative impacts with costs. Therefore, it is very important to formulate adaptation strategies that can maximize the opportunities and minimize the costs that will lead to sustainable agriculture development.

5.3.2 Climate Change Impacts on Korean Agriculture[6]

5. 3.2.1 Impacts on the Agricultural Ecosystem

An agricultural climatic area is divided into agricultural climate zones according to different agricultural climatic conditions. Average temperature of each agricultural climate zone has risen by 0.95°C for the past 35 years (1973~2007). The climate zone that has shown the lowest temperature rise is the Yeongnam inland/mountain zone , which recorded a 0.2°C rise. On the other hand, the mid-northern inland, the central inland, and the southeastern coastal zones had seen a temperature rise of 1.36~1.47°C. Especially, the average temperature of the mid-western planar zone and the Charyeong southern planar zone; the granaries of Korea rose by 1.05~1.33°C and the Taebaek highland zone, including Daegwanryeong, also experienced a temperature rise of 1.04°C.

Meanwhile, precipitation had increased by 283㎜ on average for the past 35 years (1973~2007). The Taebaek highland zone including Daegwanryeong, the Taebaek semi-highland zone, and the Yeongnam inland/mountain zone had the greatest increased rate of 452~778㎜, while the Yeongnam inland zone had the lowest increased rate of 132㎜. The increase of precipitation tends to have increased rapidly in summer but decreased slightly in winter since the 1970s.

Temperature increase by global warming has given rise to new types of blights and pests, causing damages to crops. Especially, damages to apples, peaches, grapes, and soybeans, by brown grasshoppers, are reported to have been increasing.

5.3.2.2 Impacts on Agricultural Production

For rice, the cultivation period is the basic condition for planning its production, which is decided by the climate conditions and the rice variety. Among several agricultural climate conditions, temperature is the critical factor in deciding the rice cultivation period. In general, rice is a summer crop and when the temperature rises, the area available for cultivating rice extends north and the variety and cultivation method also changes for adapting to the temperature change. For rice varieties for transplantation, the cultivation regions suitable for early-maturing variety rice will become suitable for medium-maturing variety rice and those for medium-maturing variety will become suitable for late-maturing variety. It is reported that even the mountain areas at an altitude of 600m or higher, where rice cultivation has not been possible due to low temperature, may also become suitable for cultivating some early-maturing varieties of rice.

According to the meteorological data for Korea, the proper season for earning of rice (average temperature during the ripening period: 21~23°C) was around August 15 in the 1970s but in the 2000s, it was delayed about a week to August 21. The average temperature of 21~23°C during the ripening period is favorable for the production of high-quality rice. If the average temperature during this period is higher than this range, rice cannot ripen fully. As a result, grains weigh less, contain more protein, and become less tasty and nutritious. Temperature higher than the average temperature during the ripening period results in the production of poor-quality rice.

Suitable areas for cultivating barley have been selected so as to avoid damages by severe cold during the cultivation period. At present, barley is cultivated in such areas as Donghae and Yeongduk along the east coast, Sacheon and Buseong along the south coast, and Yeonggwang and Gunsan along the west coast, but rarely cultivated in midland areas. The reason for such a shift in the suitable cultivation region for barley is because the coastal areas are less cold than the midland areas in winter, the temperature during the ripening period for barley is lower than the midland, and thus grains can ripen to their full weight. It is analyzed that, as 'mild winter' continued from 1987 to 2000 in which temperatures remained within the range between 1.5~2.5°C, the limit line for cultivating winter barley has been readjusted. According to the analysis of the average and lowest temperatures in January, for 12 years from 1987 to 1998, done for the study on winter barley cultivation regions following global warming by the National Academy of Agricultural Science (NAAS, 2000), the safe zone for winter barley cultivation has shifted far north.

Hot-temperature fruits and vegetables that require hot temperatures to grow such as watermelons, peppers and tomatoes, grow faster and have better quality including higher sugar content, as temperatures rise until it reaches the growth inhibition limit (35°C). On the contrary, for open-field vegetables that favor cool temperatures such as radish and Korean cabbage, high temperatures may result in lowered quality. With regards hot peppers, at temperature below 15°C or above 30°C, the plant pollens are not properly set to the fruit, many of the fruits drop despite being set. Strawberries also have flowering problems at high temperatures. Some vegetables like onions, green onions and lettuce may have problems at high temperatures, as it causes flowers to split. When the temperature rises, it is possible to save energy for heating greenhouses in winter. Also, it becomes possible to cultivate winter cabbages, which have been cultivated in greenhouses in the subtropical island of Jeju and in the fields of the southern coastal areas.

Climate change not only affects the growth of fruit trees but also their quality, harvest time and storage. Apple trees are perennials that can produce fruits for a long period of time over 10 years in the same place once they are planted. Therefore, the change in climatic conditions significantly affects the productivity and quality of apples. The average annual temperature in the areas in Korea where apples are cultivated is lower than 13.5°C. It is known that if the temperature goes higher than this temperature it is hard to produce good-quality apples. The regions suitable for cultivating apples should have an average annual temperature of 13°C or less and the winter temperature characteristics of midlands or basins. Due to global warming, the regions suitable for apple cultivation have shifted north and/or to the highlands. The cultivation regions for pears, peaches, grapes, and sweet persimmons have also shifted north and some areas in the south region have become unsuitable for cultivating those fruits due to high temperatures. Also, due to temperature rise in the south region, where there is not much wind, the cultivation of kiwi fruits has become popular, and in Jeju Island, subtropical fruits are being cultivated.

5.3.2.3 Impacts on the Agricultural Economy

The agro-economic model, which uses the crop production function for analyzing the economic impact of climate change, creates a problem of underestimation as it does not reflect the indirect impact of climate change such as crop conversion and adjustment of input factors for adaptation to climate change. As suggested in the Introduction of this paper, the Ricardian model was developed to solve this problem (Mendelsohn, Nordhaus, and Shaw, 1994). This model assesses the economic impact of climate change by estimating the current value of farmland price as the discounted value of future rent. Basically, it assumes that, in a long-term balanced state in which all production elements that change along with climate change, farmland price represents the quasi-rent which is the profit from utilizing the farmland.7 The Ricardian model has an advantage with regard to climate change impact assessment as it can include adaptation that cannot be accurately measured or identified. It measures the change in farmland value or revenue by taking into consideration both direct impacts, such as change in crop productivity due to climate change and indirect impacts, such as replacement effect of input production factors and change in farmland utilization, so that it is widely used for analyzing the economic impact of climate change (Kim Chang-gil and et al, 2008).

To forecast climate change, 27 cities and counties that are main production regions of Korean cabbage, radish, red pepper, garlic, apples, and pears were analyzed in addition to main rice production regions such as Gimje and Dangjin. On the assumption that climate information of all observatories affects the climate of each city and county, all observatories that had continued providing climate information between 1988~2007 were included in our analysis, which were 57 out of 79 observatories in Korea. To estimate the long-term impact of climate change on agriculture, the average of 20-year data from each observatory was used and to investigate the seasonal impact of climate change on agriculture, climate data about January, April, July, and October was applied. It was based on the assumption that the temperature in January affected the occurrence of blights and pests in April, and temperature and precipitation in July had a significant impact on crop harvests in October.

To apply the Ricardian model, socioeconomic variables and soil and geographic data were used as well as climate variables. For farmland price, the internal data from the Korea Rural Economic Institute was used. In order to prevent bias in farmland price, the average price of rice fields and dry fields only in the agricultural development regions was used. Agricultural gross income and crop income data was taken from KOSTAT data by provinces as there was no municipal data. GDP per capita in each region was also taken from the provincial data provided by KOSTAT. Data about the ratio of fields damaged by sea, wind or water, ratio of wet fields, inclination, and drainage grade was taken from Korean Soils Information.

For estimations using the Ricardian model, regression analysis was carried out by setting the algebraic linear function for farmland price with climate variable, square of climate variable, socioeconomic variable, and soil variable. The farmland price involves future investment value depending on its location in addition to future value of agricultural income from the farmland. Therefore, in order to differentiate the influences of farmland price in a specific region and of independent variables on the entire land price, weighted regression analysis approach was applied that carried out regression analysis by weighting all variables with farmland area or crop income. The regression analysis was carried out in three types: regression analysis only for the farmland price weighted with farmland area, using climate variables (Model 1); regression analysis including other variables (Model 2); and regression analysis that weighted variables with crop income (Model 3). Though there might have been differences to some extent, it appeared that all three models produced similar results for climate variables. The analytical result showed that the rise of average annual temperature had a negative impact on farmland prices but that the average annual/monthly precipitation had a positive impact. When climate variables and other independent variables were included, it appeared that both the result of weighting variables with farmland area and that of weighting them with crop income had a nonlinear relationship. When the quadratic term for the climate variable had a positive value, it meant that it was at the lowest climate condition for cultivation and thus the farmland value went up if the region was above that climate condition. On the contrary, when the quadratic term had a negative value, it meant that there was an optimum climate condition at which the farmland value was highest.

Other than climate variables, independent variables appeared to have the same sign consistently in both models. It was shown that both the per capita production and population growth rate of a region positively affected the farmland price. In other words, it can be interpreted that as income and population grows, the demand for farmlands increased resulting in the rise in farmland price. On the other hand, soil erosion and inclination, both considered negative factor in agriculture appeared to have a negative impact on the farmland price. Contrary to expectations, the ratio of frequently damaged lands had a positive impact on the farmland price. This is likely because, though 53 out of 80 cities and counties appeared to have no damaged lands, most of the cities and counties in metropolitan areas, where the farmland price was relatively high, were included in the damaged areas and so the ratio was also high.

The result of estimation based on the algebraic linear model shows that, when temperature rose by 1°C from the average annual temperature (12.4°C), the farmland price fell by approximately 14,550 thousand won/ha (Model 2) and by 19,240 thousand won/ha (Model 3). These prices correspond respectively to 5.7% and 7.5% of the average farmland prices for 80 cities and counties. Meanwhile, precipitation that had a positive impact on farmland price if it increased by 1mm from the average monthly/annual precipitation (110.8mm), i.e., if it increased 12mm a year, the farmland price appeared to rise by 330 thousand won/ha (Model 2) and 360 thousand won /ha (Model 3).

According to "A Case Study on Prediction of Climate Change Impact on the Korean Peninsula" by the Ministry of Environment, temperatures in 2020 in Korea will rise by 1.2°C and precipitation will increase by 11%. In this case, the temperature rise will result in the farmland price drop of 14,550~19,240 thousand won/ha but at the same time the precipitation increase result in the farmland price rise to 4.03~4,40 million won/ha. Therefore, the overall impact of climate change on farmland price could be evaluated as a drop of about 13.43 ~18.68 million won/ha.

In analyzing the economic impacts of climate change, it is crucial for the Ricardian approach to derive as correct a farmland price as possible under perfect competition. In a country like Korea where land area is small and population is large, the demand for land utilization is very diverse and thus it is difficult to reflect pure future value and there is a high possibility of overestimating it. For farmlands adjacent to a city, it becomes more difficult to calculate a proper land price if the demand for development increases or a speculative demand arises. Also, in Korea where the free trading of farmlands is restricted, the value of a farmland might be underestimated. The reason why the Ricardian approach is deemed useful despite the above-mentioned limitations is because it is appropriate for the market economy theory and it estimates the equilibrium assuming that all elements including the adaptation measures to climate change are optimized. In other words, the Ricardian approach has more economic implications in comparison to other economic analyses as they assess the impact of climate change under the existing conditions using the production function rather than assuming the optimized condition.

5.3.3 Climate Change Impacts on Chinese Agriculture

Climate change impacts on Chinese agriculture and livestock farming include the increase in unstable factors for agricultural production, severe drought and heat injuries in some regions, and worsening of freezing injury in early spring because of the accelerated crop growth periods due to global warming. The forage yield and quality have decreased and the loss of agriculture and stock farming by meteorological disasters has increased.

The climate change impacts on agriculture and livestock farming is expected to last into the future. It is expected that, due to climate change, the yield of three major crops in China such as wheat, rice, and corn will decrease and the productivity of Chinese agriculture will drop by 5~10% by 2030. In the late 21st century, the yield of the three major crops in China will decrease by up to 37%, affecting its long-term food security. The irrigated areas in China accounts for 47% of the entire cultivated acreage and two thirds of the food production come from the irrigated area. When moisture is reduced by 1%, the irrigated area is reduced by more than 1%, causing a serious impact on China's food production. The yield of frigid crops in most of the northern regions will also decrease considerably and the forage growth in drought or semi-drought areas will reduce as well.

Climate change will alter the production distribution and structure of Chinese agriculture. Global warming will bring about a rise in the average annual temperature of China, the rise in suitable temperature for agriculture, and the extension of growth period, shifting the cultivation areas far north. For wheat as an example, the northern limit for cultivation which is now the Great Wall will shift to Shimyang. By 2050, global warming will turn most of the present 2-crop farming areas into 3- crop farming areas and shift the present 3-crop farming areas in the Long River valley about 500km north to the Yellow River valley. The 2-crop farming areas will shift to the middle of the present 1- crop farming areas and the 1-crop farming areas will be reduced by 23.1%. Climate change that is mainly characterized by temperature rise will definitely accelerate the increase in gross food production in the northeastern region, restrain the gross food production in Hubei and the northwestern and southwestern regions to some extent, but will rarely affect the gross food production in Huadong and Jungnam regions.

Climate change will significantly increase the production costs of Chinese agriculture. Extreme climate change in China will result in an increase in torrential rain and precipitation. Especially in the 1990s, the precipitation and the rate of torrential rain had increased in the Long River valley and the southern region of the Long River, the frequency of heavy rain and flood also increased in the Yellow River and the Hoehwa valleys, and heat waves in the summer time also increased. The increase in such extreme weather conditions will increase the frequency of disasters, resulting in insecure food production and an increase in production cost.

Climate change in China is also expected to affect the crop blight and pests. According to statistics, it is expected that the agricultural production of China will have a loss of

20~25% of its gross agricultural production due to blights and pests (Government of China, 2004). Global warming will expand the activities of blights and pests that had been limited by low temperature conditions, resulting in an adverse impact on crop growth. At the same time, the greenhouse effect will lengthen the viable period of some blights and pests, aggravating the crop damages.

Under the scenario of doubling the concentration of carbon dioxide in the atmosphere, the majority of areas in China would experience climate warming and heat resources would increase. If the conditions of water, fertilizer and crop variety could meet the demands of such changes, it would be favorable for crop growing and photosynthesis. After climate warming, the areas north of the Yangtze River in China, especially in the middle-altitude and plateau areas, the crop growth season would start earlier and close later, and the potential growth season would be prolonged. However, after the climate warming, because the growth of crops is accelerated, the fertility period would be shortened generally. This would produce adverse effects on material accumulation and grain output. At the same time, the trend of drought and the deterioration of soil moisture conditions as a result of climate change would not be beneficial to wheat growth in China.

The spatial and temporal distribution of climate resources would be affected by climate warming, and the present cropping system would change accordingly. The cultivated area of a single cropping system might drop by 23%, the north boundary of a double cropping system might move northward to the middle part of the region of the present single cropping system, and the proportion of a triple cropping system might change from the present 13.5% to 36%. The northern boundary for a triple cropping system would move northward by approximately 500 km from the Yangtze River basin to the Yellow River basin.

With climate warming, a change would take place in the localities of major crop strains in China. Some crop strains currently popularized at some regions for specific climate conditions and might not be able to adapt to the changed climate conditions. New crop strains must be cultivated at appropriate time. The problem of high temperatures caused by climate warming might be offset, to a certain extent, by the adjustment of crop patterns and structure and also by making use of the temperature adaptability of crops. However, because of the impacts of other factors such as moisture content, it is hard to affirm whether the temperature increase would result in the increase of crop index, or even total yield.

Via simulations of the impacts of climate change on agriculture in China, some studies have shown the impacts of climate warming on the yield of major crops in China under $2\times CO_2$ scenario. It is found that the impacts of climate warming on the yield of spring wheat would be larger than that of winter wheat; regarding single rice, the fall of crop yields would increase gradually from the south to the north with a rate between 6~17%; yield of early rice will drop and the least drop will occur in the central south part of the Yangtze River, while in the surrounding areas, particularly in areas in the west, there would be a considerable reduction in the yield (generally between 2~5%); yield of late rice in the north-western part of the areas south of the Yangtze River would all drop significantly, while that in the south-eastern part would drop less. Spring and summer maize yields would be reduced by 2~7% and by 5~7% respectively, and irrigated and non-irrigated maize yield would be reduced by 2~6% and about 7% respectively due to global warming.

Along with climate change, occurrence of unusual disasters such as drought, flood, high temperature and freezing events might increase. The results of the simulation showed that under the assumption of no changes for the present planting system, planting varieties and production level, the total grain production might drop by about 10% due to climate change and extreme climate events during the period of 2030~2050. The production of three major crops – wheat, rice and maize – might all be reduced. Although climate change will unlikely affect China's capacity of producing grains on a self-sufficient basis, the country would require enhanced agricultural production management methods, among others, to meet the climate challenge.

5.3.4 Climate Change Impacts on Indian Agriculture

Food grain production in India has increased spectacularly due to the Green Revolution from 50 Mt in 1951 to 212 Mt in 2002, and the mean cereal productivity has increased from 500 kg/ ha to almost 1800 kg/ ha (Government of India, 2004). These increases were largely the result of area expansion, large-scale cultivation of new high yielding semi-dwarf varieties since the early 1960s, and the increased application of irrigation, fertilizers and biocides, supported by progressive government policies. Currently, Indian agriculture has 190 million ha gross sown area (142 million ha net sown area), and 40% of this is irrigated. There have been similar revolutions in the production of milk, fish, eggs, sugar, and a few other crops. India is now the largest producer of milk, fruits, cashew nuts, coconuts and tea in the world, the second largest producer of wheat, vegetables, sugar and fish, and the third largest producer of rice. This growth in agricultural production has also led to considerable surplus food stocks purchased by the government.

The droughts of 1987, 1999~2000, and of 2002~2003 could generally be managed and did not leadto severe problems of food security because of these buffer stocks. Despite this progress, food production in India, on an aggregated scale, is still considerably dependent on the rainfall quantity and its distribution. The summer monsoon (June through September) contributes to 78% of India's annual rainfall and is a major water resource. It is important to recognize that the Green Revolution was largely confined to the irrigated areas. In the past 50 years, there have been around 15 major droughts, due to which the productivity of rain fed crops in those years was affected. Limited options for other income and widespread poverty continue to threaten the livelihoods of millions of small and marginal farmers in this region.

The food security of India may be at risk once again in the future due to continued population growth. By 2050, India's population is projected to grow to 1.6 billion. This rapid and continuing increase in the population implies a greater demand for food. The demand for rice and wheat, the predominant staple foods, is expected to increase to 122MT and 103Mt, respectively, by 2020, assuming medium income growth. The demand for pulses, fruits, vegetables, milk, meat, eggs and marine products is also expected to increase very sharply. This additional food will have to be produced from the same or possibly shrinking land resource base, because there is no additional land available for cultivation. It is estimated that the average yields of rice, wheat, coarse grains, and pulses need to increase by 56, 62, 36 and 116%, respectively, by 2020. Although there is pressure to increase production in order to meet higher demands, there has lately been a significant slowdown of the growth rate in production and yield from the cultivated areas. The annual rate of growth in food production and yield peaked during the early years of the Green Revolution, but since the 1980s, it has declined since the 1980s.

The perceived gradual increase in environmental degradation, the early signs of which are becoming visible in areas that benefited largely from the Green Revolution technologies, is further compounding the problem. There is now great concern about decline in soil fertility, change in water table depth, rising salinity, resistance of harmful organisms to many pesticides, and degradation of irrigation water quality as, for example, in north-western India. Nutrient removal by crops over time has exceeded its application and consequently, farmers now have to apply more fertilizers to realize the same yield as achieved 20~30 years ago. The introduction of canal irrigation in Haryana has resulted in almost 0.5 Mha being affected by soil salinity. The rapid increase in the number of tubewells during the last three decades has resulted in an over-exploitation of groundwater in many blocks, leading to declining water tables. In some canal irrigated districts, on the other hand, the water table has risen, resulting in increased problems of salinity. Several pathogens and insect pests have also shown a tendency to increase under the intensive farming systems such as the rice wheat system. In the 21st century, one of the great challenges for Indian agriculture will be, therefore, to ensure that food production is in line with both poverty reduction and environmental preservation. The roadmap of sustainable agricultural development may also have to consider two additional important global drivers of change in agriculture in the coming decades: globalization and climate change. The on-going globalization process and multilateral trade liberalization associated with the World Trade Organization (WTO) is forcing India to make structural adjustments in the agricultural sector to increase its competitiveness and efficiency.

5.4 MITIGATION AND ADAPTATION STRATEGIES FOR THE AGRICULTURAL SECTOR

5.4.1 Countermeasures Against Climate Change

It has been known that climate change has resulted from the disruption of the energy balance for the global climatic system caused by the increase of greenhouse gases and aerosol in the atmosphere and the changes in land covering and solar radiation. Especially, it has been suggested by the results of scientific analyses that global warming is very likely to have resulted from human activities (IPCC, 2007). Countermeasures that the agricultural sector can consider in the face of risks and challenges of climate change, represented by global warming, are largely divide into the mitigation method that reduces the scale and rate of climate change by mitigating and absorbing the greenhouse gas emission and the adaptation method that admits the inevitability of global warming, understands the impacts of climate change, and minimizes the damages it could cause .

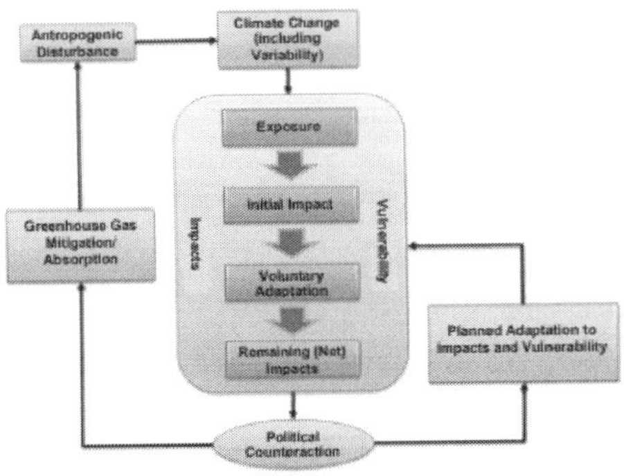

Source: IPCC (2007).

Fig. 7: Approaches to the countermeasures against climate change

Once climate change starts, components of the climate system (such as the atmosphere, hydrosphere, cryosphere, biosphere, lithosphere, etc.) are initially affected by climate change by being exposed to it and attempting to adapt themselves to that stimulus voluntarily. However, if the impact of climate change is huge, the climate system cannot handle the impact only through voluntary adaptation and thus planned adaptation that needs special measures should be attempted. If climate change still has an impact on the climate system even when the planned adaptation is in effect, it is said that there is a remaining impact. As it is difficult for systems to adapt to climate change, efforts have focused on reducing the scale of climate change through mitigation measures such as greenhouse gas reduction and absorption. Mitigation that reduces greenhouse gas emission contributes to avoiding, reducing and postponing various impacts of climate change. As climate change mitigation and adaptation are closely interrelated with each other, mitigation can be considered as belonging to adaptation measures in the long-term perspective. Therefore, adaptation to climate change is not optional but rather a compulsory countermeasure against climate change.

Mitigation measures for the agricultural sector include the improvement of cultivation methods through improved irrigation and fertilization control for the arable sector to suppress major greenhouse gases such as methane (CH4) and nitrous oxide (N2O), improvement of animal excretion treatment technologies in the livestock sector, and carbon fixing for the farmland soil.8 In relation to the countermeasures against climate change for the agricultural sector, this study focuses on the adaptation plans based on the analysis of the impacts of climate change.

5.4.2 Strategies for Climate Change Mitigation

In reality, it is almost impossible to implement the first-best solution to accurately measure the external effects of global warming and impose the economic costs to the specific greenhouse gas emitters. Under this background, the second-best solution for developing relevant policy programs and forming appropriate portfolios to approach is addressed as a practical method using a policy mix. Considering the given circumstances in a realistic manner, the means for greenhouse gas reduction are classified into economical means, regulatory means, voluntary agreement, R&D and popularization, information provision, and promotion of public awareness.

Economic means refer to policies that utilize market mechanisms, such as charges, carbon tax (or greenhouse gas tax), emission trading scheme, and subsidy.

Greenhouse gas charge, a scheme to impose a charge equivalent to the input price for the unit emission of greenhouse gas to the specific greenhouse gas emitters, is an ideal economic means to achieve efficient resource distribution using fiscal incentives. In order to put this scheme in force, greenhouse gas emissions of each emitter should be monitored and the emitter should be charged according to the actual amount of emission.

The taxes to reduce greenhouse gas emissions can be divided into carbon tax and greenhouse gas tax. Carbon tax is a system to impose taxes in proportion to the carbon content of fossil fuels used. In reality, because of the convenience of assessment and collection, carbon taxes have been put into practice in the form of a product charge. Greenhouse gas taxes differ from carbon tax in that the former is based on the amount of gas emissions while the latter is based on the input amount (or production amount) of fossil fuel which is the source of gas emissions. Though carbon tax is the same as a greenhouse gas tax, in that both can maintain the input amount of fossil fuels at efficient levels, the former is more feasible as it sets the input amount of fossil fuels as its basis for taxation. Unlike a greenhouse gas tax, which requires the taxation office to monitor the accurate amount of greenhouse gas emissions, carbon tax requires the taxation office to monitor only the shipping stage or the distribution stage of fossil fuels for taxation. As tax is input into the fossil fuel price, the input price of greenhouse gas can be reflected in the end users' decision-making. Therefore, the administrative cost for putting a carbon tax in force is relatively small.

An emission trading system is a scheme that sets emission credits based on total greenhouse gas emissions and permits the trade of the credits in the market. The demand and supply of emission credits is determined by the pricing function together with its market demand. If a certain greenhouse gas source can emit greenhouse gas at a lower cost, the emitter can reduce its emission, get as much emission credit as the reduced amount, and sell the emission credit on the emission trading market for a profit.

Subsidy is a system by which the government permits greenhouse gas emitters the right to emit greenhouse gases to a certain level and compensates them with subsidy if they give up a certain portion of the right granted to them. In order for the subsidy system to operate properly, the government should have a correct understanding of various information such as technology level, cost and reduction potential of the subsidized.

To ensure the appropriate operation of economic means that are based on the market function, reliable data and information should be collected regarding the social marginal cost of greenhouse gas emissions and the social benefits of greenhouse gas reduction.

Direct regulation is a scheme in which the government regulates emission standards using various policies and measures under law to ensure that emitters can comply and meet the required greenhouse gas emission level. An emission standard is a policy measure that specifies total greenhouse gas emission caps for each source and ensures compliance of the emitters, which is simpler and clearer than other policy measures, so that the government can introduce it and put it into force without difficulty. Examples of regulatory measures applicable to the agricultural sector include emission cap, chemical fertilizer spraying standard, manure and liquid fertilizer spraying standard and breeding density regulations.

Voluntary agreement is a non-regulatory policy measure by which firms (business establishments or farmers) and the government voluntarily agrees on greenhouse gas reduction targets, reduction plans, and the scale of government support to achieve. Voluntary agreements for greenhouse gas reduction in the agricultural sector include Good Farming Practices (GFP) and the voluntary development of resource-recycling villages by residents. Under the voluntary agreement scheme, greenhouse gas reduction targets are determined by voluntary agreements, so there is no disciplinary measure against noncompliance with the target. Therefore, it might not be an effective policy measure in achieving the national greenhouse gas reduction target.

In addition, there are R&D and popularization for GHGs reduction and capture technologies, and information provision and public awareness enhancement for greenhouse gas monitoring. Each policy measure for greenhouse gas reduction has advantages and disadvantages with respect to economic efficiency, environmental effectiveness, policy adaptation and feasibility. Therefore, in order to put any policy in force, it is desired to establish a portfolio of mixed policies based on the comprehensive evaluation of the given circumstances of each policy rather than to select and promote any one policy program. As it is not easy to pick one feasible policy mix, preliminary evaluation through policy simulation should be carried out and foreign cases where relatively successful policies are in force should be benchmarked.

5.4.3 Strategies for Climate Change Adaptation

5.4.3.1 Basic Framework for Adaptation

The concept of climate change adaptation has various definitions. According to IPCC, adaptation is defined as "adjustment in natural and human systems in response to actual or expected climatic stimuli and their effects." UNFCCC defines adaptation as "regulating process of ecological and socioeconomic systems to reduce possible damages from actual and expected climate change, that is, actions taken to help communities and ecosystems cope with changing climate conditions."

As adaptation contributes to reducing the negative risks of climate change and provides opportunities to use the climate for positive effects, it plays an important role in mitigating the impacts of climate change. Adaptation includes both actions taken to directly mitigate the damages from the climate and enhance the future adaptive capacity and actions to contribute to indirectly mitigating the damages from climate change.

Implementation of adaptation should satisfy several conditions including economic strength, technology, information, infrastructure, institutions, and equity, which are referred to as the components of adaptive capacity. Sometimes, adaptation may be implemented for free or at low cost. But in most cases, implementation for viable adaptation measures accompanies a certain amount of expenses. In addition, implementation of adaptation presupposes applicable technologies. It is also necessary for the effective implementation of adaptation measures that the necessity of adaptation measures should be acknowledged and most suitable adaptation measures should be selected through the assessment of available adaptation measures.

In relation to adaptation to climate change, adaptive capacity or adaptive capability refers to the potential of a specific system to regulate itself, reduce potential damages, or use opportunities in response to climate change, or to cope with the consequences of climate change (IPCC, 2001). It other words, it refers to the ability to effectively plan and implement adaptation and to respond to increasing risks and pressure in order to reduce the frequency and extent of the harmful consequences of climate change. Though the adaptive capacity can be obtained through a combination of abovementioned components, the relative importance of components varies with the circumstance in which adaptation is implemented and the nature of the disaster. Each component is neither independent nor exclusive of one another, but closely interrelated with one another. The components that affect adaptive capacity include the system's accessibility to wealth, information and technology; relative distribution of education and health care; and social flexibility.

Meanwhile, vulnerability refers to the risks of climate change without adaptation, such as the extent of adverse impacts from climate change including climate variability and extreme climate phenomena and the extent of being not able to cope with the adverse impacts. Therefore, vulnerability can be considered as a function of the scale and speed of climate change to which the system is exposed, and the sensitivity and adaptive capacity of the system. Here, the sensitivity of the system refers to the extent of impacts that climate disturbance has on the system. Namely, it is the system's capacity to adapt itself to changes. Adaptive capacity means the capacity of a system, region or society to adapt itself to the climate change impact, and the expansion of adaptive capacity is a substantial means of coping with the uncertainty of climate change. An assessment of vulnerability must precede the establishment of adaptation policies for climate change, and it requires the impacts of the present climate to be assessed first. The impact assessment enables one to identify what system is vulnerable in what part. Elements to be considered for the assessment of vulnerability include definite estimation of climate change, identification of the subjects of vulnerability, estimation of socioeconomic exposure to climate change, and adaptation to the expected climate changes.

Types of adaptation to climate change are divided into those reducing the sensitivity of systems significantly affected by climate change (enhancement of the storage capacity of reservoirs, cultivation of heat-resistant crops, and preparation against floods, etc.), those changing the exposure of a system to the impacts of climate change (utilization of the investment in provision against risks and Early Warning System), and those increasing the resilience of social/ecological system (resource preservation). Resilience in adaptation to climate change refers to the capacity of a community or biological system to organize or adapt itself to stresses and changes by absorbing the changes and/or obstacles. It can be interpreted as the extent of a system to withstand changes and stresses while maintaining its functions and structure.

Types of adaptation are classified according to such factors as the characteristics, intention, and point-of-time of the system. The system is divided broadly into natural systems and human systems. Adaptation is classified into autonomous and planned adaptation depending on intention; preadaptation and post-adaptation depending on the point-of-time; short-term and mid/long-term adaptation depending on the term; and adaptation for each rural household unit, community unit, and nation depending on the spatial range. It can also be divided into pre-adaptation and post-adaptation depending on the time of application and into short-term and mid/long-term adaptation depending on the period of application. In addition, it can be roughly classified into the national-level, regional-level, sector-level, and project-level adaptation depending on the scope of application.

Main groups in charge of implementing adaptation measures are broadly divided into the private sector including agricultural people and the concerned enterprises and the public sector including the central and the local governments. The private sector's goal is to maximize private profit, while the public sector pursues public value. Representative cases of application according to response time and the main sectors in charge are listed in the following table 2.

		Respon	
		Pre-adaptation	**Post-adapt**
Main sector in charge of response	Private Sector	• Utilization of the private insurance market.	• Change of crop cultivation and applicable agricultural techniques.
		• Private R&D and investment	• Regulation of the insurance market.
			• Verification of adaptation options of the minimum expense.
	Public Sector	• Utilization of Early Warning System	• Recovery from the aftermath of disasters
		• Construction of public infrastructure (irriggation systems)	• Compensation for the consequences of the impacts
		• Communication of risks Utilization of subsidies	• Insurance contract
			• Compensation system
		• Publicly available R&D	• Subsidies and supports

In reality, a wide variety of programs for the agricultural sector's adaptation to climate change are in effect depending on the national and/or regional conditions. For Canada, adaptation measures are classified into five programs: research and development (crop development, meteorological and climate information system, resource management innovation), government programs and insurance (agricultural subsidies, private insurance, resource management program), agricultural production techniques (agricultural production, land utilization, irrigation, cultivation time control), and financial management for farm households (crop insurance, crop future trading, income stabilization program, household income) (Smit and Skinner, 2002). Japan classifies the adaptation measures for the agricultural sector against climate change into 12 categories: preservation of water and soil, improvement of the soil quality, tillage activities, efficiency of water use, governmental and customary policies, R&D of new technologies, construction of infrastructure, education and public acknowledgement, land management, water resource management, human activities, and farmhouse adaptation at other levels. In the study done by Ministry of Environment to establish a master plan for adaptation to climate change, Korea classifies adaptation measures for the Korean agricultural sector into five categories: technical measures (28 measures), legal and institutional measures (7), economical measures (7), public relations and education (6), assessment (monitoring and vulnerability assessment, 14), totaling 62 detailed measures.

5.4.3.2 Priorities of Adaptation Measures to Climate Change in Korean Agriculture

To determine the priorities of adaptation measures to climate change, the Analytic Hierarchy Process (AHP) is applied.9 Solving decision problems using AHP generally involves four steps: decision hierarchy setting; pair-wise comparison of assessment criteria; weight estimation; and weight aggregation. Decision hierarchy setting is a step to set decision hierarchies for matters to be determined, by classifying each element of the matters into final objective, assessment criteria, and option. The highest hierarchy is set as the most comprehensive objective of decision-making. The next hierarchy consists of various comparable attributes that affect the objective of decision-making. The lowest hierarchy is composed of options to choose from.

To determine the priorities of adaptation measures to climate change, efficiency, effectiveness, feasibility, and acceptance are set as assessment criteria. The efficiency is to compare the cost and benefit of an adaptation policy; the effectiveness is to assess the extent of achievement of the adaptation policy objective; the feasibility is to assess whether the policy-making authorities can actually enforce the adaptation policy; and the acceptance by agricultural people is to assess whether the agricultural people would readily accept the adaptation policy.

Options for adaptation measures to climate change for the agricultural sector have been selected, total 19 items, including: R&D (breeding, production technology development, base technology development, resource management innovation, and climate information system); infrastructure management (farmland management, agricultural water management, agricultural facility management), economic means (provision of grants), legal and institutional improvement (insurance system expansion, resource management system setup, formulation of plans for each region), manpower training and education (manpower training, education/public relations), monitoring (assessment of adaptation and vulnerability), and technology and management applicable to farm households (production technology management, soil management, water management, farm household finance management) .

As climate change would occur over a considerable period of time, the roadmap for adaptation has set a target year of 2030 in three phases: short-term base buildup phase (2010~2013), mid-term takeoff phase (2014~2019), and long-term settlement phase (2020~2030). The roadmap is presented for each phase, covering seven major categories such as R&D, infrastructure management, economic means, legal and institutional improvement, public relations and education, monitoring, and technology & management applicable for individual farm household table 3.

Among major tasks for each phase, those that have to be implemented continuously are not easy to be placed in a certain phase. The development of adaptation technologies to climate change is the one that has to be implemented continuously in all phases so as to prevent and/minimize the damages caused by climate change. It is especially important to develop breeds that are in great demand and resistant to disasters and heat. This is an urgent task that should be continuously carried out to ensure that the developed breeds can be popularized to individual farm households. In the infrastructure management category, farmland management, agricultural water management, and agricultural facility management are the allotted main tasks for each phase. In particular, the popularization

Adaptaition measures	Base buildup phase (2010-2013	Take-off phase (2019)	Settlement phase (2020-2030)
R&D	-Develop new breeds that are in great demand and resistant to heat.	-Popularize the breeds adapted to global warming.	Build up the adaptaion system to global warming
	-Popularize new cultivation techonologies for fertilization and sowing.	-Provide the information about adaptaion to global warming system.	-Convert to the agricultural production system that makes the most of global warming.
	-Prepare the maps for suitable places for cultivation and crop disribution.	-Promote the crop transformation evaluation studies.	-Build up the crop transformation evaluation system.
	-Reasearches to identify physiological effects of global warming.	-Sophisticate the early warming system.	-Build up the early warming system.
Infrastructure Management	-Develop forecast models to presevent blights, pest and weeds.	-Promote the facilities to optimize the efficiency of water utilization.	-Popularize the farming simulator.
	-Develop and utilize early warning systems.	-Promote the water resource management system in prevention against natural disasters such as drought and flood.	-Promote the water resource management system in prevention agains natural disasters such as drought and flood.
	-Develop water resource management systems to prevent natural disasters including drought and flood.		
	-Popularize the technologies to reduce carbon emission from rice fields and dry fields.	-Promote the reduction of carbon emission from rice fields and dry fields.	-Promote the reduction of carbon emission from rice fields and dry fields.
	-Popularize the no-tillage farming methods.	-Expand the no-tillage farming methods.	-Settle the no-tillage farming methods.
	-Establish the standard for water-saving irrigation.	-Popularize the standard for water-saving irriagation.	-Build up TMTC system.

Adaptaition measures	Base buildup phase (2010-2013	Take-off phase (2019)	Settlement phase (2020-2030)
	-Modernize the agricultural infrastrucutre.	-Build up the automated agricultural water management.	-Expand the energy-saving fusion technology for the protected horticulture.
	-Popularize the energy-saving technology for the protected horticulture.	-Expand the energy-saving technology for the protected horticulture.	
	-Consider paying the carbon grant to the agricultural people who practice low-carbon farming methods.	-Expand the carbon grant for the low-carbon adaptaion menu methods.	-Promote the carbon grant for the low-carbon farming methods.
Economic means	-Introduce the investment incentive for water saving.	-Promote the investment incentive for water saving.	
	-Support highh-efficiency irrigation system	-Consider charging the use of agricultural water	
Legal & Institutional improvement	-Expand the insurance system for agricultural disasters.	-Promote the insuranc system for agricultural disasters.	-Promote the insurance system for agricultural disasters.
	-Expand the insurance system for damages by flood and strom.	-Promote the insurance system for damages by flood and storm.	-Promote the insurance system for damages by flood and strom.
	Operate the farm household income stabilization program.	-Settle down the programs to help farm households have stable income.	-Settle down the programs to helf farm households have stable income.
	-Establish the global warming adaptation commitee.	-Operate the global warming adaptation commitee.	-Operate the global warming adaptation commitee.
	-Introduce the system for calculation the crop damage.	-Build up the system for calculating the crop damage.	Settle down the system for caculating the crop damage and the support system.
	-Formulate the long-term development plan for rural villages.	-Settle down the long-term development plan for rural villages.	
	Operate the special task force team for main areas of production.		

Adaptaition measures	Base buildup phase (2010-2013	Take-off phase (2019)	Settlement phase (2020-2030)
Public relation & education	-Train the agricultural people specialized in risk management.	-Train the agricultural people specialized in risk management.	-Train the agricultural people specialized in risk management.
	-Train the consultants	-Utilize the consultants specialized in risk managemnt	-Complement the manual about adaptation to global warming.
	-Expand the education of farm households in the insurance for crop disaters and the risk management	-Popularize the manual about adaptation to global warming.	-Build up a systematic education system for each subject related for their adaptation to global warming.
		-Build up the adaptation education system.	
Monitoring	-Introduce the impact assessment model for producitivity forecast and biological changes.	-Utilize the impact assessment model for producitivity forecast and biological changes.	-Build up the system for assessing the environmental impact of alternative water use on crop growth.
	-Build up the agricultural ecosystem monitoring system.	-Operate the system for assessing the environmental impact on crop growth.	-Make mid/long-term forecasts of the world food demand and supply.
		-Make mid/long-term forecasts fo the world food demand and supply.	
Technology & management applicable to farm households	-Control the crop growth rate greenhouse chemicals and weedds.	-Fertilize the soil by improving the alkali soil.	-Change the places of cultivation to proper climate.
	-Cultivate the crops adapted to climate change.	-Prepare the irrigation schedule to engance the efficiency of water use.	-Fertilize the soil by improving the alkali soil.
	-Fertilize the soil by improving the alkali soil.	-Participate in the icome stabilization program	-Prepare the irrigation schedule to engance the efficiency of water use.
	-Install the water management for individual farm households.	Diversify the farm household revenues through crop diversification.	
	-Utilize the risk avoidance crop insurance.		

of no-tillage farming methods is an important task for mitigating and adapting to climate change, so that it has to be carried out continuously. In the economic means category, introduction of the low-carbon grant is a key task that needs to be done in the base buildup phase but also has to be carried on afterwards. In the legal and institutional improvement category, the agricultural disaster insurance system needs to be carried on continuously so that it can be securely settled down. Public relations and education should also be carried on with special interest in order to form consensus about adaptation to climate change. In the monitoring category, the tasks for developing a model to make mid/long-term forecasts of the world food demand and supply should be carried out in each phase. In the category of technology and management applicable individual farm households, effective programs should be developed in each phase in relation to R&D so that the developed adaptation technologies can be actually popularized to individual farm households.

5.4.4 Countermeasures for Climate Change in China

The Chinese government has the following four principles in dealing with climate change: 1) Cope with climate change to ensure sustainable development; 2) Attach equal importance to mitigation of and adaptation to climate change; 3) Convert the traditional production method and consumption pattern to cope with climate change; and 4) The whole country should be involved.

The Chinese government's countermeasures for the agricultural sector against climate change are largely divided into greenhouse gas mitigation measures and adaptation measures for climate change. The greenhouse gas mitigation measures are to continue popularizing the low-emitting multiharvesting rice varieties and half-drought type cultivation techniques; adopt scientific irrigation methods and soil-verified fertilization techniques; research and develop high-quality ruminant breeding technology and stockbreeding management technologies of scale; strengthen the management of animal excrement, wastewater and solid wastes; enhance the efficiency of methane utilization; and control the methane emission.

Adaptation measures to climate change are as follows: strengthen the measured forecast level for the extreme meteorological disasters by supplementing the measured forecast emergency action mechanism, the multi-department decision-making mechanism, and ensure comprehensive community involvement mechanism in provisions against various disasters; establish a meteorological disaster defense process by 2010 that can play a fundamental but essential role in securing the economic society; improve the comprehensive measured forecast level, defense level, and disaster mitigation capacity to cope with extreme meteorological disasters; and form 24 million ha of new grassland and clear 55 million ha of degraded, desertificated, and/or alkali grasslands by 2010, by strengthening such adaptation measures farmland construction, cultivation system adjustment, resistant variety selection and development, and biotechnology development.

The Chinese government has enacted and enforced "Agriculture Act," "Grassland Act," "Fisheries Act," "Land Management Act," "An Ordinance on Emergency Measures against Sudden Critical Animal Epidemic," "An Ordinance on Pasture Fire Prevention" as climate change adaptation policies. The government has also made efforts to establish and supplement the political and regulatory system for the agricultural sector's adaptation to climate change. In addition, they have strengthened the agricultural infrastructure construction, promoted the construction of farmland irrigation systems, expanded the irrigated agricultural area, and improved the irrigation efficiency and the general piping level for farmlands. Furthermore, they have popularized the water-saving technology for hardy crops, enhanced the agricultural disaster prevention, reduction and generation production capacity, and developed some varieties that can endure high temperature, blight, and pest.

In the future, China is planning to further expand the popularization of high-quality varieties and increase the covering rate of high-quality varieties. Also, it will strengthen the prevention of critical animal epidemics, establish and supplement the anti-epidemic measures for animals, and strengthen the measured forecast level for animal epidemics.

The Ningxia Region, one of five self-governing regions located in northwestern China, is the poorest region in China. Its weather is characterized by dry and distinctive seasons, average annual temperature of 5-9°C, and precipitation at 262mm. In the Ningxia Region, three characteristic agricultural systems of China are present: 1) utilization of rainwater; 2) utilization of rainwater and irrigation water; and 3) utilization of irrigation water. The region has a high level of poverty, has a lot of experiences in damages caused by meteorological disasters, and the region is also vulnerable to climate change. The Ningxia Region's climate had been stable but its average temperature has risen by about 0.6~1.2°C for the past 10 years and the droughts became worse in 2004~2006.

The northern Ningxia Region is the area irrigated from the river flows from the Yellow Sea. This region's main agricultural products come from corn, spring wheat, rice, and potato cultivation through intercropping, and its main livestock farming product is chicken. The central dry area has a vast pasture that uses both irrigation water and rainwater. Due to dry weather, only corn, spring wheat, and potatoes can be cultivated and some cows and sheep are raised. The southern mountain area is humid, which cultivates using rainwater, but it is still very dry. Main crops cultivated in this area are potatoes, and the main livestock farming products include beef cattle, sheep, pigs, and poultry. Because of suchregional characteristics, the Ningxia Region formulates its adaptation objectives taking into consideration how climate risk will affect each area. Furthermore, it pursues policies that can integrate its main development objectives and the objectives for adaptation to climate change. Therefore the following steps should be take: when considering the adaptation policy by the unit of region, in-depth research about the corresponding region including its natural conditions, climate system and crops cultivated in the region should be carried out; the policies differentiated for each region should be formulated based on the research results; and the existing policies should also be taken into consideration.

In response to climate change, the farming households in the Ningxia Region have taken some adaptation measures. First of all, in provision against dry weather and drought, they harrow their fields and uses film covering and/or sand cover to maintain the moisture content of the soil. Harrowing is the most widely used measure in the Ningxia Region, especially in the southern area and in Huinong and Helan in the northern area. Film covering is used in Huinong, which has a different effect depending on the crop type. For potatoes, it has been seen to increase the yield by 30%. Sand covering is used only in Haiyuan, which is known to be effective against drought.

In order to secure water, various rainwater collection measures are taken and the three areas use very different methods. To save water, irrigation is used as well, which applies mostly to the northern irrigated areas but has some difficulty in popularization as it requires supplying water to furrows without the irrigation facilities. The Ningxia Region has a high non-farming income rate of about 50% and thus has a relatively low income vulnerability to climate change. A lot of residents have moved to other areas with a better environment, and they evaluated moving as a successful adaptation measure.

Table 4. Adaptation measures for the farmers in the Ningxia Region.

Type of Adapation	Descripition
Measures to maintain the moisture content of the soil against dry weather and drought.	-Use various methods to increase the moisture content of the soil.
	-Harrow the fields and use film covering and/ or sand cover.
Rainwater collection.	-Use different rainwater collection methods.
	-The northern area rarely collects water as it has rich water resources.
	The central area collects rainwater in the underground resrvoir.
Water-saving irrigation measures.	-Water-saving irrigation is used mostly in the northern area.
Non-farming income.	-All three areas have a high non-farming income rate of about 50%.
Moving.	-Ningxia residents have move to the areas with a better environment.

5.5. LOW CARBON GREEN GROWTH STRATEGY/ROADMAP FOR THE AGRICULTURAL SECTOR

5.5.1. The Concept of Green Growth

Sustainable development means achieving sustainable economic growth reconciled with environmental protection. Since its first appearance in the Declaration of the United Nations Conference on the Human Environment in 1972, this concept was then solidly established by the World Commission on Environment and Development (WCED) in 1987, and came to serve as a basis for the Rio Declaration on Environment and Development in 1992 and its acting plan, Agenda 21. The concept of sustainable development sprang from a reflection on the economic paradigm of mainstream economists, which revolves around the 'growth first, clean up later' idea where one believes that the economy should develop first and environmental damage is dealt with later. The WCED defines sustainable development as "development that meets the present without compromising the ability of the future generation to meet their own needs" (Kim Chang-gil and Kim Jeong-ho, 2002, p.7-9).

The concept of green growth was created in order to increase the possibility for policy enforcement in sustainable development, whose concept is abstract and broad-based covering the three aspects of economic feasibility, environmental protection and social equity.10 In other words, green growth is a qualitative growth that enhances the standard of living through the attainment of ecological and economic soundness. The Korean government presents green growth as "economic growth that minimizes environmental pollution and greenhouse gas, while creating a new growth engine and jobs" (Presidential Council for Future and Vision, 2009). In addition, the relevant framework act defines green growth as "growth achieved by saving and using energy and resources efficiently to reduce climate change and damage to the environment, securing new growth engines through research and development of green technology, creating new job opportunities, and achieving harmony between the economy and the environment". As green growth is variably defined, it can be understood as a complex and open-ended concept that accepts a new civilization and an order of change. Within this context, we can understand that green growth is a concept in the making, subject to the discussions of the future.11

The operating principle of the green growth policy is to turn the vicious cycle between the environment and economic growth into a virtuous cycle through a transformation into a newly established growth pattern and economic structure. Under the new paradigm of qualitative growth, the essential factors of production are new ideas, transformational innovations, and state-of-the-art technology. The economic growth based on these drivers is expected to generate substantially intensive, qualitative growth unlike the extensive quantitative growth of the past.

Therefore, green growth continuously enhances productivity by putting green capitals (i.e., green technology and green knowledge) into the production process, thereby reducing pollution and expanding natural capital (energy and environmental resources). As a result, green growth brings about changes in production, technology and consumption patterns and ecological efficiency by taking into account environmental capacity and economic and environmental aspects of production and consumption fig. 8.

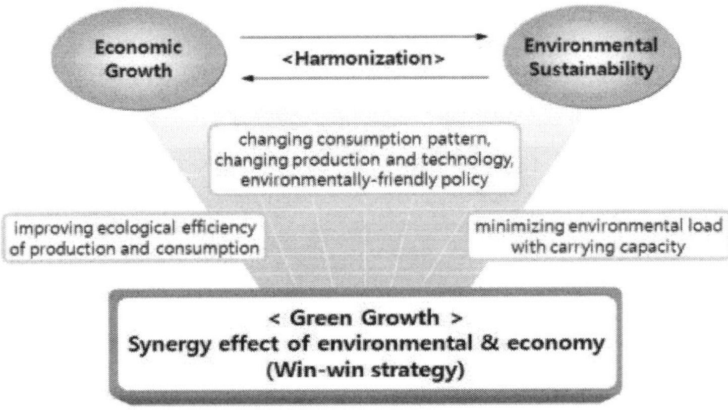

Fig. 8: Mechanism to achieve the goals of green growth

Green growth involves policy regime and social value as it aims to achieve economic growth through a low-carbon society and green industrialization. The realization of green growth is accompanied by a significant amount of economic costs and efforts, and above all, it requires us to move away from our existing patterns of living. Thus, for green growth to be materialized, economic incentives, development and the distribution of green technology, and understanding and cooperation of relevant bodies, are of the utmost importance and urgently require a paradigm shift and swift response.

Green growth in the agricultural sector is more comprehensive than sustainable agriculture, and it means a growth that guarantees environmentally sound and economically viable growth that takes into consideration environmental capacity in the agricultural ecosystem. Green growth in the agriculture sector is achieved through a shift in agricultural practice that takes into account the environmental capacity of each different region and the water system, low-carbon agriculture via greenhouse gas reduction and higher absorption capacity, and energy efficiency and savings. Green growth can be realized by shifting to a sustainable agricultural system including environmentally-friendly, low carbon agriculture. Thus, green growth in the agricultural sector can be regarded as more comprehensive than sustainable agriculture fig. 9.

Agriculture that pursues green growth can be defined as green agriculture. However, the term is not widely used. Several terms are being used in the agricultural sector with respect to the green concept. However, special attention should be paid to how they are actually related to green growth. For example, China uses the term "green food" to increase the public familiarity with the concept of environmental friendliness, and the term is relevant to green growth to some extent. In Korea, "green revolution" refers to a drastic increase in productivity through the development of high-yield rice varieties (for example, the new rice variety "Tongil" IR667), but it is hardly related to green growth. By contrast, the term "second green revolution" is being used to describe the planting of wheat, green manure crops and fodder crops on idle lands during the winter season, and it is significantly relevant to green growth in the agricultural sector as it involves energy saving and greenhouse gas mitigation.

In the meantime, green growth in the agri-food sector means a growth through a shifttowards an environmentally sound, low-carbon life cycle in the agricultural aspects of not only production but also distribution, processing and consumption.

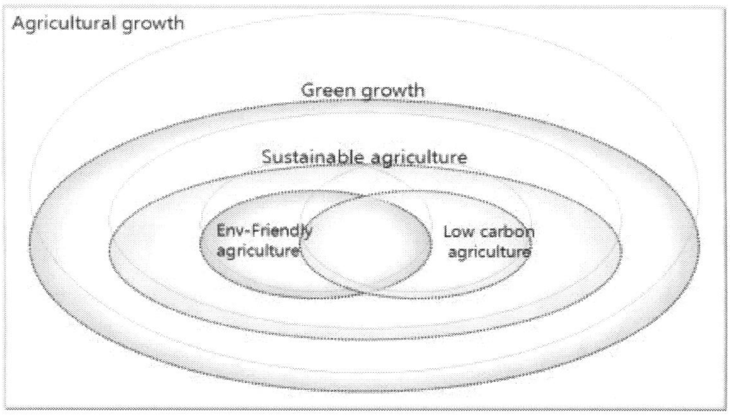

Fig. 9: Coneptual position of green growth in agriculture

5.5.2. Strategies for Green Growth in the Agricultural Sector

5.5.2.1. Basic Directions and Implementation Methods

Shifting from pre-existing conventional agricultural growth to low-carbon green growth (LCGG) requires the formulation of a vision and specific goals. An appropriate vision should reconcile agriculture with the environment, mitigate greenhouse gases and improve the agricultural ecosystem, which contributes to the lessening of negative effects of global warming and improves the living standards for present and future generations fig. 10.

As a basic direction to develop an LCGG action plan, a resource-cycling agricultural system should take its root based on the 3R cycle of "Reduced ↔ Recycled ↔ Reuse". It should be followed by the transition of the agricultural production goal from maximization to optimization. In other words, the agricultural production goal should be shifted from maximizing production to optimizing production by taking into account the local agricultural environment and greenhouse gas emitting and absorbing conditions.

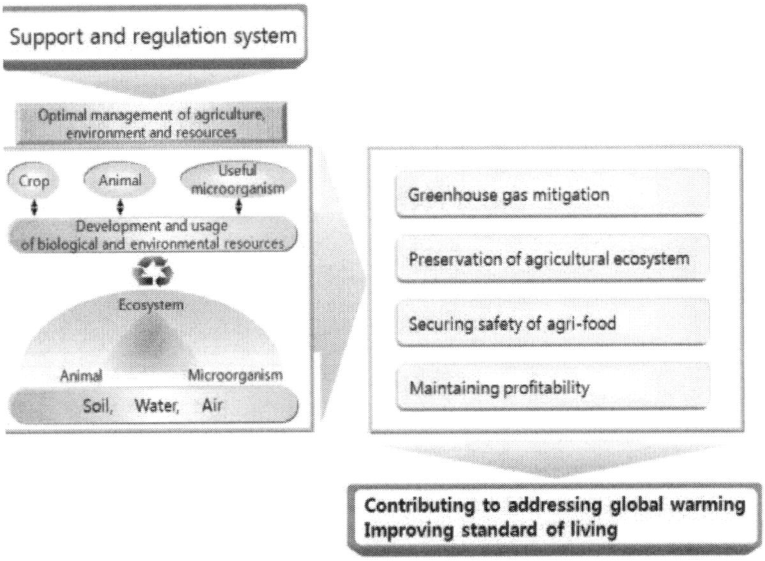

Fig. 10: Vision and goals of low carbon green growth

For the reduction of greenhouse gases and improvement of living standards, a policy combination of support, regulation and compensation should be formulated so that the right policy mix entails consolidation and coordination for environmental policies. To minimize the inconvenience and economic costs that come along with green growth, green technology should be actively developed and disseminated. Lastly, it is no less important that relevant bodies share a common understanding through information sharing, education and promotion.

5.5.2.2. Implementation Methods

As it takes a considerable amount of time to establish an agricultural system that promotes low carbon green growth with respect to obligatory greenhouse gas reduction, action plans should be devised based on a three-stage process by the target year of 2030, this includes the: groundwork establishment stage, utilization stage and settlement stage.

The groundwork establishment stage (2010~2013) involves policy promotion to foster environment-friendly agriculture and the establishment of a production base for renewable energy and a pilot project under the Kyoto mechanism to reduce greenhouse gases. Greenhouse gas absorption includes identifying the role of soil organic carbon and estimating the storage of organic carbon in the farmland. And the absorption process involves the development of a prediction system to increase the productivity of agricultural products and biota, and the development of varieties with an ability to adapt to global warming. The utilization stage (2014~2019) should pursue the development of database for greenhouse gas emissions, dissemination of greenhouse gas mitigating technologies, expansion of the use of the Kyoto mechanism and application of incentive programs aimed at enhancing the absorbing capability. The last settlement stage

(2020~2030) means the establishment of an agricultural production system that is based on low-carbon green growth and pursues an optimal policy combination of programs in various areas, including greenhouse gas mitigation, and greenhouse gas absorption and adaptation to global warming.

A roadmap should be drawn at every stage to ensure the establishment of a low-carbon green growth system by implementing green growth programs in each stage of various fields for greenhouse gas mitigation, absorption, and adaptation.

The greenhouse gas mitigation of the groundwork establishment stage involves the dissemination of greenhouse gas reduction technology, planting of bioenergy crops, and formulation of emission trading schemes and pilot projects for developing a Clean Development Mechanism (CDM). The utilization stage includes the development and dissemination of tailored technology for each region, the establishment of a bioenergy production system and an emission trading scheme. The settlement stage seeks to supplement the greenhouse gas management system and establishing a low-carbon green system.

The greenhouse gas absorption stage should lay down a foundation to identify the role of soil organic carbon, estimate the storage of organic carbon and build a utilization base. The utilization stage should introduce a direct payment for low carbon, and the settlement stage should include the establishment of an agricultural system that maximizes the ability to mitigate and absorb greenhouse gases. The greenhouse gas adaptation should -in its groundwork establishment stage- be able to develop a prediction and evaluation model for the productivity of the agricultural sector and biota, establish an eco-monitoring system, and develop crop species that can adapt to global warming.

The utilization stage should involve the development and distribution of an adaptation manual, establish an early-warning system, disseminate plant varieties with an ability to adapt to global warming, and establish educational systems.

5.5.2.3 Core Tasks for Implementing Strategies

5.5.3 Establishment of a Resource-Cycling Agricultural System

In order to achieve green growth in the agricultural sector, environmentally friendly agriculture should be built based on a solid resource-cycling system linked with related industries that are environment-friendly. As part of this plan, two environmentally friendly agricultural complexes (one that is small in size of about 10ha and the other is bigger with a size of about 1,000ha) will be utilized as a basis for the expansion of production and distribution in an environment-friendly atmosphere. It is recommendable that Regional Circulation Agriculture Support Center (tentatively named) is created and operated to help these complexes take root in an environmentally-friendly resource cycling agricultural system. Especially for the transition to an environmentally friendly agricultural system that is suitable for each region, an effective implementation plan should be formulated so that Regional-Based Maximum Nutrients Loading System could be carried out as early as 2007.

An industry of environmentally friendly machinery will be fostered to build a sound environment-friendly agriculture. Organic fertilizer and by-product fertilizer will be combined to improve the fertilizer processing standards, and an industry base should be established for the strict management of inferior organic fertilizer and post management of agricultural machinery through mandatory listings.

5.5.3.1 Integration of Agricultural Policy and Low-Carbon Eco-Friendly Policy

Green growth pursues a harmonization between agricultural activity and the environment through a paradigm shift toward a low-carbon agricultural system that mitigates or absorbs greenhouse gas. A transition to low-carbon green growth cannot be achieved only by implementing environment-friendly cultivation programs, but it requires a reorganization of the current agricultural policy system. To maximize the efficiency in using agricultural resources, while minimizing environmental pollution, an environmental evaluation test should be conducted on all agricultural policy programs so that the programs can be combined or made consistent with the low-carbon policy.

For integration of agricultural policy with environmental policy, efforts should be made to introduce a direct payment scheme for low-carbon agriculture, which is the most representative program for Environmental Cross Compliance (ECC). 12 When various kinds of low-carbon agricultural systems are practiced to encourage voluntary greenhouse gas reduction, a direct payment scheme for low-carbon agriculture, such as manual-type direct payment that provides a proper incentive through monitoring, can encourage farmers to actively participate in low-carbon green growth.

5.5.3.2 Activation of Climate Smart Agriculture

A new initiative that calls for agriculture to be part of the solutions to climate change, and not part of the problems, has been launched at the COP1613 climate negotiations from November 29 to December 10, 2010 in Cancun, Mexico.14 It advocates getting the right policies and programs in place that will increase farm productivity and incomes, make agriculture more resilient to variations in climate and make the sector part of the solution to climate change by sequestering more carbon into the soil and biomass. There is a growing recognition that agriculture in developing countries must become "climate smart" to cope with the combined challenges of feeding a warmer, more heavily populated world. The Food and Agriculture Organization (FAO) suggested food security and climate change should be addressed together, by transforming agriculture and adopting practices that are "climate-smart" to eradicate hunger from the world. 15 Climate smart means agriculture that sustainably increases productivity and resilience to environmental pressures, while at the same time reduces greenhouse gas emissions or removes them from the atmosphere. The FAO stressed that a variety of climate-smart practices already exist and are being used in some places, providing examples that could be more widely implemented in developing countries. Agriculture in developing countries must become 'climate-smart' in order to cope with the combined challenge of feeding a warmer, more heavily populated world. There are a number of areas where changes in the food production sector are required. Farming must become more resilient to disruptive events like floods and droughts through improving agricultural water and soil management. The vulnerability of farming communities to climate-related disasters must be reduced, and better warning and insurance systems to help them cope with climate-related problems need to be established. In this note, agriculture has to find ways to reduce its environmental impacts -including lowering its own greenhouse gas emissions- without compromising food security and rural development. The FAO's report goes on to highlight examples from around the world of how farmers are already moving to tackle these issues and adopt new, climate-smart practices In order to transform into climate smart agriculture, agriculture must produce more food, waste less, and make it easier for farmers to get their produce to consumers. In addition, farming must do a better job of managing natural resources like water, land and forests, soil nutrients and genetic resources to be more resilient to natural disasters. The report also argues that greater coherence among agriculture, food security and climate change policy-making is urgently needed. Policies in all three of these areas impact smallholder production systems and a lack of coherence can prevent them from capturing synergies. Improving mechanisms for getting data, science and knowhow to farmers so they can adapt is another area in need of attention. In particular, key requirements for an enabling policy measures to promote climate-smart smallholder agricultural transformation is greater coherence, coordination, and integration between climate change, agricultural development and food security policy process. In addition, adequate investment in national climate-smart agricultural formulation, research and extension, including capacity building, is important in supporting action by farmers (FAO, 2010).

The most significant achievement of the COP16 was the establishment of the Green Climate Fund. Its purpose is to finance poorer countries to develop less pollution-intensive energy sources, and adapt to a changing global climate as natural disasters increase in frequency and intensity, and infrastructures are taxed to cope with more severe natural conditions. Developed nations are committed to contributing up to $100 billion dollars annually to the fund by 2020, resulting in a massive resource that poorer nations will be able to draw upon. The new fund will be managed under the auspices of the UN rather than the World Bank, although the bank will still be involved as the interim trustee of the fund subject to a review three years after the launch of the fund.[16]

5.5.3.3 Development and Dissemination of Low-Carbon Green Technology

Green technology refers to technology that minimizes the use of materials and energy, decreases environmental load and weakens entropy through the circulation and utilization of renewable materials and energy for the purpose of establishing a low-carbon paradigm. Therefore, green technology pursues the minimal use of energy and materials and promotes material circulation and dynamic balance in the realm of nature and renewable energy technology. In addition, green technology entails an adjustment of input and production to the extent that the natural ecosystem can handle.

Green technology with aforementioned characteristics is manifested in the form of four distinctive technologies: greenhouse gas mitigation technology, energy efficiency technology, clean energy technology and new environmental technology.

First, greenhouse gas mitigation technology in agriculture includes mitigation technology for reducing methane and nitrous oxide in farmlands, soil organic carbon storage technology, for improving intestinal fermentation of ruminant animals and animal manure treatment facility, and technology for biomass utilization and fossil fuel reduction. Most of these technologies are at commercialization stage thanks to the technology development effort so far, which is a significant technological achievement. Nevertheless, greenhouse gas mitigation technology is not widely used in the rural scene due to a low level of technology acceptance by farmers.

Second, energy efficiency technology includes technologies improving heat recovery ventilators and heat exchangers. Technology development in this area has gone to such a great extent that heat exchanger improving technology for warm air heaters and exhaust heat recovery facilities for warm air heaters are already at the distribution stage. Technology development and formulation of complementary plans for achieving economic efficiency should continue so that energy efficiency technology can be easily accessible in rural areas.

Third, clean energy technology refers to a technology that utilizes clean energy sources, such as geothermal, solar power (photovoltaic), wind power, and multifold thermal covers and water filter protection curtains. Clean energy technology also involves the development of production models for cellulose crops to produce bio-ethanol, such as canola, sweet potato for ethanol and C4 (non-food crops), and Korean cellulose ethanol. Fourth, new environmental technology includes bio-crop protection agents using natural materials, biological pesticide such as environmentally friendly microorganisms and natural enemies, development of urban building-type plan systems for crop production through the convergence of agriculture and cutting-edge technologies, such as nanotechnology (NT), biotechnology (BT), information technology (IT), environment technology (ET) and crop production technology using LED. More efforts should be made for the commercialization of new convergence green technology, including the production of rice bran products using supercritical fluid (materials put onto temperature and pressure above critical level) and the introduction of bio-refineries for rice.[17]

Green technology of the future is actively involved in the development of state-of-the-art technologies, such as NT, BT, IT, and ET, leading to the expectation that convergence technology will be expanded in the agricultural sector. And more attention and support need to be paid to the installment and expansion of plant factories and vertical farms that use such convergence technologies.

5.5.3.4. Development of Policy Programs Based on Carbon Information

In order to effectively pursue green growth, the success of which largely depends on the greenhouse gas reduction in each sector, it is essential to develop policy programs that actively utilize carbon information. One of the major policy options is carbon labeling. It seeks to achieve the reduction of market-oriented greenhouse gases by quantifying carbon emissions during the entire life cycle of a product from production to transportation, distribution and disposal. It allows this information to be available to consumers and induce them to lead a low-carbon, greener life. Korea introduced a carbon reduction labeling system of its own in February 2009. In the agri-food sector, its implementation is currently limited to processed products such as precooked rice and tofu, but it is necessary to expand the system to agricultural commodities and other food-related areas through appropriate measures including the establishment of GHGs inventory along the steps of production, processing and consumption.

Different types of carbon emission trading are currently in place where carbon emission rights - the right to emit greenhouse gases- are commercialized and traded. The trading market types can be categorized into "allowance-based market" and "project-based market" depending on what origin the right is based upon, and into "mandatory carbon market" and "voluntary carbon market" depending on whether the reduction is compulsory by the Kyoto Protocol.

Clean Development Mechanism (CDM), a regime of the project-based market, can be considered for introduction in the agricultural sector. The utilization of CDM is still in its early stages in the domestic agricultural sector, but has a huge potential to contribute to greenhouse gas reduction and green growth if potential business items are explored and identified by benchmarking overseas CDM operation cases. The launch of CDM projects should be more actively considered including the establishment of internationally-approved animal waste biogas plants and bioenergy projects using agricultural byproducts. In order to boost CDM projects in the domestic agricultural sector, feasibility studies should be conducted first on projects with a higher possibility of greenhouse gas reduction, such as methane reduction in livestock waste disposal facilities and utilization of biomass and bioenergy.

The second option to consider in regard to emission trading system (cap and trade scheme) is to trade the emissions reduced in the agricultural sector with players in the non-agricultural sectors. The cap and trade scheme, the implementation tool of the Kyoto Protocol, is operated or test-operated in the EU, Britain, Canada, Denmark and Japan. Each country committed to the Protocol allocates emission allowances to domestic sectors and companies to achieve the mandatory reduction targets under the Protocol.

A company that fails to reach the target can purchase emission rights from other sectors. It is necessary to develop appropriate programs for early reduction or absorption of greenhouse gases under a new market-oriented policy that can encourage farmers to participate in carbon reduction efforts.

5.5.3.4 Utilization of Green Finance in the Agricultural Sector

Green finance stems from "environmental finance" or "sustainable finance" used in foreign countries, but in Korea it implies something more comprehensive and economy-oriented. Green finance is a concept that incorporates indirect support for green industries by including the process of environmental risk management, etc. Green finance in Korea has been actively planned and pursued mostly by state-run financial institutions. Non-governmental financial institutions have recently started to engage in green finance by devising green products and systems in support of green industries, but there have been little tangible results.

An example of green finance in the agricultural sector is "Green World Installment Deposit" launched by the National Agricultural Cooperative Federation, which applies up to 0.6% of preferential interest rates to the deposit if the subscriber participates in low-carbon, green growth related activities. As another example of green finance, LIG Insurance Company introduced a special policy for losses related to eco-friendly agricultural products. Under the policy, if pesticide residues are detected in eco-friendly products purchased by consumers, the producer of such products is compensated for the costs to redeem the consumer confidence on the concerned products and to prevent future occurrence of such cases (e.g., direct compensation to consumers). Measures to promote green finance in the agricultural sector include financial support for the introduction of green energy technology in capital-intensive areas such as protected livestock and horticulture; financial assistance for agricultural production facilities utilizing green technologies for energy saving and higher efficiency in production as well as for packaging and transportation in the process of distribution; and creation of agricultural investment funds to provide loan programs with preferential rates for businesses distributing green agricultural technologies.

5.5.3.5. Education and Communication for Green Growth

The shift to a low-carbon green growth regime will require substantial investment and support for education and training programs aimed at promoting active participation in the green growth effort of relevant players such as farmers, businesses and policy makers. Actions on the spot for low-carbon green growth in agriculture are led by the Agricultural Research and Extension Services of each province, the Agricultural Technology Centers in municipalities, local producer groups, and leading farmers. Therefore, structured education and communication programs targeting core leaders in the sector should be one of the priorities.

5.6. AT A GLANCE AND POLICY IMPLICATIONS

Scientific diagnosis and assessment of the impact of climate change on the agricultural sector is essential for formulating the vision of future agriculture and the direction of agricultural administration. Specifically, it can provide useful information for formulating the long-term agricultural development plan for each region and the adaptive measures for farming households.

Scientific diagnosis and assessment of the impact of climate change on the agricultural sector is essential for formulating the vision of future agriculture and the direction of agricultural administration. Specifically, it can provide useful information for formulating the long-term agricultural development plan for each region and the adaptive measures for farming households.

This paper was carried out in order to suggest scientific and phase-by-phase counterstrategies to prevent against climate change through the diagnosis of climate change phenomena and in-depth analysis of climate change impacts on the agricultural sector. The countermeasures for the agricultural sector against climate change have mostly focused on greenhouse gas mitigation. However, more interest and policy support should be directed to adaptation measures in consideration of the inevitability of global warming and the characteristics of climate-dependent agriculture. In particular, it is necessary to understand that the countermeasures for the agricultural sector against climate change are to minimize the risks of climate change and utilize it as an opportunity. For this, proper education and training programs for agriculture workers, public officials and the personnel from the related agencies should be developed and put into practice so that they can properly cope with climate change. It is also urgent to expand the adaptation measures of the agricultural sector that have been limited to research and development, to more active policy programs including popularization of technology and adaptation manuals to agriculture workers and proper incentive programs. As shown by the result of questionnaire research from farmers, they are highly interested in adaptation to climate change and very willing to participate in carbon reduction program, it is necessary to develop technologies and customized programs that they can apply in the field. The improvement of agricultural production infrastructure has focused on the farmlands for rice farming to become self-sufficient in food production, such as securing water resources in preparation against drought, preventing floods, and arranging for mechanized farming. Now, more scientific measures in water and facility management should be formulated and implemented in preparation against unusual weather including localized torrential rain and typhoon. In ensuring the effective implementation of the agricultural sector's adaptation strategies, it is necessary to divide roles properly between the concerned bodies such as the government, farmers, researchers, and other related institutions and to build up the integrated management system for comprehensive planning and implementation for those adaptation strategies.

In the age of the so-called "energy-climate era," a new development strategy is required to replace the conventional growth strategy which is faced with many limitations. In preparing for the future, low-carbon green growth has become the main stream as an inevitable core task to be performed home and abroad, and it is anticipated that there will be a lot of discussions in relation to the preparation of a green growth strategy to cope with global warming for a considerable period of time. As the agricultural sector takes up a very low portion of the total amount of domestic greenhouse gas emissions, with approximately 3%, there is a possibility that proper attention may not be paid to the administration of greenhouse gas reduction activities in the agricultural sector. In this regard, we need to benchmark advanced countries such as the U.S., Europe and Japan where the agricultural sector is given significant attention in terms of implementing green growth. The agricultural sector has a significant potential to contribute not only to the administration of national greenhouse gas reduction but also to the national development of green industries in the future.

Given that the growth paradigm of agriculture so far has been based on a productivity-oriented quantitative approach, green growth implies a significant paradigm shift into a qualitative approach which takes both productivity and ecology into account. In light of this shift, Korea's agricultural sector should leave the old paradigm behind and pro-actively seek measures to play a part in resolving global warming and achieve a balance between agricultural development and environmental protection. In order to accomplish green growth in the agricultural sector, we should create an innovative way to turn inconvenience into a growth engine by leaving existing convenience and inertia behind, and by achieving a shift in thinking among relevant parties, to ensure that inconvenience and hazard can be properly managed. For this to happen, an amicable atmosphere should be created with a bold paradigm shift, where the suggestion of various ideas and active discussion can take place. First of all, it is urgent that we come up with an implementation strategy that allows us to maintain the unique characteristics of agriculture as a green industry, and thereby eventually achieve green growth by actively developing public functions, such as atmospheric purification and environmental protection through agricultural production innovation and clean technology. It is particularly necessary to establish green governance where all farmers, relevant organizations and policy makers concerned can work together, where a strong will to implement green growth and an effective execution system are required to accomplish green growth. However, the policy to promote environmentally friendly agriculture itself is not enough to ensure an assured transition toward a low-carbon agricultural system, but reorganization of the overall agricultural system is needed. Above all, agricultural policy and low-carbon environmental policy should be properly integrated so that the concept of green growth in the overall agricultural sector takes root. In order to maximize the policy effectiveness through a proper combination of policy instruments in various relevant sectors, a green innovation system should be established where policymakers, researchers, relevant organizations, farmers and other relevant bodies can have proper understanding of green growth and share their roles. In addition to that, a systematic stage-by-stage strategy to develop technology should be devised and implemented on a steady basis so that green technology reduction or absorbing of greenhouse gases in the agricultural sector can be utilized as a growth engine. When green growth in the agricultural sector is successfully implemented, agriculture will solidify its position not only as a green industry that manages national land in an environmentally sustainable manner but also as a life industry that supplies safe agricultural products and manages national greenhouse gas emissions.

Systematic researches into the impacts of climate change on the agricultural sector and the formation of counterstrategies to fight against climate change have to be done as interdisciplinary research among agricultural science, ecology, agricultural engineering, hydrology, meteorology, and agricultural economics. For a more reliable analysis of climate change impacts, efforts need to be made to develop integrated models consolidating the estimates produced by climate change scenarios, the simulation analysis based on agricultural characteristics, and the economic analysis reflecting the socioeconomic factors of agriculture. Furthermore, future studies should carry out a more detailed analysis for the economic and policy effects of each adaptation measures for the agricultural sector.

The multiple challenges the East Asian Region is facing in terms of climate change, degradation of ecosystems, food insecurity require an integrated approach. The challenges and many of the solutions are understood. Now is the time to get down to earth and take urgent action on solving those issues towards a sustainable, inclusive and resource efficient path. The notion that greens growth and climate-smart agriculture can help to solve urgent issues has been the basis for developing and starting the roadmap for the East Asian Region. The roadmap for action is part of an ongoing process to identify, stimulate and broaden actions. At the same time, it is a stepping stone to further initiate and broaden the partnerships and its activities with inclusive engagement by all stakeholders, such as the private sector, governments, scientists, non-governmental organizations, and farmers.

6

GLOBAL WARMING AND CLIMATE CHANGE

6.1 INTRODUCTION: AN OVERVIEW

In the year 2060 my grandchildren will be approaching seventy; what will their world be like? Indeed, what will it be like during the seventy years or so of their normal life span? Many new things have happened in the last seventy years that could not have been predicted in the 1930s. The pace of change is such that even more novelty can be expected in the next seventy. It is fairly certain that the world will be even more crowded and more connected. Will the increasing scale of human activities affect the environment? In particular, will the world be warmer? How is its climate likely to change?

Before studying future climate changes, what can be said about climate changes in the past? In the more distant past there have been very large changes. The last million years has seen a succession of major ice ages interspersed with warmer periods. The last of these ice ages began to come to an end about 20 000 years ago and we are now in what is called an interglacial period. Chapter 4 will focus on these times far back in the past. But have there been changes in the very much shorter period of living memory – over the past few decades?

Variations in day-to-day weather are occurring all the time; they are very much part of our lives. The climate of a region is its average weather over a period that may be a few months, a season or a few years. Variations in climate are also very familiar to us. We describe summers as wet or dry, winters as mild, cold or stormy. In the British Isles, as in many parts of the world, no season is the same as the last or indeed the same as any previous season, nor will it be repeated in detail next time round. Most of these variations we take for granted; they add a lot of interest to our lives. Those we particularly notice are the extreme situations and the climate disasters. Most of the worst disasters in the world are, in fact, weather- or climate-related. Our news media are constantly bringing them to our notice as they occur in different parts of the world – tropical cyclones (called hurricanes or typhoons), wind-storms, floods, tornadoes and droughts whose effects occur more slowly, but which are probably the most damaging disasters of all.

6.2 LAST DECADES OF THE TWENTIETH CENTURY

The 1980s and 1990s were unusually warm. Globally speaking, the decades have been the warmest since accurate records began somewhat over a hundred years ago and these unusually warm years are continuing into the twenty-first century. In terms of global average near-surface air temperature, the year 1998 was the warmest in the instrumental record and the nine warmest years in that record have occurred since 1990. The period has also been remarkable (just how remarkable will be considered later) for the frequency and intensity of extremes of weather and climate. For example, periods of unusually strong winds have been experienced in western Europe. During the early hours of the morning of 16 October 1987, over fifteen million trees were blown down in southeast England and the London area. The storm also hit Northern France, Belgium and The Netherlands with ferocious intensity; it turned out to be the worst storm experienced in the area since 1703. Storm-force winds of similar or even greater intensity but covering a greater area of western Europe have struck since – on four occasions in 1990 and three occasions in December 1999.

But those storms in Europe were mild by comparison with the much more intense and damaging storms other parts of the world have experienced during these years. About eighty hurricanes and typhoons – other names for tropical cyclones – occur around the tropical oceans each year, familiar enough to be given names. Hurricane Gilbert that caused devastation on the island of Jamaica and the coast of Mexico in 1988, Typhoon Mireille that hit Japan in 1991, Hurricane Andrew that caused a great deal of damage in Florida and other regions of the southern United States in 1992 and Hurricane Mitch that caused great devastation in Honduras and other countries of central America in 1998 are notable recent examples. Low-lying areas such as Bangladesh are particularly vulnerable to the storm surges associated with tropical cyclones; the combined effect of intensely low atmospheric pressure, extremely strong winds and high tides causes a surge of water which can reach far inland. In one of the worst such disasters in the twentieth century over 250 000 people were drowned in Bangladesh in 1970. The people of that country experienced another storm of similar proportions in 1999 as did the neighbouring Indian state of Orissa also in 1999, and smaller surges are a regular occurrence in that region.

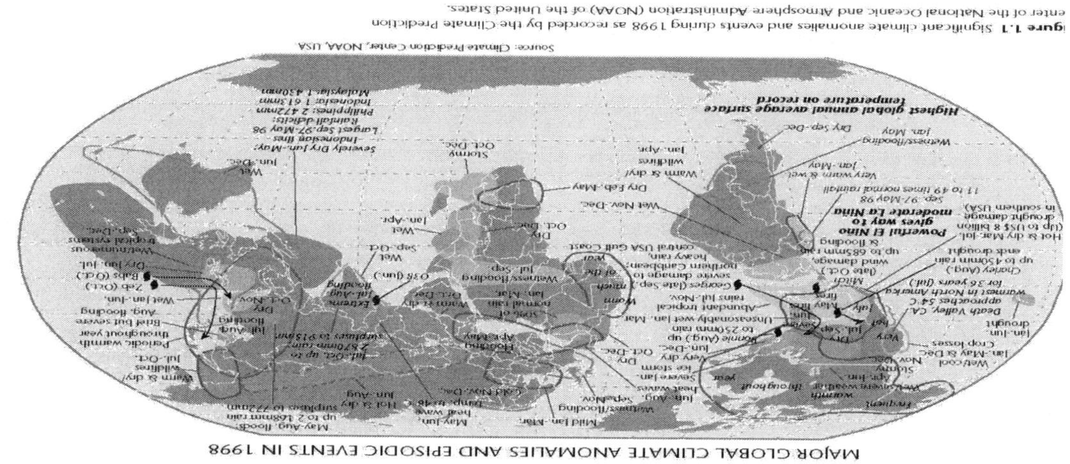

Fig. 1: Significant climate anomalies and events during 1998 as recorded by the Climate Prediction Center of the National Oceanic and Atmosphere Administration (NOAA) of the United States.

Familiar enough to be given names. Hurricane Gilbert that caused devastation on the island of Jamaica and the coast of Mexico in 1988, Typhoon Mireille that hit Japan in 1991, Hurricane Andrew that caused a great deal of damage in Florida and other regions of the southern United States in 1992 and Hurricane Mitch that caused great devastation in Honduras and other countries of central America in 1998 are notable recent examples. Low-lying areas such as Bangladesh are particularly vulnerable to the storm surges associated with tropical cyclones; the combined effect of intensely low atmospheric pressure, extremely strong winds and high tides causes a surge of water which can reach far inland. In one of the worst such disasters in the twentieth century over 250 000 people were drowned in Bangladesh in 1970. The people of that country experienced another storm of similar proportions in 1999 as did the neighbouring Indian state of Orissa also in 1999, and smaller surges are a regular occurrence in that region.

The increase in storm intensity during recent years has been tracked by the insurance industry, which has been hit hard by recent disasters. Until the mid 1980s, it was widely thought that windstorms or hurricanes with insured losses exceeding one billion (thousand million) US dollars were only possible, if at all, in the United States. But the gales that hit western Europe in October 1987 heralded a series of windstorm disasters which make losses of ten billion dollars seem commonplace.

Hurricane Andrew, for instance, left in its wake insured losses estimated at nearly twenty-one billion dollars (1999 prices) with estimated total economic losses of nearly thirty-seven billion dollars. Figure 2 shows the costs of weather-related disasters2 over the past fifty years as calculated by the insurance industry. It shows an increase in economic losses in such events by a factor of over 10 in real terms between the 1950s and the 1990s. Some of this increase can be attributed to the growth in population in particularly vulnerable areas and to other social or economic factors; the world community has undoubtedly become more vulnerable to disasters. However, a significant part of it has also arisen from the increased storminess in the late 1980s and 1990s compared with the 1950s.

Windstorms or hurricanes are by no means the only weather and climate extremes that cause disasters. Floods due to unusually intense or prolonged rainfall or droughts because of long periods of reduced rainfall (or its complete absence) can be even more devastating to human life and property. These events occur frequently in many parts of the world especially in the tropics and sub-tropics. There have been notable examples during the last two decades. Let me mention a few of the floods. In 1988, the highest flood levels ever recorded occurred in Bangladesh, eighty per cent of the entire country was affected; China experienced devastating floods affecting many millions of people in 1991, 1994–5 and 1998; in 1993, flood waters rose to levels higher than ever recorded in the region of the Mississippi and Missouri rivers in the United States, flooding an area equivalent in size to one of the Great Lakes; major floods in Venezuela in 1999 led to a large landslide and left 30 000 people dead, and two widespread floods in Mozambique occurred within a year in 2000–1 leaving over half a million homeless. Droughts during these years have been particularly intense and prolonged in areas of Africa, both north and south. It is in Africa especially that they bear on the most vulnerable in the world, who have little resilience to major disasters. Figure.3 shows that in the 1980s droughts accounted for more deaths in Africa than all other disasters added together and illustrates the scale of the problem.

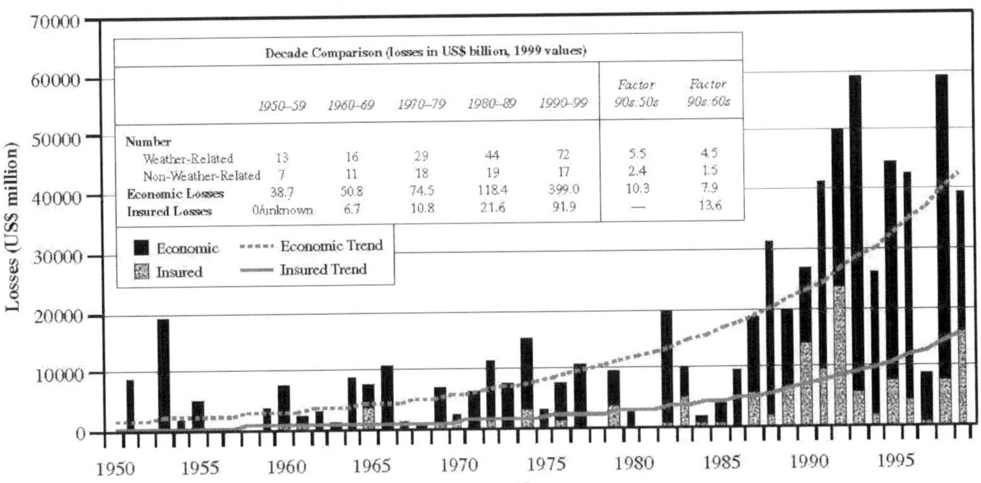

Fig. 2: The total economic costs and the insured costs of catastrophic weather events for the second half of the twentieth century as recorded by the Munich Re insurance

company. Both costs show a rapid upward trend in recent decades. The number of non-weather-related disasters is included for comparison.

6.3 EL NI˜NO EVENTS

Rainfall patterns which lead to floods and droughts especially in tropical and semi-tropical areas are strongly influenced by the surface temperature of the oceans around the world, particularly the pattern of ocean surface temperature in the Pacific off the coast of South America . About every three to five years a large area of warmer water appears and persists for a year or more. Because they usually occur around Christmas these are known as El Ni˜no ('the boy child') events.3 They have been well known for centuries to the countries along the coast of South America because of their devastating effect on the fishing industry; the warm top waters of the ocean prevent the nutrients from lower, colder levels required by the fish from reaching the surface.

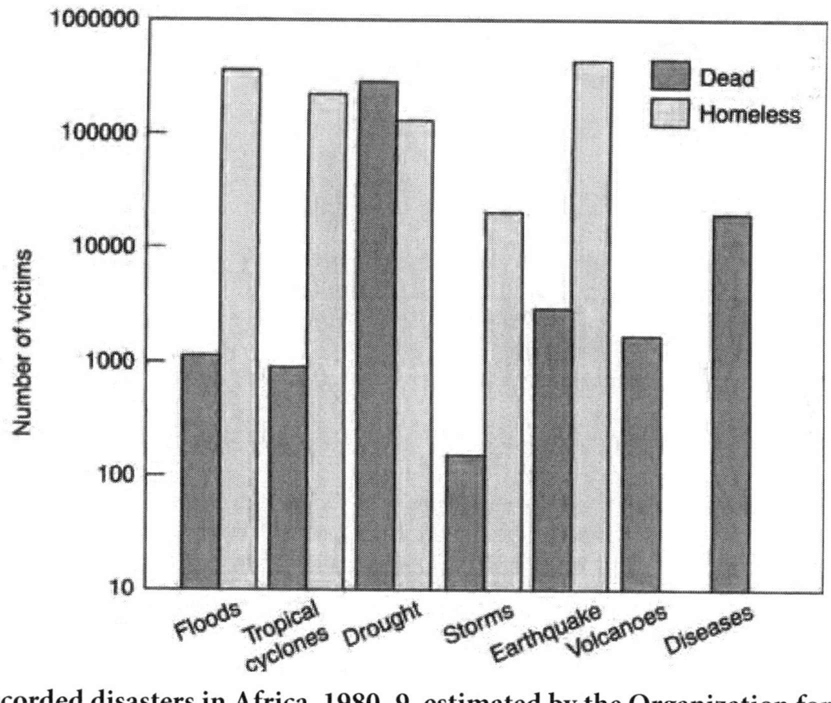

Fig. 3: Recorded disasters in Africa, 1980–9, estimated by the Organization for African Unity.

A particularly intense El Ni˜no, the second most intense in the twentieth century, occurred in 1982–3; the anomalous highs in ocean surface temperature compared to the average reached 7 °C. Droughts and floods somewhere in almost all the continents were associated with that El Ni˜no (Figure.4). Like many events associated with weather and climate, El Ni˜nos often differ very much in their detailed character; that has been particularly the case with the El Ni˜no events of the 1990s. For instance, the El Ni˜no event that began in 1990 and reached maturity early in 1992, apart from some weakening in mid 1992, continued to be dominated by the warm phase until 1995. The exceptional floods in the central United States and in the Andes, and droughts in Australia and Africa are probably linked with this unusually protracted El Ni˜no. This, the longest El Ni˜no of the twentieth century, was followed in 1997–8 by the century's most intense El Ni˜no that brought exceptional floods to China and to the Indian sub-continent and drought to Indonesia – that in turn brought extensive forest fires creating an exceptional blanket of thick smog that was experienced over a thousand miles away (Figure 1). Studies with computer models provide a scientific basis for links between the El Ni˜no and these extreme weather events; they also give some confidence that useful forecasts of such disasters will in due course be possible. A scientific question that is being urgently addressed is the possible link between the character and intensity of El Ni˜no events and global warming due to human-induced climate change.

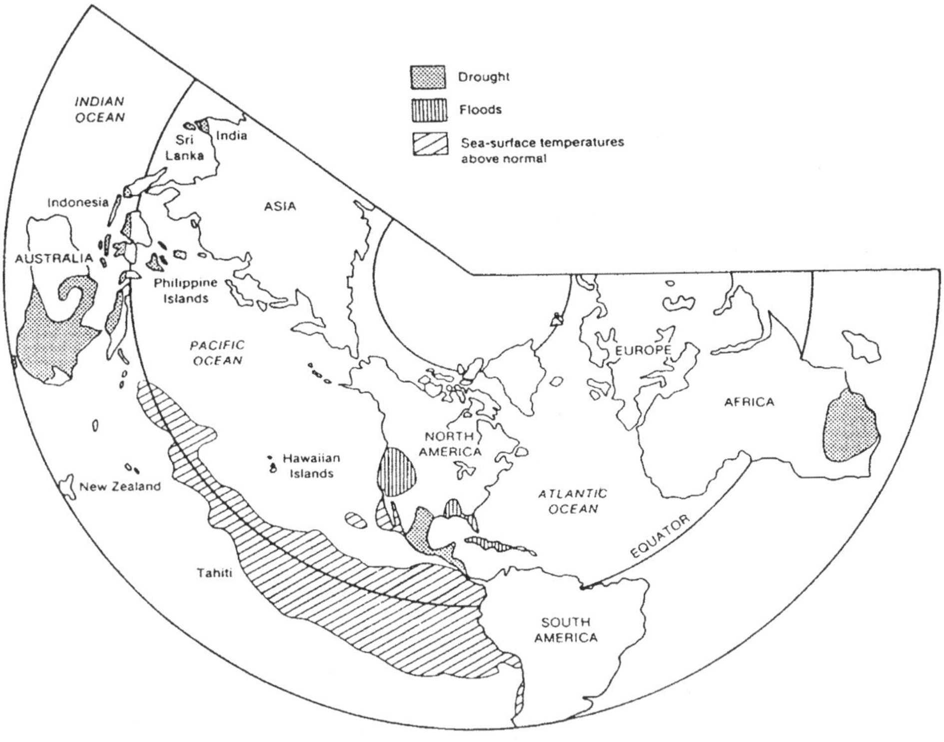

Fig. 4: Regions where droughts and floods occurred associated with the 1982–3 El Ni˜no.

6.4 THE EFFECT OF VOLCANIC ERUPTIONS ON TEMPERATURE EXTREMES

Volcanoes inject enormous quantities of dust and gases into the upper atmosphere. Large amounts of sulphur dioxide are included, which through photochemical reactions using the Sun's energy are transformed to sulphuric acid and sulphate particles. Typically these particles remain in the stratosphere (the region of atmosphere above about 10 km in

altitude) for several years before they fall into the lower atmosphere and are quickly washed out by rainfall. During this period they disperse around the whole globe and cut out some of the radiation from the Sun, thus tending to cool the lower atmosphere. One of the largest volcanic eruptions in the twentieth century was that from Mount Pinatubo in the Philippines on 12 June 1991 which injected about twenty million tonnes of sulphur dioxide into the stratosphere together with enormous amounts of dust. This stratospheric dust caused spectacular sunsets around the world for many months following the eruption. The amount of radiation from the Sun reaching the lower atmosphere fell by about two per cent. Global average temperatures lower by about a quarter of a degree Celsius were experienced for the following two years. There is also evidence that some of the unusual weather patterns of 1991 and 1992, for instance unusually cold winters in the Middle East and mild winters in western Europe, were linked
with effects of the volcanic dust.

6.5 VULNERABLE TO CHANGE

Over the centuries different human communities have adapted to their particular climate; any large change to the average climate tends to bring stress of one kind or another. It is particularly the extreme climate events and climate disasters which emphasise the importance of climate to our lives and which demonstrate to countries around the world their vulnerability to climate change – a vulnerability which is enhanced by rapidly increasing demands on resources. But the question must be asked: how remarkable are these events? Do they point to a changing climate due to human activities? Do they provide evidence for global warming because of the increased carbon dioxide and other greenhouse gases being emitted into the atmosphere by burning fossil fuels?

Here a note of caution must be sounded. The range of normal natural climate variation is large. Climate extremes are nothing new. Climate records are continually being broken. In fact, a month without a broken record somewhere would itself be something of a record! Changes in climate that indicate a genuine long-term trend can only be identified after many years.

However, we know for sure that, because of human activities especially the burning of fossil fuels, carbon dioxide in the atmosphere has been increasing over the past two hundred years and more substantially over the past fifty years. To identify climate change related to this carbon dioxide increase, we need to look for trends in global warming over similar lengths of time. They are long compared with both the memories of a generation and the period for which accurate and detailed records exist. Although, therefore, it can be ascertained that there was more storminess, for instance, in the region of the north Atlantic during the 1980s and 1990s than in the previous three decades, it is difficult to know just how exceptional those decades were compared with other periods in previous centuries. There is even more difficulty in tracking detailed climate trends in many other parts of the world, owing to the lack of adequate records; further, trends in the frequency of rare events are not easy to detect.

The generally cold period worldwide during the 1960s and early 1970s caused speculation that the world was heading for an ice age. A British television programme about climate change called 'The ice age cometh' was prepared in the early 1970s and widely screened – but the cold trend soon came to an end. We must not be misled by our relatively short memories.

What is important is continually to make careful comparisons between practical observations of the climate and its changes and what scientific knowledge leads us to expect. During the last few years, as the occurrence of extreme events has made the public much more aware of environmental issues, scientists in their turn have become more sure about just what human activities are doing to the climate. Later chapters will look in detail at the science of global warming and at the climate changes that we can expect, as well as investigating how these changes fit in with the recent climate record. Here, however, is a brief outline of our current scientific understanding.

6.6 THE PROBLEM OF GLOBAL WARMING

Human activities of all kinds whether in industry, in the field (e.g. deforestation) or concerned with transport or the home are resulting in emissions of increasing quantities of gases, in particular the gas carbon dioxide, into the atmosphere. Every year these emissions currently add to the carbon already present in atmospheric carbon dioxide a further seven thousand million tonnes, much of which is likely to remain there for a period of a hundred years or more. Because carbon dioxide is a good absorber of heat radiation coming from the Earth's surface, increased carbon dioxide acts like a blanket over the surface, keeping it warmer than it would otherwise be. With the increased temperature the amount of water vapour in the atmosphere also increases, providing more blanketing and causing it to be even warmer. Being kept warmer may sound appealing to those of us who live in cool climates. However, an increase in global temperature will lead to global climate change. If the change were small and occurred slowly enough we would almost certainly be able to adapt to it. However, with rapid expansion taking place in the world's industry the change is unlikely to be either small or slow. The estimate I present in later chapters is that, in the absence of efforts to curb the rise in the emissions of carbon dioxide, the global average temperature will rise by about a third of a degree Celsius every ten years – or about three degrees in a century.

This may not sound very much, especially when it is compared with normal temperature variations from day to night or between one day and the next. But it is not the temperature at one place but the temperature averaged over the whole globe. The predicted rate of change of three degrees a century is probably faster than the global average temperature has changed at any time over the past ten thousand years. And as there is a difference in global average temperature of only about five or six degrees between the coldest part of an ice age and the warm periods in between ice ages (see Figure 4.4), we can see that a few degrees in this global average can represent a big change in climate. It is to this change and especially to the very rapid rate of change that many ecosystems and human communities (especially those in developing countries) will find it difficult to adapt.

Not all the climate changes will in the end be adverse. While some parts of the world experience more frequent or more severe droughts, floods or significant sea level rise, in other places crop yields may increase due to the fertilising effect of carbon dioxide. Other places, perhaps for instance in the sub-arctic, may become more habitable. Even there, though, the likely rate of change will cause problems: large damage to buildings will occur in regions of melting permafrost, and trees in subarctic forests like trees elsewhere will need time to adapt to new climatic regimes. Scientists are confident about the fact of global warming and climate change due to human activities. However, substantial uncertainty remains about just how large the warming will be and what will be the patterns of change in different parts of the world. Although some indications can be given, scientists cannot yet say in precise detail which regions will be most affected. Intensive research is needed to improve the confidence in scientific predictions.

6.7 ADAPTATION AND MITIGATION

An integrated view of anthropogenic climate change is presented in Figure.5 where a complete cycle of cause and effect is shown. Begin in the lower right-hand corner where economic activity, both large and small scale, whether in developed or developing countries, results in emissions of greenhouse gases (of which carbon dioxide is the most important) and aerosols. Moving in a clockwise direction around the diagram, these emissions lead to changes in atmospheric concentrations of important constituents that alter the energy input and output of the climate system and hence cause changes in the climate. These climate changes impact both humans and natural ecosystems altering patterns of resource availability and affecting human livelihood and health. These impacts in their turn affect human development in all its aspects. An anticlockwise arrow represents other effects of development on human communities and natural systems, for instance changes in land use that lead to deforestation and loss of biodiversity. Figure.5 also shows how both causes and effects can be changed through *adaptation* and *mitigation*. In general adaptation is aimed at reducing the effects and mitigation is aimed at reducing the causes of climate change, in particular the emissions of the gases that give rise to it.

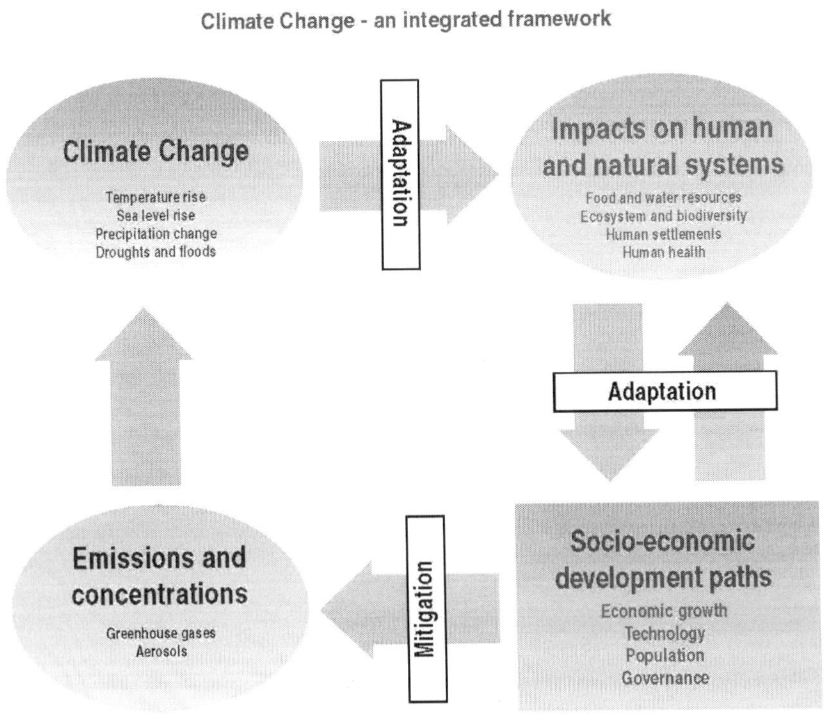

Fig. 5: Climate change – an integrating framework; see text for explanation.

6.8 UNCERTAINTY AND RESPONSE

Predictions of the future climate are surrounded with considerable uncertainty that arises from our imperfect knowledge both of the science of climate change and of the future scale of the human activities that are its cause. Politicians and others making decisions are therefore faced with the need to weigh all aspects of uncertainty against the desirability and the cost of the various actions that can be taken in response to the threat of climate change. Some mitigating action can be taken easily at relatively little cost (or even at a net saving of cost), for instance the development of programmes to conserve and save energy, and many schemes for reducing deforestation and encouraging the planting of trees. Other actions such as a large shift to energy sources that are free from significant carbon dioxide emissions (for example, renewable sources – biomass, hydro, wind, or solar energy) both in the developed and the developing countries of the world will take some time. Because however of the long timescales that are involved in the development of new energy infrastructure and in the response of the climate to emissions of gases like carbon dioxide, there is an urgency to begin these actions now.

7

LAND VALUE CHANGES AND CARBON SEQUESTRATION: AS AN ECOSYSTEM

7.1 INTRODUCTION: AN OVERVIEW

> *Land use change plays a very significant role in climate change mitigation and adaptation, and a land value database could provide both direct and indirect benefits. Land prices or values integrate supply and demand information and could also provide better information on future land-base conditions for climate change analysts. For example, for forest ecosystems, they could provide a thorough and unified description of anticipated change in the valuation by society.*

Global climate change induced by anthropogenic release of greenhouse gases (GHGs), including carbon dioxide (CO_2), may modify the growth and geographic distribution of forests, as well as productivity of competing land uses. Impacts on agriculture and forests could arise from increases in atmospheric CO_2 concentrations, change in temperature regimes, and variations in patterns of rainfall over the course of a year. Shifts in global climatic conditions could affect agricultural crop and forest growing conditions, and hence land values for those uses. Climate change policies may in turn affect land and resource markets, thereby modifying land values, land use, and forest cover distribution. A broad examination of such changes necessitates assessing multiple markets, including those for carbon sequestration as an ecosystem service, especially if policies arise to promote carbon sequestration as a mitigation activity.

Carbon credits and carbon markets are often key components of proposed national and international attempts to mitigate the growth in concentrations of GHGs. At the same time as policy is causing land-use changes, climate change itself is expected to cause changes to environmental conditions, influencing the characteristics and growth of forests in rural and urban settings, as well as agricultural production and that of other competing land uses. The focus of this paper is how climate change may change land values at broad scales, both from the direct impact from future altered climate conditions and from policies formulated to address climate change and increase the supply of tradable carbon credits. We summarize findings from U.S. studies, to illustrate examples of land value changes in response to climate change and associated policies. Understanding land use changes and land values in the face of climate change is important for creating efficient strategies for the mitigation of climate change at local, regional, and global levels.

7.2 CHANGES IN LAND VALUES

Land resources such as forest land can produce a future flow of recurring net returns or land rents. Land values typically represent the present value of such net returns, dependent on estimating the future net returns and the choice of an appropriate capitalization rate. Theoretical models of land-use change are grounded in the landowner's dynamic optimization problem. A land- owner is assumed to choose the land use that maximizes the net present value (NPV) of the stream of rents to his or her land. A basic model posits that a change in land use occurs whenever the NPV of the rents to an alternative land use exceeds the NPV of the rents to the current land use by more than the cost of conversion.1 Over time, changes in supply and demand for land affect relative land rents. The rents accruing to any particular land use are determined by a host of factors such as land productivity, location, and demographics (e.g., population or income changes). Usefully, land rents reflect the interaction of these determinants of land values, which are frequently difficult to observe. Many empirical analyses of land-use change use the rents accruing to different land uses to model and forecast land-use transitions. Given the infinite time horizon, our lack of perfect knowledge and foresight complicates the determination of future rents.

To the extent that we are interested in how the allocation of land uses may change with possible carbon revenues, we are also interested in how land values as a key measure may be affected by carbon markets.

Some empirical studies also incorporate market values of land, which when expressed as a stream of rents are conceptually similar to net rents. However, market values may incorporate expectations about the future differently than do calculated net returns. First, market values may incorporate expectations of changes in land use. For example, forest-land prices anticipate future development close to urbanizing areas.

The market value of land can be thought of as having two components:

- The value of present assets (e.g., structures or standing crops or timber) and
- The speculative value of future land uses and assets that incorporates expectations about a wide range of variables including commodity prices, development patterns, government policies such as for carbon sequestration, and regional population and income levels. Inconsistencies between real estate values and

rents for the current land use can be indicative of an expected land-use change. Second, the time horizon for which expectations are formed is likely to be longer for market values.

Projected shifts in land use reflect economic and soil quality conditions in the case of forest and agriculture, whereas conversion of forest land to development is influenced significantly in many cases by location. Increases in population and the demand for land can give rise to incentives to bring less fertile lands into use. For example, with expanded development, growing scarcity of forest land acts to raise timber prices. If timber prices and associated timber-related incomes rise enough, it may be profitable for some owners to intensify timber management on some lands and also afforest additional land. At the same time, changes in transportation costs may affect the extent of the area within which forest products can be profitably produced. Such changes affect land values in forestry.

Urban and developed uses typically sit on top of the economic hierarchy of land uses, with rents often at least an order of magnitude higher than those for forest land. It is reported that the ratio of average value of land in urban use compared to forest use is approximately 87 in the Southeast and 111 in the Pacific Northwest Westside region. The highest forest values on a county basis in the Pacific Northwest Westside are about 0.04 the lowest urban values and 0.007 the highest urban value. The ratios in the Southeast are roughly similar; however, overall, the Pacific Northwest Westside has much larger land values in forest use and urban use. The Pacific Northwest Westside has 14 counties with land values at least $200,000 per acre ($494,200 per ha) in urban use and $2,000 per acre ($4,942 per ha) in forest use, and the Southeast has none. Such higher rents mean that determinants of land-use transitions in many cases are demand-side factors pertaining to developed uses, such as population and income. Thus, in interface areas such as metro or urbanizing locations, the economic hierarchy of land uses suggests that, in land markets, development-related land-use factors tend to strongly dominate forestry-related ones.

7.3 ECOSYSTEM SERVICE MARKETS

In this section we discuss the implications of climate change for ecosystem service provision and examine how changes in ecosystem service provision affect land values. Defining ecosystem services— broadly defined, ecosystem services are the benefits people derive from the natural processes that sustain ecosystems. They can be generally atalogued into four broad areas: supporting, regulating, provisioning, and cultural. Supporting services include basic ecosystem functions like soil formation, and provisioning services are important sources of food and fiber. Regulating services help control climate change through carbon sequestration. Cultural services include recreation and education. So ecosystem services comprise both goods and processes.

Although the concept of ecosystem services is inherently based on the value or importance to humans, the underlying ecological structure and processes involve complexities that complicate quantifying these services at different scales. In some cases, the linkage between structure, processes, and resulting services is fairly straightforward. For example, the degree to which a specific plant community can support a given wildlife population can be determined directly by measuring community attributes such as species composition, height, and age. Other services, such as improving water quality by converting nitrate to nitrogen gas through denitrification, are controlled by more complex interactions among multiple eco- system attributes that are more difficult to measure (for example, carbon, redox status, denitrifier population, and temperature). These relationships are also altered by temporal and condition gradients, which result in dynamic processes and significant variability across and within different ecosystems, making them difficult to measure and quantify at large spatial scales. Ecological production function models based on biophysical inputs are often used to produce spatial estimates of specific services.

An ecosystem service differs from an ecosystem function because a service takes the human context into account. For example, consider a wetland that pro- vides, among other benefits, flood control. The level of this ecosystem function is the wetland's capacity to absorb rising flood waters and gradually release them over time. The level of ecosystem service is the benefit of that flood control; thus it depends on what lies downstream. A wetland that contributes to flood control for a town or city provides a higher level of service than a wetland protecting undeveloped land. Location is important.

Fig. 1: Efforts to reduce the divergence between the private provision of ecosystem services and socially optimal amounts has involved payments for ecosystem services.

Theoretically, ecosystem services tend to be under supplied by the private sec- tor because they often represent a positive externality. Although a landowner could manage to increase the provision of an ecosystem service, little economic incentive exists to do so because the benefits of these management practices do not accrue solely to the landowner and little increased revenue may result from increased provision of that service. Ultimately, the provision of ecosystem services flowing from private lands is less than what is desirable to society collectively. Payments for ecosystem services (PES) constitute one strategy to increase the provision of ecosystem services via an economic incentive. Examples include government programs (e.g., the U.S. Conservation Reserve Program [CRP]), eco-labeling and certification programs (if the rents flow to farmers), and purchase of conservation easements (e.g., by The Nature Conservancy for increasingly prevalent land trusts). The provision of ecosystem services from private lands could also be increased through regulations (e.g., best management practices for forestry). In this context, regulations mandate desirable management practices, and PES programs incentivize desirable management practices.

The goal of PES programs is to increase, on the margin, the provision of eco- system services by paying landowners to adopt beneficial management practices. Carbon sequestration is a prominent example in this context. Sequestering carbon should mitigate negative climate change impacts for society at large. In the absence of some sort of carbon market, landowners will sequester too little carbon because they do not reap the full benefits of sequestration. A carbon market or allowance system incentivizes private land sequestration by making sequestration pay. Given the opportunity to increase revenues, many landowners may change behavior to increase carbon sequestration to levels that are more socially desirable.

The level of ecosystem service provision from a parcel or a region depends on site characteristics (e.g., soil, typical climate conditions—average rainfall, typical growing season, etc.), stochastic shocks (e.g., variable weather patterns, natural disasters), and land management practices. Altering management practices should alter the provision of services, holding other factors constant. Thus, the relation- ship between ecosystem service and ecosystem function varies with management practices.

Effects of changing climate— In general, changing climate conditions can be expected to result in changes to site characteristics and stochastic shocks—changes to ecosystem function. Thus, for a given site, the average conditions will change as will the frequency and severity of departures from that average. The implications for ecosystem service provision depend on management response. With static management, ecosystem service and function should move together. However, human adaptation to changing conditions can mitigate the ecosystem service impacts. For example, it is showed that climate change impacts on forestry in North America would be significantly altered by management practices including salvaging dead trees and speeding up the transition between species. In fact, their results suggest gains for the U.S. forestry sector from climate change. Thus, climate change interacts with human behavior, determining the outcome in terms of ecosystem services in the process. Although there is a general scientific consensus that climate change will be costly overall, the impacts on particular sectors or services may be positive in some regions.

Changes in land values from ecosystem services provision— But how does ecosystem service provision affect land value, and thus how will climate-induced changes in ecosystem service provision affect land value? First, climate change affects land value through changes in productivity. In forestry and agriculture, site productivity drives land values and rental rates. So as climate change affects productivity, land value changes. Over time, land tends to move to the highest value use. Thus, if we see land moving from forestry to agriculture, we can surmise that returns to agriculture currently outstrip returns to forestry.

However, returns to agriculture and forestry do not depend solely on productivity of traditional products. Returns also depend on payments landowners receive from other sources (e.g., government conservation programs, conservation groups, carbon markets). Payments for ecosystem service schemes enhance the returns to a particular land use or management strategy, rendering that strategy relatively more attractive. Recent years have seen significant expansions of PES schemes. Government expenditures have increased steadily in both the United States and the European Union, and the involvement of conservation groups and institutional actors (such as the Chicago Climate Exchange) is also on the rise. Payment for ecosystem services programs essentially creates markets for services that previously had no markets. As more and more markets come online, this should push land values up. Indeed, empirical evidence shows that the CRP, a PES pro- gram that pays farmers to remove environmentally sensitive land from production, has resulted in higher values for both agricultural and developed land. Simultaneously, the share of land value attributable to ecosystem services should rise, assuming all other things being held constant.

Growing concern over climate change (and other environmental problems) is a major driver of PES programs, which could affect land value. Carbon sequestration is a particularly important ecosystem service in this context. With a carbon-trading scheme in place, landowners will have the option of managing to enhance carbon sequestration and selling this enhanced sequestration potential in the form of carbon credits. Agriculture and forestry differ in terms of carbon sequestration potential, so the introduction of a carbon market affects relative returns, land value, and land use.

Of course, carbon sequestration rates, similar to other ecosystem services, will vary with climate change. The direction of this movement is important. If carbon sequestration potential rises with increased CO_2 concentration, we have a positive feedback loop. Higher CO_2 concentrations could lead to enhanced sequestration rates. Holding land use constant, the effect of sequestration on climate is self-limiting, with a limit on how long the positive feedback would be significant or exist. If, on the other hand, sequestration potential falls with increased CO_2 concentration, we have a negative feedback loop. Higher concentrations lead to lower sequestration, so concentrations continue to rise.

In either case, the market response will enhance the feedback effects. If carbon sequestration potential rises with CO_2 concentration, the return from using a given parcel to sequester carbon also rises. On the margin, some land moves into carbon sequestration, dampening climate change further. If sequestration falls as concentration increases, the return from using a parcel to sequester carbon also falls. Thus, on the margin, some land will move to other uses, contributing to further increases in CO2 concentration.

Given the economic hierarchy of land uses, the addition of carbon-related revenues associated with that ecosystem service are not likely to cause major changes in land allocation between developed uses and forestry in interface areas or urbanizing areas. However, adding carbon-related revenues to the future stream of net returns for forestry may boost benefits enough to change the allocation of land between forestry and agriculture in some regions. Agricultural land is afforested because expected incomes per hectare under forest use rise above the expected returns per hectare associated with continuing agricultural use. Increasing financial incentives to plant trees, where plantations store multiple metric tons of forest carbon per hectare, would boost forest incomes in some cases well beyond land rental rates from agriculture. As an example, afforested cropland in mixed hardwood forest in the Corn Belt region is estimated to sequester approximately 60 metric tons of forest ecosystem carbon per ha by age 20. A $15 price per metric ton of CO2 is equivalent to about $55 per metric ton of carbon. At this price, the carbon payments from the hardwood stand would amount to $8,154 per hectare ($3,300 per acre) as a lump sum at age 20. As an equal annual payment over the same 20-year period, that would be equivalent to about $274 per hectare ($111 per acre) per year (at a 4 percent real discount rate).

The $15 price for CO_2 could result in several hundred thousand more hectares of afforestation in the Corn Belt region as compared to business as usual. If that price per metric ton is doubled to $30, then about 2.5 million more hectares could be afforested with carbon markets. Other ecosystem services also provide benefits, and some values could be estimated using techniques such as the benefit transfer method. However, unless the amounts are very large, they are not likely to tip the land allocation balance between forest and developed uses. Further, unless there is some sort of market or PES scheme, enhanced ecosystem services may enhance landowner utility, but may not be as likely to be capitalized in land value. One example of another service is nitrogen uptake, sometimes discussed alongside carbon sequestration. In addition to benefits, analysts need estimates of costs for providing ecosystem services, such as converting agricultural land to wetlands.

Considering both benefits and costs allows for net benefits to be estimated, including examining the present values of benefits and costs. Although costs of a mitigation activity may be significant, values of the marketable commodities along with the potential ecosystem services values could be even greater. Thus, depending on the assumptions of the benefits to be included, the return on investment in the mitigation activity could be significant.

Fig. 2: Afforestation on erodible or other environmentally sensitive agricultural land can enhance ecosystem services and co-benefits, such as enhanced wildlife habitat, in addition to contributing to climate change mitigation.

Explicitly recognizing the complex relationships between ecosystem structure, processes, and services is critical to understanding the potential ancillary effects of carbon sequestration strategies. Any change, either anthropogenic or naturally occurring, that affects structural components such as plant community composition or processes such as nutrient cycling, will affect the quality, quantity, and types of services produced from that ecosystem. Quantification of the effect is a difficult task, as some services, such as biodiversity, can be both a cause of ecosystem functioning and a response to changing management activity. Thus, effects of carbon- specific components are hard to separate. Another problem is that the responses of multiple services to specific carbon-related management activities have not been well studied. It concluded that policies aimed at increasing carbon sequestration did not necessarily increase species conservation and that highly targeted policies were not necessarily better than more general policies. The study by Nelson et al. (2008) demonstrates the likelihood that many of the possible management activities and sequestration strategies will affect ecosystem services of direct importance to landowners and managers. For example, afforestation designed to increase carbon sequestration will alter migratory bird habitat depending on the location and species composition of the forest, so ecological tradeoffs will be involved.

Understanding prospects for carbon markets— The promise of ecosystem service markets greatly depends on the particular circumstances of program implementation, including what services are to be traded and whether they are amenable to trading, the ability to enact and enforce regulation sufficient to induce trading, and how expected program results compare with those likely to arise from other conservation policy approaches. In the case of tradable forest carbon credits, still needed from an institutional viewpoint is a cap on the amount of carbon emissions. In an unregulated world, carbon emissions exemplify the problem with externalities and public goods. Car- bon emitters can carry out production without fully internalizing the social costs of the carbon they emit. All producers contribute to total atmospheric carbon and the resulting negative externality of global climate change. The lack of property rights to the atmosphere provides producers little incentive not to pollute, because they need not pay for the right to emit carbon.

With a regulatory approach, one proposal is to reduce total emissions by establishing a uniform cap on the emissions that individual producers can emit. Thus, the total cap is the uniform cap times the number of producers. A uniform cap is not the least-cost option for achieving the total cap because different producers face different marginal pollution control costs. The least-cost or cost-effective option equalizes marginal control cost across producers. The intent of creating a carbon market is to provide a compliance option for heavy emitters—the ability to purchase allowances to maintain production without having to reduce their carbon emissions. Increasing the flexibility of compliance options for emitters in this way equalizes marginal cost and lowers the cost of achieving the total cap. Over time, the regulatory agency could ratchet down the overall cap on emissions to gradually reduce total carbon emissions.

In this carbon market, forest owners could conceivably participate by selling offsets in the form of tradable carbon credits, increasing the supply of emission or damage allowances in a market. The forest carbon owners are performing activities that offset damage caused by others. Landowners might be given the right to sell emissions allowances based on forest management or afforestation activities that increase forest carbon storage on their lands above some predetermined baseline.

Next, we summarize aspects of land-use changes, land markets, and land values based on modeling studies. The opportunity for exurban and urban development's to influence the supply of terrestrial carbon sequestration is outside the scope of this chapter, and interested readers are referred to examples of development area projections. In this chapter, we focus on land-use change within a climate change context, although land-use change can have many implications for a landscape and society. Land-use changes can also affect product prices, wildlife habitat, and many other parts of ecosystems and the interconnected economic system.

7.4 LAND VALUE PROJECTIONS USING THE FASOM-GHG MODELING SYSTEM

Global questions involving forestry's and agriculture's potential contributions to climate change mitigation are framed within a context of increased demands for cropland, forage, and wood products to help feed and house an additional 3 billion people globally by 2050, increased land demand for bioenergy production, and millions of hectares of land needed to house another 125 million U.S. residents by midcentury. Here we illustrate how the Forest and Agricultural Sector Optimization Model—Greenhouse Gas (FASOM-GHG) modeling system is used to examine such linked forestry and agricultural issues.

Examples of components in FASOM-GHG's integrated mitigation analysis include:

- Simultaneous consideration of the agricultural (including both crops and livestock) and forestry sectors;
- Discount rate;
- Modeling of impacts on traditional industries;
- Dynamic modeling of forest stand growth and investment;
- GHG accounting, including non-CO_2 gases;
- GHG (e.g., CO_2) prices; (g) modeling conversion of forest and agricultural lands to developed uses (e.g., urban); (h) examination of bioelectricity as part of bioenergy analyses; and
- Land-use changes between the forestry and agricultural sectors, among other components.

Examining land use change and climate change can involve looking at effects of climate change on forestry and agricultural activities, adaptation activities, mitigation activities, and the interplay among all three. Within the narrower mitigation modeling, such modeling should account both for the GHG impacts of substituting bioenergy for fossil fuels and any GHG changes owing to land-use change—so we should get the net GHG impact of any change to bioenergy production. Increasingly, modeling could point out exactly the kinds of perverse GHG incentive results described in some of the literature. For example, some types of subsidies could lead to reduced carbon sequestration as an ecosystem service. Including both the agricultural and forest sectors, modelers can directly recognize the tradeoffs between the needs to reduce GHG emissions and expand food production. Models such as the Forest and Agricultural Optimization Model demonstrate the forest environmental impacts of different types of policies. Tradeoffs with food and environment will occur, and we can use models such as FASOM-GHG to investigate such tradeoffs along the path to reducing net GHGs.

In the next section, we examine studies that project how future climate condi- tions may impact land values, as well as how policies proposed in response to climate change may affect forest-land values. This involves using models that link expected forest changes from climate change to the inputs and parameters used in forest sector economic models. A linked model of the forest and agricultural sectors simulates land-use competition between the two sectors, with resultant impacts on land allocation and land values in the sectors.

7.5 MODELING CHANGES FROM CLIMATE CHANGE

Most studies thus far have assumed that the total area of land in agriculture and the total area of land in forests will remain about the same even as climate changes.

For example, scientists valued the impact of climate change as a large-scale ecological change in the U.S. timber market, without a mechanism for accounting for any productivity-related shifts between forestry and agriculture. Although the total area of land in agriculture and forestry may indeed remain relatively constant over time, climate change could alter the distribution of land uses over time. Given that forest areas could be affected and the potential human and social consequences of these impacts, it is important to consider and assess the implications of climate change on the distribution of U.S. land uses.

Differential impacts of climate change in agriculture and forestry could lead to land-use shifts as one possible adaptation strategy by landowners. For example, if climate change results in relatively higher agricultural productivity per hectare, some hectares may be converted from forests to agricultural use. Such changes would alter the supply of products to national and international markets, changing the prices of forest products and the economic well-being of both producers and consumers. Conversely, if climate change renders forestry more attractive than agriculture in terms of yields and production costs, land could shift from agricul- ture to forestry as these two sectors adjust. Given that the agricultural and forestry sectors sometimes compete for the same land, shifts in productivity of agricultural land could affect the ultimate distribution of forest land, and vice versa.

Using four climate change scenarios from the national assessment of climate change, and found that less forested area was projected under four climate scenarios relative to the base case (no climate change). Furthermore, less cropland and more pasture land were projected to convert to forests under all scenarios. In general, climate change was projected to increase the overall supply of timber, causing timber prices to fall and reducing forest-land values. The net conversion of forest land to agricultural use was prompted by aggregate increases in forest productivity relative to agriculture, with price signals in markets leading to adjustments. Although climate change is likely to affect the margin between forestry and agriculture differently in specific locations, aggregate productivity changes in forestry appear larger overall than aggregate productivity changes in agriculture.

Fig. 3: Impacts of climate change on agricultural yields

Other studies have examined impacts of climate change on agricultural yields. For example, scientists estimated impacts of climate change on crop yields, considering the importance of nonlinear temperature effects. However, they did not consider any land use shifts involving forestry that might arise from impacts on productivity in the two sectors.

To our knowledge, no study to date has isolated the impacts on forest-land val- ues from carbon sequestration owing to climate change. The earlier version of the FASOM model) did not have carbon-related prices in the objective function. Scientists' did note changes in economic welfare by sector associated with the different climate change scenarios, with changes relatively small compared to those associated with policy scenarios.

Economic impacts of climate change on land use distribution could also involve human migration patterns and impact the amount and pattern of urban and developed areas by region. Regional patterns of growth and decline in the United States show shifts in population and property value to more vulnerable areas, and concerns about climate change and severe weather events could alter coastal settlement patterns. A large amount of uncertainty surrounds any such movement of population and land use impacts, and is outside the scope of this paper.

7.6 MODELING CHANGES IN LAND VALUES WITH POLICIES AND CARBON MARKETS

Forests are cost-competitive in their ability to play some role in U.S. carbon sequestration activities. Cost studies of forest carbon sequestration over more than a dozen years have estimated costs of between about $10 and $150 per metric ton of carbon. Some estimates point to forests being able to sequester between approximately 250 and 500 million metric tons in the United States per year, mirroring the growth of the forest and storage of carbon in the accumulating woody biomass. It is pointed out that the cost effectiveness of forest-based sequestration depends on the secondary benefits of sequestration (e.g., habitat creation and watershed improvement).

Next, we look at the use of the FASOM-GHG model to investigate changes in values for forestry, agriculture, and tradable carbon credits with carbon-related policies. The FASOM-GHG modeling framework can also be used to examine how expectations of land-base shifts between the agriculture and forest sectors impact potential policy effectiveness. Because market equilibrium is determined with endogenous interaction between the forest and agriculture sectors, varying levels of future land use patterns can be explored by the introduction of constraints.

It is used the FASOM-GHG modeling system to investigate effects of carbon dioxide equivalent (CO_2e) prices of between \$0 and \$50 per metric ton. These carbon-related values are added to the objective function in a symmetric fashion. A symmetric system, also called a carbon subsidy-tax system, pays forest owners for increments and "taxes" them for any decrements in carbon stock. Hypothetical policies involving carbon-related prices could cause landowners to undertake changes in forest management as well as afforest additional agricultural land. Although outside the scope of this chapter, climate change policies may involve mitigation actions that promote reduced deforestation.

The CO_2 pricing scenarios included a CO_2 price of zero, \$25/metric ton, and \$50/ metric ton. The \$25/metric ton price approximates levels estimated, who estimated that costs of mitigation actions in forestry and agriculture would range from \$15 to \$25/ metric ton of CO2 mitigated. Their analysis using FASOM model runs indicated that between 10 and 25 percent of current U.S. GHG emissions could be offset through a combination of actions in forestry and agriculture, including reduced cropland tillage, afforestation, improved forest management, improved nutrient management, manure management, and bioenergy production. They also simulated a scenario with \$50/metric ton to investigate effects of a higher CO_2 price.

It is reported that soil expectation values (SEV), as one proxy for effect on land values, on average almost doubled with a \$25 CO_2 price compared to the base case with no carbon price. These are "bare" forest-land values. Moving to a \$50 CO2 price increased the SEV values by more than 2.5 times compared to the base case.

Although these are substantial increases, the increases in land value are a magnitude of order smaller than the differential between forest land and developed use values reported. For example, in the Southeast, SEV timberland values in the base run were approximately \$500 to \$750 per hectare (approximately \$200 to \$300 per acre) on average. Doubling or tripling such land values would leave them short of \$2,500 per hectare (\$1,000 per acre), whereas scientist reported average urban values in the Southeast of more than \$90,000 per hectare (\$36,000 per acre), an average of more than 30 to 40 times as much.

Therefore, although carbon-related revenues are not likely to appreciably affect the land competition between forestry and developed uses in urbanizing areas, such carbon revenues can conceptually cause switches in land use between forestry and agriculture. As shown in figure 4, with afforestation area sensitive to CO_2 prices, projected net change in forest area involving land exchanges with agriculture (afforestation area minus area deforested to agriculture) summed over the first 50 years of the projection was positive with a \$20 CO2 price. In contrast, net change levels without a CO2 price are negative. Afforestation dominates deforestation to agriculture early in the projection with \$25 CO_2, but then is exceeded by such deforestation for most of the projection period. The degree of land use change associated with a particular carbon price will vary by region.

For example, scientists showed that the Pacific Northwest region is relatively less responsive to incentive payment levels for afforestation than the Southeast/Appalachia region.

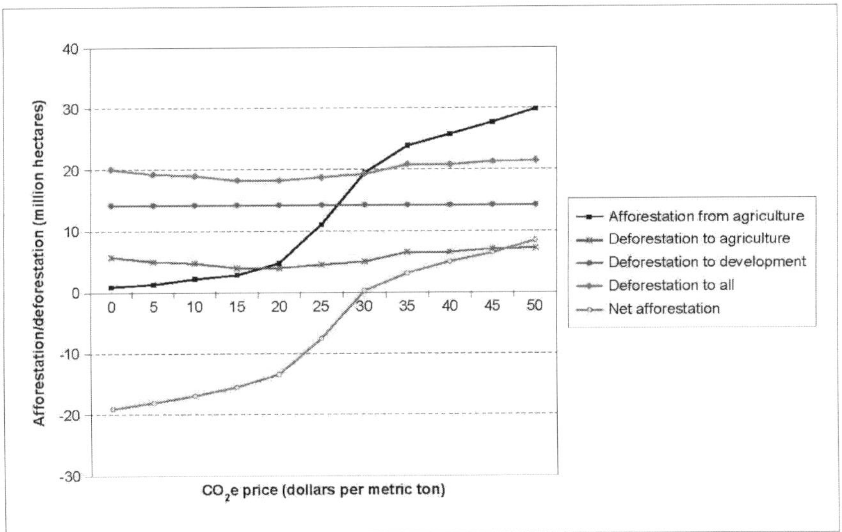

Fig. 4: Net projected afforestation under different carbon dioxide prices.

Projected area of afforestation follows a generally declining trend over time in the national study. Afforestation area with \$50/metric ton CO_2 averages 0.92 million hectares for the first 25 years of the projection, and then drops to 0.31 million hectares over the next 25 years. The highest CO_2 prices boost forest area by 55 percent compared to the base case, especially in the South and North. Doubling the CO2 price from \$25 to \$50 per metric ton boosts forest area in 2050 by 17 percent.

Deforestation for conversion to agricultural use is sensitive to the CO2 price assumption. Deforestation to agriculture is reduced to about one-fifth the base area amount with \$25 CO_2 prices. With \$50 CO_2, the amount of deforestation to agriculture drops essentially to zero over the first two projection decades. However, even with \$50 CO_2 prices, the deforestation for agricultural use does increase later in the projection period, given the expanded forest area and changes in land prices across the two sectors.

Altering the assumption of no loss of forest to development has a smaller effect than having CO_2 payments for landowners. Compared to the base level assumption, avoiding deforestation to urban and developed uses can boost forest area in 2050 by about 9 percent, about half the increase in moving from \$25 to \$50 CO_2 prices. The projected base level of forest converted to development is approximately 15 million hectares over 50 years, drawn from exogenous projections for the 2010 Resources Planning Act (RPA) Land Base Assessment by the USDA Forest Service. The largest projected losses are in the South and Northeast, consistent with recent historical trends.

7.7 EVIDENCE OF CARBON MARKETS

The section above pertains to hypothetical policies for carbon sequestration. However, what is actually going on with carbon markets in the United States? Currently, there is no binding global agreement to reduce carbon emissions, and no associated carbon market. Climate change is a topic replete with uncertainty and the progress of carbon markets would require a regulatory cap for significant offset values to arise. Ecosystem service markets are one approach among several for ecosystem protection and part of the future course of such markets will depend on recognition of whether greater global ecosystem protection is necessary. The Europeans have had a cap-and-trade market running since 2005. The first few years were experimental. Initially, there was a lot of price volatility and a fundamental imbalance between supply of tradable carbon credits and demand for them. With a more stable and clear regulatory framework, the European carbon trading experiment continues. A mature trading program will provide valuable data and give economists the opportunity to observe interaction between carbon trading policy and a real economy. Future transactions in forest carbon credits could constitute a market of billions of dollars annually and involve potentially large transfers of funds between players in our economy.

In terms of actual transactions, buyers of ecosystem services can already purchase carbon credits on the Chicago Climate Exchange (CCX) to offset activities that emit GHGs. In turn, the CCX buys credits from projects that offset the accumulation of GHGs in the atmosphere. Many forestry projects do just that, and some are already generating income for forest landowners. One example is carbon currently being traded on the CCX. Figure 2-2 shows that carbon prices in recent times have fallen below \$1 per metric ton.

A number of organizations are angling to position forest landowners to be able to trade carbon sequestered on their land on the CCX. In October 2007, California Air Resources Board adopted the first standards in the United States for forest- generated, CO_2 emissions reduction projects. Agra Gate Climate Credits Corp. is now able to provide a Forestry Offset contract on the CCX to forest owners. F&W Forestry Services, Inc. is in a similar position. The Georgia Carbon Sequestration Registry assists in the application of land management practices to sequester carbon and creates new economic opportunities for landowners. The Texas Forest Service received approval as an Authorized Verifier of Forestry Offset Projects for the CCX. FORECON Eco Market Solutions, LLC announced its successful development, presentation, and approval from the CCX for the first managed forestry carbon offset project for a timber investment management organization. The pilot project covered about 4050 ha (10,000 acres), mostly in Pennsylvania.

However, challenges include the fact that standards lack any sort of national uniformity. There are no compliance-based forest programs in the United States, and monitoring actual implementation of carbon offset programs is challenging. It is suggested that some people are buying into projects that are never completed, or paying for ones that would have been done anyway. Some green schemes range from selling protection for existing trees to the promise of planting new ones that never thrive. In some cases, the offsets have consequences that their purchasers never foresaw. Scientists characterized carbon offsets as the environmental equivalent of financial derivatives: complex, unregulated, unchecked, and in some cases not worth their price.

Implementation issues also include possible leakage effects, which could result in governments spending large amounts of money while gaining little or no net gain in carbon reductions. Thus, although it is possible to increase carbon sequestration in forests through afforestation or forest management activities, the net effects on overall carbon sequestration may not be as large as anticipated because land markets in areas where forests are unprotected may respond by moving some forests back into agriculture (i.e., deforestation) for example, found an approximate one-to-one correspondence between hectares converted to forests from agriculture and hectares converted the opposite direction, suggesting the need for a comprehensive policy that recognizes leakage. Murray et al. (2004) found similarly large leakage effects for some regions of the United States, and Sohngen and Brown (2004) found smaller, though still potentially substantial, leakage effects for tropical regions. More recent efforts suggest that efficient policies with flux constraints or carbon pricing could provide net sequestration, and that these would increase timber supply both in the short run and long run.

It is also important for any carbon credit (offset) to prove a concept called additionality. The concept of additionality addresses the question of whether the project would have happened anyway, even in the absence of revenue from carbon credits. Only carbon credits from projects that are additional to the business-as-usual scenario represent a net environmental benefit. Thus, assessing additionality requires specifying some sort of baseline— what would have happened without carbon markets. Carbon projects that yield strong financial returns even in the absence of revenue from carbon credits, that are compelled by regulations, or that represent common practice in an industry are usually not considered additional, although a full determination of additionality requires specialist review.

7.8 LAND VALUE DATA NEEDS

Forest-use valuation is increasingly complicated, as is our understanding and modeling of our environment and economy. The increasing recognition of forest resource values other than timber, such as for tradable carbon credits, is part of a larger societal concern about sustainability of land to provide the goods and ser- vices that we as a society demand. Given the fixed amount of land, we need to ask what prices we are willing to pay for those goods and services and how that will affect land values. We need improved databases pertaining to land markets that provide evidence on revealed behavior about what people are willing to actually pay for a bundle of rights necessary to gain access to land that can provide goods and services into perpetuity.

On the empirical side, analyses of land-use change over the past two decades have tended to express land value in terms of the stream of rents, or net returns, accruing to a particular land use. Three general types of data used to estimate net returns are (1) land coverage data, (2) production data, and (3) land value or price data. Land coverage data identify a particular land use at a point in time and may also provide a measure of land quality or suitability for alternative uses (e.g., soil classification). Production data identify the quantities of output associated with a particular land use for a specific geographic area and period. An example of a need for improved production data is afforestation yields on former cropland and pasture- land in all regions. Price data identify input and output prices for the production and sale of goods related to a particular land use, such as for forest carbon sequestration. Efforts to improve land valuation for climate change studies would likely benefit most from attention to the latter category. Increased interest in carbon markets has resulted from growing awareness about climate change and the need for mitigation activities, many of which could involve terrestrial carbon sequestration. Ecosystem service markets lately have become a popular topic among environmental policymakers, and carbon sequestration has received a lot of attention. However, with emerging or nascent markets, many questions remain to be answered. For example, will demand for various ecosystem services be comple- mentary or competitive, and how may pricing of new services affect land use, food prices, and the prospects for biofuels or other bioenergy production?

One information need is for monitoring of emerging forest carbon markets and the relationship to land values. Emerging markets may involve speculative elements and often include nontraditional markets for forest goods and services. Valuation information for these emerging markets may be notably more limited than for commodity-related forest products. Further, forest carbon sequestration opportuneties exist on both private and public lands and often are a joint or byproduct of forest management for other objectives.

One example of a database as a candidate to be built upon is Lubowski's (2002) construction of a nationwide database of net return estimates for several major land uses on private lands. It should be noted that the construction of such estimates can be labor intensive and may require a large degree of extrapolation or inference. His database was constructed to support national land-use modeling, and was later used in other studies.

In general, gaps exist in land value data, such as for forest-land values based on actual transactions generally available to the public. For forest-land values, as estimated the weighted net present value (NPV) of saw timber from different forest types at the county level. The area of each forest type was identified using the USDA Forest Service's Forest Inventory and Analysis data. Timber yields for different forest types were based on existing forest-land estimates, and need to be augmented for afforestation possibilities. State-level stumpage prices were obtained from state and federal agencies or from private companies. Where state-level stumpage prices for private forest were unavailable, sales prices from state or national forests were used.

Carbon markets are emerging in some areas, and further expansion could affect forest-based mitigation opportunities.

Unavailability of nationwide stumpage price data is an impediment to estimating rents. Although Timber Mart-South has historical records of stumpage prices, its coverage is limited to 11 Southern States. In constructing nationwide rents, it proceeded state by state to gather reported stumpage prices by surveying various state agencies and private sources. Hence, price data are likely to be inconsistent in terms of methodology and quality (varying both regionally and over time). Furthermore, developing a nationwide database of stumpage prices in this manner is likely to be labor intensive.

Other data sources may be able to complement, approximate, or improve upon more established sources at the regional, if not national, level. Some of the sources identified below are real estate values. Although real estate values will typically be similar to stream of rent values, they cannot be considered perfect substitutes. Real estate values incorporate expectations about future land uses that may differ from the current use. For example, the real estate value of forestry or agricultural land on the fringe of urban development may be well in excess of the NPV of indefinite forestry or agricultural rents because the land value is "ripening" in accordance with buyers' expectations about its impending change to developed land. In this sense, disparities between market land values and rent values may indicate an impending land use change.

Since 1997, USDA's National Agricultural Statistics Service (NASS) as a public source has published agricultural land value and rental rate data at the state level, but 2009 marks the first time NASS published the information at the county level. For 2009, NASS reports mean county-level cash rents per acre for cropland (irrigated and no irrigated) and pastureland. The population target for the data is all farms and ranches, and information is collected primarily by telephone interview. The county-level rents are available online in tabular and map formats, providing a readily available proxy for cropland net returns at the county level. State-level cash rents are available from 1994 to present. However, similar data are not available for forest land (Fig. 5).

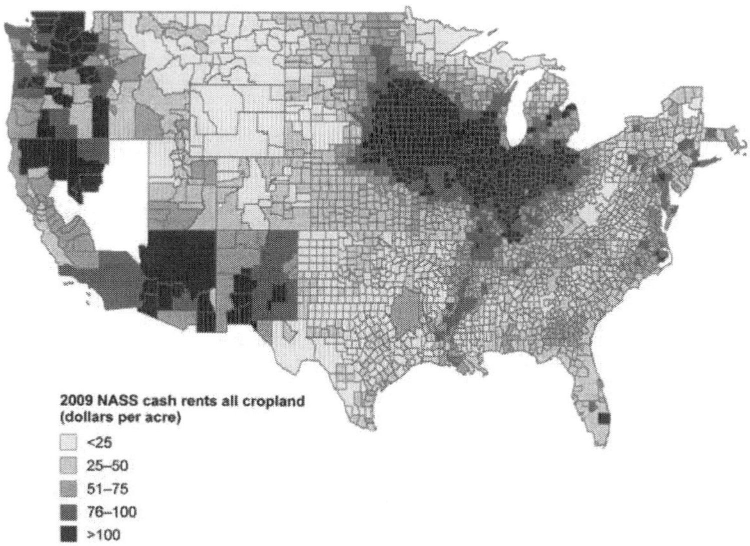

Fig. 5: 2009 National Agricultural Statistical Service (NASS) cash rents, across all cropland. Note that NASS cash rents are based on a survey of farmers. Data are unavailable for parts of Nevada, California, and New England, which appear white in the map.

Conceptually, the cash rent data are the same as the net returns to cropland calculated. To obtain a single rent figure for cropland, irrigated and nonirrigated rents should be weighted proportionally to those lands' acreages. The cash rent data could supplement future updates to the crop rent database by serving as a comparative check or by providing an easy way to expand the database.

In addition to cash rent data, NASS publishes regional and state estimates of average real estate value per acre for irrigated, nonirrigated, and pastured agricultural land. These real estate values could be used to compute annual rents. The disparity between average real estate value and crop rents could be used as a measure of development pressure.

7.9 AT A GLANCE

Land prices and other measures of values reflect society's valuation of land for different uses. Modeling changes in land values from climate change and climate change policies can provide valuable information to landowners, managers, and policymakers as society grapples with how to deal with climate change. One facet is how land values may change with increased demand for tradable carbon credits. Transactions in forest carbon credits could constitute a market of billions of dollars annually and involve potentially large transfers of funds between players in our economy.

Forest carbon credits could add significantly to the possible streams of income from timberland. In addition to the growth and harvest of timber, the possibility of selling carbon credits would affect investments in timberland. Forest lands in general sequester (bind up from the atmosphere) substantial amounts of carbon, which in markets could be translated into carbon credits. Revenues from carbon credits would tend to boost forest-land values, especially in comparison to some agricultural alternatives in some regions. One example of a key activity for which better forest and carbon yield data are needed is afforestation of former agricultural lands. In general, we need to better understand how land productivity for differ- ent uses may change under future climate change in the various regions. Climate change is being monitored, with some 24 separate indicators (USEPA 2010) showing how climate change affects the health and environment of U.S. citizens. That could be usefully augmented with related indicators about how land productivity is being affected.

Ecosystem service provision can be expected to change with climate. Productivity will change with climate, affecting land value and land use. Additionally, concern over environmental degradation and climate change has increasingly led to the establishment of conservation programs, which can augment land value. A central ecosystem service in the climate change context is carbon sequestration.

The direction of the relationship between CO_2 concentration and sequestration determines whether we have a negative or positive feedback loop. In both cases, incorporating the effect on land values tends to reinforce the loop.

If global carbon markets emerge, a comprehensive assessment of the carbon storage and GHG fluxes within the Nation's ecosystems would aid in the evaluation of potential policy and mitigation actions. Such an assessment is inherently complex, as it is geographically broad; covers many different ecosystems, pools, and flux types; and is influenced by many present and future potential conditions (such as climate change, population change, land cover change, ecosystem disturbances, land management activities). If policies adopted to address climate change influence the extent, composition, and management of future forests as well as competing land uses, effects are likely to differ by region. In addition, the nature of the altered climate could vary substantially by region, so that comparative advantages in forest and agricultural production across regions could be affected along with land values.

The policy development process must be supported with integrated, systems analysis. Continued investment in developing and refining the models that provide this type of analysis is critical. The over-arching nature of the climate change issue, which potentially could affect all facets of our lives and work activities, means our reliance on integrated models and analysis will continue to increase as policymakers grapple with a confusing array of choices and tradeoffs. Climate change may substantially affect baseline emissions as well as the cost and effectiveness of mitigation options. Although there has been increasing recognition of these linkages, there remains a need for additional model development and analyses to improve our understanding of the implications for mitigation, adaptation, and risk management policy. Demands for bioenergy feedstocks, carbon sequestration, food, fiber, and ecosystem benefits from our land base are increasing. At the same time, the land base faces enormous potential changes in productivity and risks of extreme events. Thus, there are significant needs to improve our understanding of potential effects and to model the complex interactions between competing demands under changing conditions. Risk management has an important role in influencing responses to climate impacts and in developing appropriate mitigation and adaptation policy. There is a need to improve our understanding of how people will change their land use and production practices and the implications for agriculture, forests, and natural ecosystems as they respond to changing risks within the context of policies influencing management choice.

As ecosystem services become scarcer owing to population growth and rising global standards of living (consumption), interest grows in devising markets and other means for procuring these benefits. Research to support carbon policy analyses includes compensation amounts needed as incentives for landowners to participate in tradable carbon credit markets. Land-use change is a key part of global change in that deforestation, urban sprawl, agriculture, and other human influences have substantially altered and fragmented our global landscape. Such disturbances of the land can change the global atmospheric concentration of CO_2 , the principal heat-trapping gas, as well as affect local, regional, and global climate by changing the energy balance on Earth's surface. Research questions relate to the relationship between land market activity, land management, and carbon sequestration on the global landscape. For example, given that land management can notably alter future levels of forest carbon sequestration (e.g., Adams et al. 1998), interplay between land markets and markets for products and services can influence the path of carbon sequestration over time. Thus, how might variations in landowner behavior by region, including development of different densities and types, influence the land market outcomes?

Land use change plays a very significant role in climate change mitigation and adaptation, and a land value database could provide both direct and indirect benefits. Land prices or values integrate supply and demand information and could also provide better information on future land-base conditions for climate change analysts. For example, for forest ecosystems, they could provide a thorough and unified description of anticipated change in the valuation by society.

8

CLIMATE CHANGE AND MENTAL HEALTH

8.1 INTRODUCTION: AN OVERVIEW

Climate change affects mental health in a variety of direct, indirect, and overarching pathways—disproportionately affecting those most marginalized. The mental health implications of climate change can result in mental problems and illness as well as affirmative psychosocial outcomes. While the timing and triggers associated with climate change and mental health may vary, making it challenging to establish the manifold links between climate change and mental health, the opportunities of attributing mental health to climate change support climate mitigation as well as mental health action and psychosocial resiliency.

The authors argue the following three points: firstly, while attribution of mental health outcomes to specific climate change risks remains challenging, there are a number of opportunities available to advance the field of mental health and climate change with more empirical research in this domain; secondly, the risks and impacts of climate change on mental health are already rapidly accelerating, resulting in a number of direct, indirect, and overarching effects that disproportionally affect those who are most marginalized; and, thirdly, interventions to address climate change and mental health need to be coordinated and rooted in active hope in order to tackle the problem in a holistic manner. This discussion paper concludes with recommendations for priority actions to address the mental health consequences of climate change.

It is well understood that human health is threatened by the impacts of climate change. In the 2017 Lancet Countdown on Climate Change and Health, authors state: "The human symptoms of climate change are unequivocal and potentially irreversible—affecting the health of populations ar ound the world today". Climate change is no longer a looming threat but rather a destructive reality with dire predictions for the future. The World Health Organization (WHO) estimates an increase of 250,000 excess deaths per year between 2030 and 2050 due to the "well understood impacts of climate change".

Impacts include heat-related morbidity and mortality, increases in vector-borne diseases (e.g. dengue fever, malaria), increased respiratory illness, and morbidity and mortality due to extreme weather events. The lesser-known, and often overlooked, effects of climate change include the risks and impacts to mental health—the focus of this article.

Mental health refers not just to mental illness, mental problems, and mental disorders, but also includes states of mental wellness, emotional resilience and psychosocial wellbeing. Psychosocial wellbeing is the interplay between social and psychological conditions that shape human welfare; a broad term which encompasses the states of being mentally healthy, experiencing mental problems, and mental illness.

Investigating the current state of evidence and knowledge about the climate change impacts to mental health, this article pays particular attention to the inequitable impacts of climate change on the mental health of marginalized and vulnerable populations. We argue the following three points: firstly, while attribution of mental health outcomes to specific climate change risks remains challenging, there are a number of opportunities available to advance the field of mental health and climate change with more empirical research in this domain; secondly, the risks and impacts of climate change on mental health are already rapidly accelerating, resulting in a number of direct, indirect, and overarching effects that disproportionally affect those who are most marginalized; and finally, interventions to address climate change and mental health need to be coordinated and rooted in active hope in order to tackle the problem in a holistic manner. This paper explores each of these facets and concludes with recommendations to enhance the state of knowledge and actions on climate change and mental health. Before diving into the topic area, a brief overview of climate change and health effects are noted in the sections below.

8.2 CLIMATE CHANGE AND HEALTH

An extensive body of research continues to strengthen knowledge about the impact of climate change on physical health, including for example, a rise in vector-borne, water and food-borne diseases; an increase in acute and chronic respiratory conditions (including asthma and allergies); and, heat-related and extreme weather-related morbidity and mortality. Indirect health implications that are increasingly recognised in global reports on climate change and health include illness related to food and water safety, under-nutrition related to food insecurity, malignant melanoma from UV exposure, and chronic kidney disease from dehydration. In late 2017, the Lancet released its first, full tracking report on climate change and health. In this report, there is an explicit request for more information on, and actions to address, the "often-unseen" impacts of climate change on human health, notably, the mental health consequences of climate change.

8.3 CLIMATE CHANGE AND MENTAL HEALTH

The expanding research literature on climate change and mental health includes increasing evidence that extreme weather events—which are more frequent, intense, and complex under a changing climate—can trigger post-traumatic stress disorder (PTSD), major depressive disorder (MDD), anxiety, depression, complicated grief, survivor guilt, vicarious trauma, recovery fatigue, substance abuse, and suicidal ideation. Incremental climate changes, such as rising temperatures, rising sea levels, and episodic drought, can change natural landscapes, disrupt food and water resources, change agricultural conditions, change land use and habitation, weaken infrastructure and give rise to financial and relationship stress, increase risks of violence and aggression, and displacement of entire communities. The overarching threats of a changing climate, can also incite despair and hopelessness as actions to address the 'wicked problem' of climate change seem intangible or insignificant in comparison to the scale and magnitude of the threats. Paradoxically, these same disastrous circumstances may also inspire altruism, compassion, optimism, and foster a sense of meaning and personal growth (otherwise referred to as post-traumatic growth) as people band together to salvage, rebuild, and console amongst the chaos and loss of a changing climate.

8.4 CLIMATE CHANGE AND HEALTH INEQUITY

It is well understood that climate change augments existing inequalities, rendering those most marginalized at greater peril to the health consequences of a changing climate. In fact, the first key message from the Lancet's Countdown on Climate Change and Health report emphasizes the disproportionate impact climate change has on the world's most marginalized people and the consequential impacts this has on these populations if social and environment justice concerns are not addressed. As state: "By undermining the social and environmental determinants that underpin good health, climate change exacerbates social, economic, and demographic inequalities, with the impacts eventually felt by all populations". Those who are at greatest risk to the effects of climate change are those who are most marginalized based on socially and environmentally mediated factors, such as socioeconomic status, culture, gender, race, employment, and education. Marginalized groups who tend to be the most affected by the mental and physical health implications of climate change are: Indigenous peoples, children, seniors, women, people with low-socioeconomic status, outdoor labourers, racialized people, immigrants, and people with preexisting health conditions. Importantly, these marginalized groups are not homogenous. People may experience intersections of marginalization based on a variety of the above social indicators.

8.5 EXPLORING THE RELATIONSHIP BETWEEN MENTAL HEALTH AND CLIMATE CHANGE

An updated overview of recent evidence on the mental health implications of climate change is timely given the ongoing, rapid expansion of research in the broad field of health and climate change as well as increasing public concern about climate change trends and risks. Since 2007, media reports on climate change and health have increased by 78% and the academic literature on climate and health issues has tripled. There is also increasing public and academic recognition of the extent to which rising global temperatures threaten planetary and human health. While public awareness about the health implications of climate change continues to grow, the topic of mental health is frequently absent from this discourse. In some ways, this reflects the global discourse, where, in comparison to physical health, mental health *in general* has been neglected.

Globally, the prevalence of mental health issues is extremely high even without considering the added mental health consequences of a changing climate. Based on a 10-year systematic analysis of global burden of disease from 1990 to 2010, Murray et al. find that mental illness comprises 7.8% of the global burden of disease. Mental and behavioural disorders also account for the greatest global burden of years lived with a disability (YLDs). It contend that these figures are actually much higher if co-morbidities related to mental illness are considered within burden of illness studies and if a more accurate definition of mental illness is used, a definition that includes health behaviours like self-harm and suicide. The failure of global investment in mental health care to address the consequences and impacts of rapidly escalating levels of mental illness has been described as a "global tragedy" reflecting a long "legacy of the neglect and marginalization of mental health". Similarly, authors from the 2016 Lancet report on sustainable development and global mental health describe the state of mental health as the "most neglected of all human health conditions" and a "failure of humanity. The inattention to mental health is of particular concern in the field of climate change and mental health given the evidence that psychological impacts from any form of disaster exceed physical injury by 40–1, and that since 2000 the frequency of climate change-related weather disasters has increased by 46%. Crucially, it is the most marginalized who are especially vulnerable to climate change's impacts on mental health. As McMichael notes, climate change acts as a health "threat amplifier", compounding existing social injustices. There is, therefore, a strong case for continuing to explore and communicate research and policy learning about the relationship between climate change and mental health—especially as the topic area pertains to health equity.

8.6 EXPLORING THE CHALLENGES AND OPPORTUNITIES OF ATTRIBUTION

The lack of attention to the topic of climate change and mental health is often imputed to the challenges of attribution. Attribution in this case is the scientific association between greenhouse gas emissions and meteorological change on the one hand, and between climate change-related meteorological change and mental health effects on the other. There is now an increasingly strong body of literature which highlights the causal linkages between climate change and extreme weather events). One of the key messages within this literature is that while we cannot say with certainty that any one specific extreme weather event is directly caused by climate change; we do know that because of climate change, extreme weather is more generally on the rise, making extreme weather events more frequent, intense, and complex. In other words, climate change therefore 'loads the dice' for more weather extremes.

Within the disaster mental health literature, the links between extreme weather events and mental health effects are well established. However, many of these studies tend not to connect extreme weather to a changing climate—instead referring to extreme weather events as natural disasters rather than events linked to anthropogenic climate change. Studies within this domain tend to focus on mental health outcomes of specific hazard events positioning each hazard as an isolated incident unconnected to the wicked problem of climate change. The risk of overlooking or minimizing the role of climate change within these hazardous events is that this creates a reactive culture of emergency response that inhibits appropriate and effective adaptation planning and preparation for complex emergencies that a changing climate can create.

An additional concern is that much of the disaster mental health research has traditionally focussed on three distinct phases (the emergency and crisis stage, the post-impact stage, and the rehabilitation and recovery phase), while little attention has been paid to psychosocial phenomena that can occur during the pre-disaster phase. Such phenomena include, for example, heightened anxiety levels, feelings of impending doom, hopelessness, and fatalism that can be triggered by approaching extreme events or associated weather warnings; and which may also be amplified due to the perceived risk of subacute, environmental changes like rising temperatures and episodic droughts.

8.6.1 Key Challenges of Attributing Climate Change to Mental Health

Attribution related to climate change and mental health can be challenging for four key reasons: firstly, there is a risk of pathologising common transitory distress responses to abnormal events and underdiagnosing mental health effects of a changing climate; secondly, there is a wide array of potential climate change and mental health outcomes related to a changing climate; thirdly, there is substantial scope with respect to the timing of the climate change effects on mental health, thus causal links become harder to determine; and finally, attribution related to climate change and mental health is not well understood because of the complex interaction between mental health and other social determinants of health.

There is a simultaneous risk of pathologising 'normal' responses to a changing climate and of underdiagnosing the real mental health effects of a changing climate. Pathologising 'normal' responses to disaster situations may result in a failure to differentiate between mild transitional distress or grief and more severe, persistent mental health problems. Both overinflating or underinflating mental health outcomes associated with climate change can lead to erroneous prevalence estimates and skewed assumptions about mental health service needs. A further consideration noted by Whaley in the aftermath of Hurricane Katrina is that in some cases medical professionals did not assess pre-existing mental health conditions and, therefore, attributed disaster trauma as a typical stress response, or alternatively diagnosed patients with stress response when in fact there were much larger mental health issues related to the effects of the Hurricane that went undiagnosed. Crucially important to consider is that pre-existing mental health conditions or problems can be exacerbated or even triggered by changes in climate.

As noted earlier, mental health includes states of mental wellness as well as mental problems and disorders. With this in mind, the current application of tools to assess mental health have some limitations. Researchers tend to conceptualize mental health solely as mental illness and mental problems, administering surveys using validated instruments that assess mental health problems and issues like: generalized anxiety disorder (using the general anxiety disorder, GAD-2 questionnaire), PTSD (PCL-6 checklist), and psychological distress (via the general health questionnaire, GHQ-12) following an extreme weather event. Few empirical studies that use these survey methods capture positive psychological consequences of extreme weather events, like feelings of compassion, altruism, sense of meaning, post-traumatic growth, or even increased acceptance of climate change and engagement with climate mitigation. This information can elucidate the complexity of mental health impacts from a changing climate and also help to understand any predisposing factors that may influence positive mental health outcomes and build psychosocial resilience.

Timing of psychosocial implications from climate related hazards poses another challenge. Scholars have discovered wide-ranging timeframes for the psychosocial impacts to manifest. Azuma et al. find that the incidence of psychological disorders (including PTSD) tended to be most significant within 6 months after a flood. Kessler et al. conducted interviews with Hurricane Katrina survivors 5–8 months post-event and 1-year post-event, these authors found an increase in mental health disorders as time progressed. For example, PTSD increased from 14.9% at 5–8 months to 20.9% after 1 year. Anderson et al. suggest that psychosocial impacts tend to peak within the first-year, post-extreme weather event. On the other hand, found that residents who experience significant flooding self-report long-term psychosocial impacts (namely anxiety when it rains) from 2.5 to 5 years post-flooding. A recent news article suggests that over 7000 people who experienced Hurricane Katrina in 2005 are still receiving mental health care for trauma associated with the Hurricane. Noting that psychosocial health outcomes can have latent effects, or that these outcomes can occur as a result of sequelae, knowing how and when to study the climate change-related impacts and psychosocial outcomes becomes increasingly challenging—especially if the aim is to demonstrate the magnitude and attribution of effects.

Distilling the precise impact of climate change on mental health can be difficult to separate from other social determinants. As Watts et al. note, measuring the impacts of climate change on mental health is challenging not only because of attribution but also because of the "complicated nature of mental health, which embraces a diverse array of outcomes (e.g. anxiety and mood disorders), many of which co-occur and all of which vary with contexts and during lifetimes. Mental health impacts are often products of long and complex causal pathways, many of which can be traced back to distal but potent root causes, such as famine, war, and poverty, of which climate change is an accelerator". Mental health, like physical health, is shaped by social and ecological factors that can influence—and often amplify—other determinants of health, like a changing climate.

8.6.2 Opportunities of Attributing Mental Health to Climate Change

It is important to locate climate change within the discourse on mental health because the frequency, intensity, duration, and complexity of climate change effects is on the rise and thus climate-related mental health outcomes are also increasing—adding to the already burgeoning burden of mental illness and mental problems globally. Acknowledging the mental health consequences of climate change helps the mental health community to discern and anticipate patterns of mental illness, like for example PTSD following extreme weather events. Also, an understanding of the unequal impacts of climate change on marginalized groups supports public health prevention strategies that seek to protect those most susceptible to mental illness and mental problems.

There is a risk that climate-related psychosocial consequences may become diluted in the high prevalence of mental health disorders globally; therefore, there is a need for additional research within the specific domain of climate change and mental health. If there is a better understanding of the linkages between climate change and mental health, there are more opportunities to understand and address climate change and mental health via actions rooted in climate change mitigation and adaptation that support psychosocial resilience. Specifically, the field requires more empirical research on the mental health consequences of climate change, especially as this research relates to marginalized communities and the risks and impacts associated with chronic climate change-related hazards and consequences (like sea-level rise, rising temperatures, and ecological degradation). To address the overarching interplay of social and environmental determinants of health that can magnify climate change-related risks on mental health, a health equity approach to this area of study is required. Secondly, there is a need to better understand climate change-related hazards within the context of mental health sequelae. Research in this area could help to explore the complexity of climate change and mental health attribution by recognizing the role of predisposing mental health conditions while also taking into account the perceived and actual risks and impacts related to a changing climate. This type of research can support a better understanding of the triggers and timing of climate change-related mental health effects as well as support policy and program development for mental health resources.

Attributing mental health outcomes to climate change also presents opportunities to assess, build, and strengthen mental health systems. In 2015, the World Health

Organization (WHO) set forth the *framework for building climate resilient health systems* . This framework provides guidance for health professionals to predict, prevent, and prepare for climate change-related shocks with the ultimate aim of protecting population-level health by increasing health systems' capacity to cope, adapt, sustain, and strengthen in the wake of a changing climate. While the framework overlooks the intricacies of mental health systems—like the current and prevailing lack of mental health infrastructure, funding, and resourcing globally—it does provide the necessary guidance to build mental health systems resiliency. Articulating climate change as a determinant of mental health not only brings awareness of the broad consequences of climate change on health but also supports the enhancement of mental health systems.

In sum, the key benefits of understanding the linkages between climate change and mental health include: enhanced knowledge of patterns of illness; an added emphasis on the global call to action to reduce and address climate change risks and impacts; in-depth knowledge of the risks and impacts to marginalized communities; and, better planning for mental health response and mental health systems resiliency.

8.7 CURRENT RISKS AND IMPACTS OF CLIMATE CHANGE ON MENTAL HEALTH

It is challenging for people to recognise changes in climate because these changes appear distant or abstract—especially because climate is often confused or lost in perceptions about weather or seasonal change. The influential sociologist Anthony Giddens refers to this space and time distancing of the climate change problem as the Giddens Paradox. The Giddens Paradox states that: "since the dangers posed by global warming aren't tangible, immediate or visible in the course of day-to-day life, many will sit on their hands and do nothing of a concrete nature about them. Yet waiting until such dangers become visible and acute—in the shape of catastrophes that are irrefutably the result of climate change—before being stirred to serious action will be too late" . Marshall contends that part of the time and space distancing of the climate change problem, and thus the reluctance to act, is reinforced by the Western political discourse on climate change as a future-facing problem that intentionally overlooks the centuries of industrialization, fossil fuel consumption, and land degradation that contribute to anthropogenic climate change. Marshall calls for a reckoning with this discourse by noting:

"Climate change is a future problem. But it is also a past problem and a present problem. It is better thought of as a developing process of long-term deterioration, called, by some psychologists, a "creeping problem." The lack of a definite beginning, end, or deadline requires that we create our own timeline. Not surprisingly, we do so in ways that remove the compulsion to act. We allow just enough history to make it seem familiar but not enough to create a responsibility for our past emissions. We make it just current enough to accept that we need to do something about it but put it just too far in the future to require immediate action".

Noting the Giddens Paradox and the reckoning that Marshall asks us to have with the "creeping problem" of climate change, it becomes important to confront the current mental health consequences related to climate change that are happening now. To do so, it is useful to explore the conceptual framework of climate change and mental health developed. These authors organize climate change-related hazards into three categories: acute (flooding, hurricanes, etc.), sub-acute (pervasive drought), and chronic (rising sea-level, increasing temperatures). These climate change-related hazards lead to a variety of direct, indirect, and overarching psychosocial consequences that are occurring now—disproportionately affecting those most marginalized.

Direct psychosocial consequences of climate change include trauma related to extreme weather events, like floods, hurricanes, wildfires, and heat waves. Indirect mental health consequences of climate change occur through social, economic, and environmental disruptions (e.g. famine, civil conflict, displacement, and migration) related to a changing climate. The overarching psychosocial consequences of climate change relate to the long-term emotional distress caused by awareness of the threats and impacts of climate change on the current and future wellbeing of the earth and its inhabitants. The multidimensional climate change and mental health pathway leads to a variety of unequal psychosocial consequences explored below.

8.7.1 Direct Mental Health Consequences of Climate Change

There is now an extensive and rapidly expanding body of research exploring the current mental health consequences of climate change-related extreme weather events. Extreme heat events and humidity have been noted to increase hospital admissions for mood and behavioural disorders, including schizophrenia, mania, and neurotic disorders. Scholars in the field note that heat-related mental health morbidity tends to occur most often in people with impaired thermoregulation, namely people with pre-existing mental health illness and problems, people taking prescription medications (specifically lithium, neuroleptic and anticholinergic drugs), and those with substance abuse (alcohol and drugs) problems. Extreme heat is also linked with an increased risk of wildfires, which also directly impact mental health. Bryant *et al.* mapped the psychological outcomes of the Black Saturday bushfires in Victoria, Australia; in communities most at risk to the impacts of bushfires, these authors found incidences of PTSD, psychological distress, and depression related to the fires.

In a study of 30 locations in England and Wales, reascher conducted interview surveys with residents affected by flooding. They found that psychological impacts were more commonly reported than physical effects]. One study researching the psychosocial impacts following Hurricane Katrina estimates that 20–35% of survivors experienced some form of mental health issue following the disaster. Also reported a 31.2% prevalence of anxiety-mood disorders amongst Hurricane Katrina survivors, while Rhodes and Chan found that nearly half (47.7%) of marginalised community members of New Orleans (mainly low-income, African American women) showed probable signs of PTSD after Hurricane Katrina.

While PTSD is often reported as one of the most severe mental health impacts related to acute climate change-related disasters, there have also been increasing reports of suicide and suicidal ideation following extreme weather events. One Italian study that found higher rates of suicide in northern communities with greater climate variability. Scientist highlight the risk of homicide-suicides after extreme weather events by noting the doubling of these incidents following Hurricane Andrew in 1992 in Miami-Dade County. There is also observed evidence of increased suicidal thoughts (from 2.8 to 6.4%) and plans to commit suicide (from 1.0 to 2.5%) 18-months after an extreme weather event. Notably, however, the overall evidence linking changing climate and suicide is still inconclusive. Studies on suicidality in natural disaster contexts, for example, vary considerably in study methodology and timeframes considered, with recent reviews indicating divergent trends in suicidality rates following exposure to extreme events, ranging from an initial decline, to neutral effects, all the way to a delayed increase in suicidality.

On a deeper level, the psychological responses of communities and individuals to disasters are complex and varied and do not necessarily simply result in more mental illnesses. Rebecca Solnit, in A Paradise Built in Hell, usefully describes the complicated psychosocial consequences that can arise after an extreme weather event as, "that sense of immersion in the moment and solidarity with others caused by the rupture in everyday life, an emotion graver than happiness but deeply positive. We don't even have a language for this emotion, in which the wonderful comes wrapped in the terrible, joy in sorrow, courage in fear. We cannot welcome disaster, but we can value the responses, both practical and psychological". Exploring the complexity of psychological responses in the book, Climate change and human well-being, as discuss the full spectrum of psychosocial consequences of climate change-related events ranging from mental illness to more positive experiences like 'Post Traumatic Growth' (PTG), empathy, compassion, altruism, and emotional resilience.

8.7.2 Indirect Mental Health Consequences of Climate Change

The indirect mental health consequences of climate change can occur as a result of damages to physical and social infrastructure, physical health effects, food and water shortages, conflict, and displacement from acute, subacute, and chronic climactic changes. One of the most well-documented climate hazards that indirectly influences mental health is drought. Long-term droughts affect food and water supplies and can subsequently affect the economic and mental wellbeing of land-based workers, most often impacting those living in rural and remote communities. In a quantitative analysis of drought and distress in Australia over a 7-year period, authors found that rural dwellers experience more distress due to the droughts than their urban counterparts]. In a systematic review of the literature, authors note the most prominent causal pathway linking drought and mental health is via the economic effects from land degradation. These effects are most prominent amongst farmers whose economic livelihoods depend on environmental conditions. Exemplifying this, a 2008 study in New South Wales, Australia reports that nearly three quarters of farmers report stress related to persistent drought. Some authors also suggest that income insecurity related to drought increases the risk for suicide among farmers.

Long-term drought has also been increasingly linked to conflict and forced migration, which can influence psychosocial outcomes like the propensity for stress, PTSD, anxiety, and trauma. The Institute for Environment and Human Security of the United Nations University estimates that migration due to climate change may vary drastically, citing estimates of between 25 million to 1 billion by 2050, with 200 million as the most frequently cited estimate]. The rise in the number of 'climate migrants' has been identified as a significant risk by an increasing number of defence and security experts. Gleick postulates that the civil conflict in Syria can be traced to the agricultural failures in 2006–2009 and the returning drought in 2011. In 2011, over 1.5 million Syrians moved from rural, agricultural areas to urban areas seeking refuge from the pervasive drought, failed agriculture, and lack of food and water. Pervasive ecological degradation, poor policy response to water and food insecurity, and ongoing tensions between rural and urban community members, have arguably all contributed to civil unrest and ongoing conflict in Syria. According to the United Nations, the number of displaced Syrians has reached over 5 million people in the past 5 years. Migration from a war-torn country to a host country where culture, language, and lifestyle may be vastly different may also contribute to psychosocial malaise as displaced migrants can face stressors associated with xenophobia and racism from people in their new host country. Conversely as Siriwardhana and Stewart note, displacement may also support psychosocial resilience by fostering hope and belonging for refugees in host countries where they feel welcomed, safe, and experience better living conditions.

At the community level the indirect mental health consequences of climate change are understudied. These consequences may include things like a diminishment in community cohesion, the loss of community identity, threats to a sense of continuity and sense of belonging as people are forced to move in and out of communities because of environmental stressors, and an undermining of cultural integrity if people have to leave their homelands. Migration challenges the identity, sovereignty and heritage of people who have to leave their homelands. It also challenges the integrity and continuity of people's traditional ways of life. Threats to community health also include an increased likelihood of criminal behaviour, violence and aggression as community members experience various stressors related to climate change.

8.7.3 Overarching Psychosocial Consequences of Global Climate Change

Awareness of the looming threats and current risks and impacts of climate change presents challenges to emotional and social wellbeing. Since early 2007, environmental philosopher Glenn Albrecht and colleagues have taken note of emotional distress related to the awareness of the overarching problem humans face as a result of global climate change. As suggest that this awareness contributes to 'psychoterratic syndromes'. Psychoterratic syndromes include phenomena such as 'ecoanxiety', 'ecoparalysis', and 'solastalgia'. 'Ecoanxiety' refers to the anxiety people face from constantly being surrounded by the wicked and threatening problems associated with a changing climate. 'Ecoparalysis' refers to the complex feelings of not being able to take effective action to significantly mitigate climate change risks. 'Solastalgia' refers to "the distress and isolation caused by the gradual removal of solace from the present state of one's home environment". The term 'solastalgia' is also commonly referred to throughout much of the literature on climate change and mental health to articulate the feelings associated with displacement following a climate change-related extreme weather event. This new vocabulary provides the language to explore some of the broader mental health implications of escalating climate change risks.

For many people, climate change is experienced by way of vicarious threats or as an existential threat to civilisation. People may experience vicarious threats when they receive weather warnings related to future disaster seasons or when they hear about environmental stressors experienced by people in other places. For many people, this is largely how climate change is experienced—not as a direct threat, but as a global threat, often distant in time and place, or as a threat to our very way of life. Qualitative research finds evidence of some people being deeply affected by feelings of loss, helplessness, and frustration as they engage with the problems of global climate change.

8.8 PRIORITY ACTIONS TO ADDRESS CLIMATE CHANGE AND MENTAL HEALTH

Acting on the health consequences of climate change requires actions rooted in both mitigation and adaptation at all levels—from global to local—and from all sectors and individuals. Climate change mitigation refers to overarching efforts to reduce greenhouse gas emissions and enhance carbon sinks to slow the speed, scale, and magnitude of climate change. Key climate change mitigation priorities include reducing energy demand (through reduced consumption and increased energy efficiency); a swift and equitable transition from fossil fuels to renewable energy; reducing emissions from agriculture and forestry; and strengthening land-based emissions sequestration. Climate change adaptation refers to interventions that respond to the effects of climate change by adjusting, moderating, and coping with the risks and impacts of climate change. Adaptation is ultimately affected by the capacity to adapt, which is the ability and willingness to respond to climate change mediated by individual and collective agency. Adaptive capacity is determined by things like: governance, economics, infrastructure, technology, information and skills, institutions, and equity. Examples of adaptation interventions that address climate change and health include: surveillance and monitoring of disease burdens related to climate change and health; education (e.g. public health promotion of the risks of vector-borne illness), and capacity building (e.g. psychological first-aid, and surge capacities at hospitals and health care facilities); preparing for extreme weather events; and re-locating entire communities to geographic areas where sea-level rise and frequent extreme weather events are less-likely to occur.

Within international approaches to combat climate change there is a significant focus not only on mitigation but also adaptation. From the Paris Accord to the Lancet Countdown on Climate Change, to the Planetary Manifesto and climate action marches—policy makers, academics, and the general population are taking steps to mitigate and adapt to the current threats and impacts to preserve a future for the next generation. These actions, however, often fail to address the gap between stated goals of emissions reduction commitments and the speed of actions required to keep global warming well below 1.5–2 °C.

With a specific focus on mental health and climate change, there are a number of global programs in place that indirectly address the topic area—like for example measures to enhance and protect mental health in the Sustainable Development Goals 2016–2030; efforts by the Movement for Global Mental Health to increase the holistic conceptualization of health to incorporate mental health; the Sendai Framework, a 15-year disaster risk reduction program; and, the United Nations Human Settlement Program that promotes sustainable urban development. There is a need, however, to harness health and mental health related synergies amongst these global agreements since none of these in and of themselves will likely be sufficient to address the future risks and impacts of climate change.

Coordinated, collaborative efforts to address the mental health implications of climate change not only require policy frameworks but also concrete actions on behalf of mental health practitioners. Such concrete actions may include: communicating about climate change and mental health in a way that helps people to see that it is relevant and salient to them; advocacy for greenhouse gas reductions in health care facilities and engagement in efforts to reduce the environmental footprint of the health care sector; and, engaging in adaptation measures like preparing for and responding to extreme events.

8.8.1 Psychological Adaptation

Psychological adaptation requires a set of responses, it requires an acknowledgement of the grave threats posed by climate change and the profoundly consequential global crisis. It requires coping strategies to manage the feelings and thoughts that arise so that people can face up to, and come to terms with, these threats and consequences rather than avoiding the creeping problem of climate change. It also requires behavioural and psychological engagement, in which people change and adjust their behaviour and lifestyle in order to reduce the threat and protect themselves.

Active hope—something Macy and Johnstone champion—supports psychological adaptation. Active hope is required to move hopeful intentions from a passive state where waiting for someone else to take-on the task of addressing the climate change problem is replaced with an active process of climate change mitigation and adaptation behaviours. The key point here is that hope alone cannot provide sufficient protection from the escalating risks of climate change. This active process occurs when the reality of the problem is acknowledged as is the magnitude of the problem, intentions to address the problem are set, and engaged actions take place. While these three steps may oversimplify the complexity of acting in the face of bureaucracy, climate denialism, or downright avoidance and ignorance of the magnitude of the problem area, these three steps are indeed the pivot points of transformation. These pivot points, however, need to be upheld by global political will and policy commitments that tackle the problem at the appropriate scale and speed. To do so, public awareness of the severity, magnitude and range of health impacts—current and projected—is required to pressure governments and communities to act now. Also, discernible interventions are needed to demonstrate a tangible path forward to respond to the risks and impacts we face in a changing climate. Examples of these types of interventions are explored below.

8.8.2 Adaptation Measures

Adaptation measures that address the psychosocial impacts of climate change come in a variety of forms, i.e. policies, practices, behavioral interventions, community-based interventions, specific training, and pharmacotherapeutics. Some general approaches to address climate change-related mental health problems or illnesses include: primary care interventions, individual and group-based therapy, cognitive based interventions (including cognitive based therapy, cognitive restructuring, and, stress inoculation training), and crisis counselling. More broadly, emotional resiliency may be sustained by engaging with art, literature, and spirituality. In addition to the above, the list below contains some specific priority adaptation mechanisms that ought to be considered to support population-level mental health in a changing climate:

- Policy responses: improving access and funding to mental health care;
- Surveillance and monitoring: administering epidemiological surveys after extreme weather events, and monitoring emergency department visits during heat waves and following extreme weather events;
- Practice: the application of a stepped-care approach to mental health that is often used in disaster mental health to support different levels of interventions depending on the timing of the disaster and the level of distress;
- Preparation and response: climate change adaptation/resilience planning in the mental health system;

- Community-based interventions: climate change resilience plans that address psychosocial wellbeing; and,
- Special training for care providers and first responders: e.g. psychological first aid.

Other innovative approaches to addressing mental health and wellbeing in a changing climate writ large include experiencing and preserving nature. It is suggest that environmental preservation provides people with a sense of stewardship and personal investment that can help people overcome feelings of hopelessness, anxiety, and ecoparalysis. Also suggest: "if people feel a deep connection to places, wilderness, and other species, then threats to these others are much more likely to be viewed as personal issues". Other research on the restorative benefits of natural environments and settings has found that biodiversity in natural environments is important for human health and wellbeing and has a particularly positive effect on mood, attention and cognition. A common practice in Japan to reduce stress and anxiety is the practice of shinrin-yoku, otherwise referred to as forest bathing. In a study found that forest bathing resulted in decreased cortisol levels, pulse rates, and negative feelings and significantly increased positive feelings. Research on people's interactions with nationally important ecosystems, like World Heritage Areas for example, highlights positive impacts including quality of life, a sense of place and belonging, self-identity, restoration and inspiration.

While there are a number of interventions to support psychosocial wellbeing within a changing climate, it is important to highlight that many of these interventions are still quite nascent and administered in ad hoc fashion, and these interventions are mainly accessible in developed countries. Sustainable mental health care in developed and developing nations is urgently needed as the realities of climate change become more and more apparent—especially for those most marginalized. Further, there are research needs in this domain where the efficacy and accessibility of mental health interventions-related to climate change are assessed.

8.9 IMPACT OF CLIMATE CHANGE: AT A GLANCE

Climate change affects mental health in a variety of direct, indirect, and overarching pathways—disproportionately affecting those most marginalized. The mental health implications of climate change can result in mental problems and illness as well as affirmative psychosocial outcomes. While the timing and triggers associated with climate change and mental health may vary, making it challenging to establish the manifold links between climate change and mental health, the opportunities of attributing mental health to climate change support climate mitigation as well as mental health action and psychosocial resiliency. Global commitments, like the Paris Accord, the SDGs, and the Sendai Framework are needed to help advance global mental health and climate action; however, coordination amongst these commitments is required—as are concrete actions on behalf of health practitioners—if the issue of mental health and a changing climate is to be efficiently and holistically addressed. Further, a reckoning with social, environmental, and climate injustice is needed if actions to address climate change and mental health are to be rooted in health equity. Transformative action—where inequities are addressed, active hope is demonstrated, and communities are mobilized—is the defining opportunity of the twenty-first century to address the climate change impacts on mental health.

9

GLOBAL ENVIRONMENTAL CHANGE AND HEALTH

Human actions are changing many of the world's natural environmental systems, including the climate system. These systems are intrinsic to life processes and fundamental to human health, and their disruption and depletion make it more difficult to tackle health inequalities. Indeed, we will not achieve the UN millennium development health goals if environmental destruction continues. Health professionals have a vital contributory role in preventing and reducing the health effects of global environmental change.

9.1 PROBLEMS OF FOCUS

In 2000 the United Nations set out eight development goals to improve the lives of the world's disadvantaged populations. The goals seek reductions in poverty, illiteracy, sex inequality, malnutrition, child deaths, maternal mortality, and major infections as well creation of environmental stability and a global partnership for development.2 One problem of this itemisation of goals is that it separates environmental considerations from health considerations. Poverty cannot be eliminated while environmental degradation exacerbates malnutrition, disease, and injury. Food supplies need continuing soil fertility, climatic stability, freshwater supplies, and ecological support (such as pollination). Infectious diseases cannot be stabilised in circumstances of climatic instability, refugee flows, and impoverishment.

The seventh millennium development goal also takes a limited view of environmental sustainability, focusing primarily on traditional localised physical, chemical, and microbial hazards. Those hazards, which are associated with industrialisation, urbanisation, and agriculture in lower income countries, remain important as they impinge most on poor and vulnerable communities. Exposure to indoor air pollution, for example, varies substantially between rich and poor in urban and rural populations. And the World

Health Organization estimates that a quarter of the global burden of disease, including over one third of childhood burden, is due to modifiable factors in air, water, soil, and food.This estimated environment related burden is much greater in low income than high income countries overall (25% versus 17% of deaths—and widening further to a twofold difference in percentages between the highest and lowest risk countries). Heavy metals and chemical residues contaminate local foods, urban air pollution causes premature deaths, and waterborne enteric pathogens kill two million children annually.

These relatively localised environmental health hazards, though, are mostly remediable. Meanwhile, a larger scale, less remediable, and potentially irreversible category of environmental health hazard is emerging. Human pressures on the natural environment, reflecting global population growth and intensified economic activities, are now so great that many of the world's biophysical and ecological systems are being impaired. Examples of these global environmental changes include climate change, freshwater shortages, loss of biodiversity (with consequent changes to functioning of ecosystems), and exhaustion of fisheries. These changes are unprecedented in scale, and the resultant risks to population health need urgent response by health professionals and the health sector at large.

9.2 WHO WILL BE AFFECTED

The health effects of global environmental change will vary between countries. Loss of healthy life years in low income African countries, for example, is predicted to be 500 times that in Europe. The fourth assessment report of the Intergovernmental Panel on Climate Change concluded that adverse health effects are much more likely in low income countries and vulnerable subpopulations.[7] These disparities may well increase in coming decades, not only because of regional differences in the intensity of environmental changes (such as water shortages and soil erosion), but also because of exacerbations of differentials in economic conditions, levels of social and human capital, political power, and local environmental dependency.

These differential health risks also reflect the wider issue of access to global and local "public goods." Most of the world's arable land has now been privatised; stocks of wild species (fish, animals, and wild plants) are declining as population pressures and commercial activities intensify; and freshwater is increasingly becoming subject to market pricing. Social policies should therefore pay particular attention to the health inequalities that flow from unequal access to environmental fundamentals.

Availability of safe drinking water illustrates the point about access to what, historically, was common property: 1.1 billion people lack safe drinking water, and 2.6 billion lack basic sanitation. Beyond diarrhoeal disease, water related health risks also arise from chemical contamination—such as arsenic as a cause of skin pigmentation, hyperkeratosis, cardiovascular disease, neuropathy, and cancer.

9.3 ROLE OF SOCIAL CONDITIONS

The relation of environmental impoverishment to health risks and inequalities is complex. Environmental degradation impairs health, while health deficits (for example, malnutrition or depletion of the workforce from AIDS) can amplify environmental mismanagement. This causes inequalities in both health endangering exposures and health outcomes.

India provides a good example of the complexity of these relations. The country's average life expectancy is relatively low but is expected to improve with industrialisation and modernisation. Industrialisation is contributing to the rapid increase of coal burning in India, and the resultant addition to global emissions and climate change amplifies health risks worldwide. These health risks will affect the world's most vulnerable populations.

The risks to population health from environmental change have far reaching implications for prevention strategies (**Fig. 1)** Global changes result in loss of natural resources. Resolution of these risks therefore requires a different approach from that used for the more familiar challenges presented by time limited and reversible local environmental contamination.

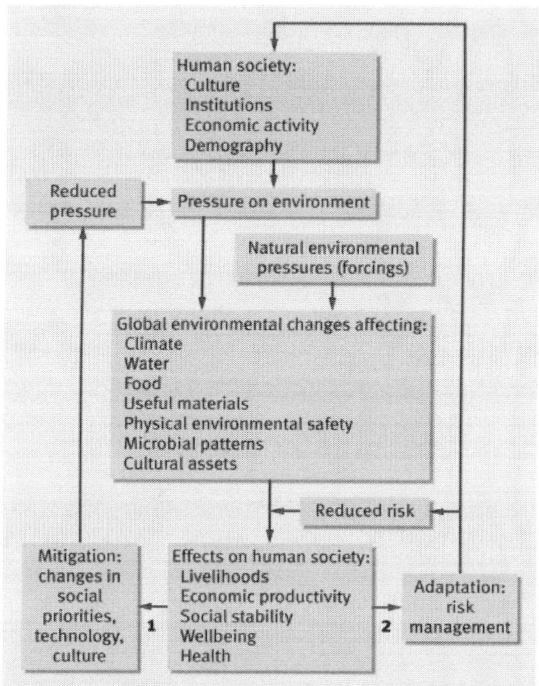

Fig. 1: The risks to population health from environmental change

Relations between human induced global environmental changes affect health and social policy responses. True primary prevention (path 1) reduces or eliminates the human pressures on environment. A more defensive type of prevention is attained through adaptive interventions to lessen risk (path 2), particularly in vulnerable communities.

9.4 CLIMATE CHANGE AND HEALTH

Human induced global climate change is now an acknowledged reality. We have taken a long time to recognise the resultant health risks, current and future, and their unequal effects around the world, but the topic is now attracting much attention. Risks to health will arise by direct and indirect pathways and will reflect changes in both average climate conditions and in climatic variability. The main risks are:

- Effects of heat waves and other extreme events (cyclones, floods, storms, wildfires)
- Changes in patterns of infectious disease
- Effects on food yields
- Effects on freshwater supplies
- Impaired functioning of ecosystems (for example, wetlands as water filters)
- Displacement of vulnerable populations (for example, low lying island and coastal populations)
- Loss of livelihoods

Extreme weather events, infection, and malnutrition will have the greatest health effects in poor and vulnerable populations (box 1). In sub-Saharan Africa over 110 million people currently live in regions prone to malaria epidemics. Climate change could add 20-70 million to this figure by the 2080s (assuming no population increase, and including forecast malaria reductions in West Africa from drying). Any such increase would exacerbate poverty and make it harder to achieve and sustain health improvements.

Box 1 Africa and climate change

Africa is very vulnerable to climate change because of other environmental and social stresses. The economy depends critically on agriculture, which accounts for two thirds of the workforce and up to half of household incomes and food.

- Climate models predict regional increases in mean temperatures of several degrees centigrade by 2100, a decline in summer rainfall in southern and northern Africa and some increase in west and east Africa. Drying, plus the demands of population growth and economic development, will exacerbate regional water scarcity
- Falls in crop yields due to 1-2°C warming by 2050 would add an estimated 12 million additional Africans to the 200 million currently undernourished[18]
- Extreme events such as flooding change will affect food availability by damaging roads, storage, and markets—floods in 2000 in Mozambique damaged about 10% of farmland and 90% of irrigation, displaced two million people, and affected up to 1.5 million livelihoods (mostly in poor rural areas)

- Livestock viral diseases such as east coast fever, foot and mouth disease, blue tongue virus, Rift valley fever are climate sensitive. Regional increases in temperature and rainfall could affect tsetse fly habitat and hence trypanosomiasis in livestock
- Climate change and agricultural downturn in Africa may force populations to move, generating conflicts over territory. Pastoralists forced to search for grazing land because of wells drying up may partly explain the Darfur crisis in Sudan

Some links between climate change and human health are complex. For example, the predicted drying in sub-Saharan Africa could increase the incidence of HIV infection, as impoverished rural farming families move to cities where conditions foster sex work and unsafe sex.

The recent report of the Global Environmental Change and Human Health project gives a good summary of the major categories of current and predicted health effects of global environmental changes other than climate change.

9.5 ROLES FOR DOCTORS AND OTHER HEALTH PROFESSIONALS

The spectrum of potential strategies to reduce health risks is wide, commensurate with the diversity of threats to health posed by climate change and other global environmental changes. Local policies and actions, both to mitigate environmental change at source and to adapt to existing and unavoidable risks to health, will often need support from health attuned policies at provincial, national, and international levels. For example, community programmes to mosquito-proof houses will need to be reinforced by improvements in the national surveillance of infectious diseases and in outbreak warning systems.

Doctors and other health professionals have particular knowledge, opportunity, and, often, political leverage that can help ensure—through advocacy or direct participation—that preventive actions are taken. Actions include promoting public understanding, monitoring and reporting the health effects of environmental change, and proposing and advocating local adaptive responses (box 2).

Box 2: How health professionals can promote adaptive strategies

- Public education, especially through healthcare settings such as doctors' waiting rooms and hospital clinics
- Preventive programmes eg, vaccines, mosquito control, food hygiene and inspection, nutritional supplementation
- Health care (especially mental health and primary care) for communities affected by environmental adversity
- Surveillance of disease (especially infectious disease) and key risk factors

9.6 FORECASTING FUTURE HEALTH RISKS FROM PROJECTED CLIMATE CHANGE

- Forecasting future health risks and gains from mitigation and adaptation strategies
- Health sector workforce training and in-career development
- Strategies that extend beyond health sector:
- Early warning systems for impending extreme weather (eg, heat waves, storms)
- Neighbourhood support schemes to protect the most vulnerable people
- Climate-proofed housing design, urban planning, water catchment, and farming practices
- Disaster preparedness, including capacity of the health system

Various websites list and discuss actions for doctors to take, both individually and collectively (box 3). For example, the US Centers for Disease Control and Prevention lists 11 functions for the public health system and practitioners for responding to climate change. And Doctors for the Environment Australia has run a successful, continuing, national campaign of patient education by distributing posters and pamphlets for use in doctors' waiting rooms.

Box 3 Environmental websites helpful for doctors and other health professionals

- Centers for Disease Control and Prevention (US) (www.cdc.gov/nceh/climatechange)
- Doctors for the Environment Australia (www.dea.org.au)
- Global environmental change and human health project (www.essp.org/en/joint-projects/health.html)
- Intergovernmental Panel on Climate Change (www.ipcc.ch)
- International Society of Doctors for the Environment (www.isde.org)
- Medact UK (www.medact.org/env_climate_change.php)

9.7 ADAPTIVE STRATEGIES TO LESSEN HEALTH RISKS

Many local actions can be taken to reduce the vulnerability of communities and populations. These will vary considerably between different regions of the world, and in relation to prevailing socioeconomic conditions and available resources. During Australia's recent prolonged drought (2001-7), some rural health doctors reported that fostering and supporting communal activities (community choirs, social gatherings, financial advisory networks, etc) increased local resilience against depression associated with loss of livelihood.

Climate change and other large scale environmental changes are unlikely to cause entirely new diseases (although they may contribute to the emergence of new strains of viruses and other microbes that can infect humans). Rather, they will alter the incidence, range, and seasonality of many existing health disorders. Hence, existing healthcare and public health systems should provide an appropriate starting point for adaptive strategies to lessen health effects.

9.8 PREVENTIVE ACTION

Although adaptive strategies will minimise the effects of climate change, the greater public health preventive challenge lies in stopping the process of climate change. This requires bold and far sighted policy decisions at national and international levels, entailing much greater emissions cuts than were being proposed a decade ago. Scientists have concluded that we need to prevent atmospheric carbon dioxide concentrations exceeding 450-500 ppm to avoid the serious, perhaps irreversible, damage to many natural systems and ecological processes that a global average temperature increase of 2-3°C would cause. Thisrequires early radical action as today's concentrations are approaching 390 ppm (compared with 280 ppm before industrialisation). Health professionals, acting through citizens' or professional organisations, have both the opportunity and responsibility to contribute to resolving this momentous issue. Improving awareness of the problem is the first step. Since 1993, doctors from 14 countries (including six low income countries) have had a central role in the Intergovernmental Panel on Climate Change's assessment of the health effects of climate change. We should also add this topic, including its relevance to health professional activity, to the medical curriculum.

The health sector, meanwhile, must minimise greenhouse gas emissions from its own infrastructure, especially hospitals. Health researchers should act to minimise greenhouse gas emissions from their own studies.

9.9 CONCLUSION

The Stern report, in 2006, highlighted the potentially great damage to the world's economic system from unconstrained climate change. The greater risk, however, is to the vitality and health of all species, including humans, if current trends continue to weaken the earth's life support systems. The health professions have a crucial role in promoting public understanding of this fundamental association and health protecting responses to it.

9.10 SUMMARY POINTS

- The biophysical and ecological systems of the natural environment are fundamental to human health.
- Mounting human pressures on the environment are disrupting and depleting these systems.

- The resulting health risks will particularly affect vulnerable and poorly resourced populations.
- Adaptive strategies can lessen current and impending health risks.
- Health professionals are well placed to contribute to adaptive and preventive strategies.

Notes

Contributors and sources:

AMcM has contributed to the assessment of health effects for the Intergovernmental Panel on Climate Change, and has been an environmental health adviser to WHO, the World Bank, and the UN Environment Program. He co-chairs the International Council of Science research project on global change and human health. SF has worked in public health nutrition and health promotion research at national and international level for 15 years. She has worked closely with ministries of health, education, agriculture, and social affairs in healthy public policy development. AN is an environmental scientist and has contributed widely to global environmental change research in Africa and to the Africa chapter for the fourth assessment report of the Intergovernmental Panel on Climate Change (2007). He is a senior programme specialist on climate change with the International Development Research Centre. CC is a senior environmental epidemiologist at WHO. He has contributed substantially to recent scientific reviews and policy development work on climate change and global environmental change and their health effects.

10

CLIMATE CHANGE: TWENTY-FIRST CENTURY AND BEYOND

10.1 EMISSION SCENARIOS

The increase in greenhouse gases is by far the largest of the factors which can lead to climate change during the twenty-first century. The likely climate changes for a range of scenarios of greenhouse gas emissions have been described in terms of global average temperature and in terms of regional change of temperature and precipitation and the occurrence of extremes. The rate of change is likely to be larger than any the Earth has seen at any time during the past 10 000 years. The changes that are likely to have the greatest impact will be changes in the frequencies, intensities and locations of climate extremes, especially droughts and floods. Sufficient fossil fuel reserves are available to provide for continuing growth in fossil fuel emissions of carbon dioxide well into the twenty- second century. If this occurred the climate change could be very large indeed and have unpredictable features or 'surprises.

A principal reason for the development of climate models is to learn about the detail of the likely climate change this century and beyond. Because model simulations into the future depend on assumptions regarding future anthropogenic emissions of greenhouse gases, which in turn depend on assumptions about many factors involving human be- haviour, it has been thought inappropriate and possibly misleading to call the simulations of future climate so far ahead 'predictions. They are therefore generally called 'projections' to emphasize that what is being done is to explore likely future climates which arise from a range of assumptions regarding human activities.

A starting point for any projections of likely climate change into the future is a set of descriptions of likely future global emissions of green- house gases. These will depend on a variety of assumptions regarding human behavior and activities, including population, economic growth, energy use and the sources of energy generation.

A wide range of scenarios was developed by the IPCC in a Special Report on Emission Scenarios (SRES)1 in preparation for its 2001 Report (see box below). It is these scenarios that have been used in developing the projections of future climate presented in this chapter. In addition, because it has been widely used in modelling studies, results are also presented using a scenario (IS 92a) taken from a set developed by the IPCC in 1992 and widely described as representative of 'business-as-usual'.

The SRES scenarios include estimates of greenhouse gas emissions resulting from all sources including land-use change. Estimates in the different scenarios begin from the current values for land-use change including deforestation (see Table 1). Assumptions in different scenarios vary, from continued deforestation, although reducing as less forest remains available for clearance, to substantial afforestation leading to an increased carbon sink.

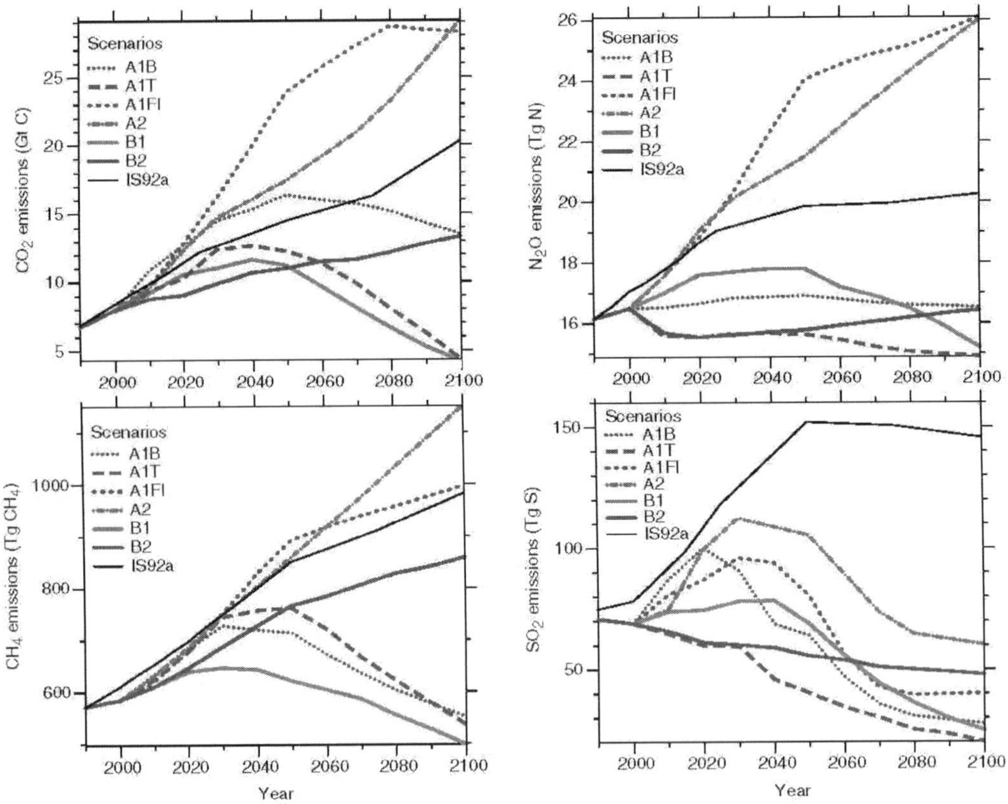

Fig. 1: Anthropogenic emissions of carbon dioxide, methane, nitrous oxide and Sulphur dioxide for the six illustrative SRES scenarios, A1B, A2, B1 and B2, A1F1 and A1T. For comparison the IS 92a scenario is also shown.

The next stage in the development of projections of climate change is to turn the emission profiles of greenhouse gases into greenhouse gas concentrations (Figure 2) and then into radiative forcing. For the carbon dioxide concentration scenarios these uncertainties, espe- cially those concerning the magnitude of the climate feedback from the terrestrial biosphere, amount to a range of about −10% to +30% in 2100 for each profile.

For most scenarios, emissions and concentrations of the main green- house gases increase during the twenty-first century. However, despite the increases projected in fossil fuel burning – very large increases in some cases – emissions of Sulphur dioxide (Figure 1) and hence the concentrations of sulphate particles are expected to fall substantially because of the spread of policies to abate the damaging consequences of air pollution and 'acid-rain' deposition to both humans and ecosystems. The influence of sulphate particles in tending to reduce the warming due to in- creased greenhouse gases is therefore now projected to be much less than for projections made in the mid-1990s. The other anthropogenic sources of particles in the atmosphere listed in Figure 8 will also contribute small amounts of positive or negative radiative forcing during the twenty-first century.

10.2 MODEL PROJECTIONS

Results which come from the most sophisticated coupled atmosphere– ocean models of the kind described in the last chapter provide fundamental information on which to base climate projections. However, because they are so demanding on computer time only a limited number of results from such models are available. Many studies have also therefore been carried out with simpler models. Some of these, while possessing a full description of atmospheric processes, only include a simplified description of the ocean; these can be useful in exploring regional change. Others, sometimes called energy balance models, drastically simplify the dynamics and physics of both atmosphere and ocean and are useful in exploring changes in the global average response with widely different emission scenarios. Results from simplified models need to be carefully compared with those from the best coupled atmosphere–ocean models and the simplified models 'tuned' so that, for the particular parameters for which they are being employed, agreement with the more complete models is as close as possible. The projections presented in the next sections depend on results from all these kinds of models.

In order to assist comparison between models, experiments with many models have been run with the atmospheric concentration of carbon dioxide doubled from its pre-industrial level of 280 ppm. The global aver- age temperature rise under steady conditions of doubled carbon dioxide

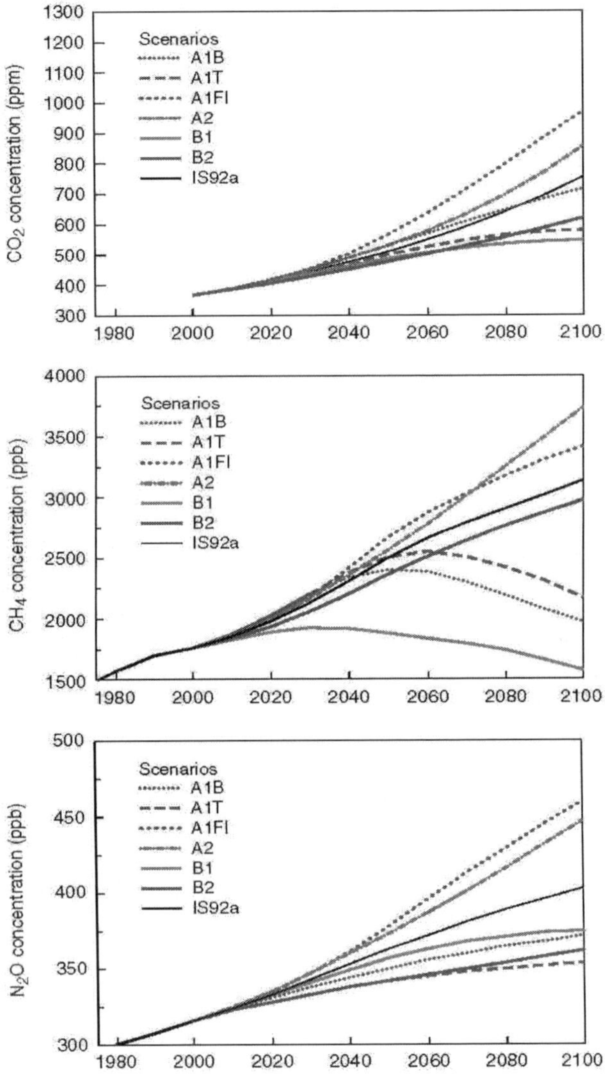

Fig. 2: Atmospheric concentrations of carbon dioxide, methane and nitrous oxide resulting from the six illustrative SRES scenarios and from the IS 92a scenario. Uncertainties for each profile, especially those due to possible carbon feedbacks, have been estimated as from about −10% to +30% in 2100.

Table 1 Radiative forcing (W m^{-2}), globally averaged, for greenhouse gases from the year 1750 to 2000 and from SRES scenarios to 2050 and 2100

Greenhouse gas	Year		A1B	A1T	A1FI	A2	B1	B2	I S 92a
CO_2	2000	1.46							
	2050		3.36	3.08	3.70	3.36	2.92	2.83	3.12
	2100		4.94	3.85	6.61	5.88	3.52	4.19	4.94
CH_4	2000	0.48							
	2050		0.70	0.73	0.78	0.75	0.52	0.68	0.73
	2100		0.56	0.62	0.99	1.07	0.41	0.87	0.91
N_2O	2000	0.15							
	2050		0.25	0.23	0.33	0.32	0.27	0.23	0.29
	2100		0.31	0.26	0.55	0.51	0.32	0.29	0.40
O_3(trop)	2000	0.35							
	2050		0.59	0.72	1.01	0.78	0.39	0.63	0.67
	2100		0.50	0.46	1.24	1.22	0.19	0.78	0.90
Halocarbons	2000	0.34							
	2050	0.49							
	2100	0.57							

Radiative forcing of climate change. In Houghton, Climate Change 2001,. Data selected from Tables.1 and 4. For 2050 and 210w0 for the halocarbons, all scenarios make the same as- sumptions concentration has become known as the climate sensitivity. The Inter- governmental Panel on Climate Change (IPCC) in its 1990 Report gave a 'best estimate' of 2.5 C for the climate sensitivity; it also considered that it was unlikely to lie outside the range of 1.5 C to 4.5 C, a range that encompasses the results of the best coupled atmosphere–ocean general circulation models (AOGCMs). The IPCC 1995 and 2001 Reports have confirmed these values

Simple climate models

To carry out more simulations under differ- ent future emission profiles of greenhouse gases or of aerosols or to explore the sensitivity of fu- ture change to different parameters (for instance, parameters describing the feedbacks in the atmo- sphere which largely define the climate sensitiv- ity), extensive use has been made of simple climate models. These simpler models are 'tuned' so as to agree closely with the results of the more complex AOGCMs in cases where they can be compared.. The most radical simplification in the simpler mod- els is to remove one or more of the dimensions so that the quantities of interest are averaged over lat- itude circles (in two-dimensional

10.3 PROJECTIONS OF GLOBAL AVERAGE TEMPERATURE

When information of the kind illustrated in Figures 1, 2 and 4(a) is incorporated into simple or more complex models, projections of climate change can be made. As we have seen in earlier chapters, a useful proxy for climate change that has been widely used is the change in global average temperature. The projected rise in global average temperature due to the increase in greenhouse gases and aerosols from pre-industrial times is illustrated in Figure 4(b). It shows an increase of about 0.6 C up to the year 2000 and an increase ranging from about 2 C to about 6 C by 2100 – that wide range resulting from the very large uncertainty regarding future emissions and also from the uncertainty that remains regarding the feed- backs associated with the climate response to the changing atmospheric composition.

models) or over the whole globe (in one-dimensional models). Such models can, of course, only simulate latitudinal or global averages – they can provide no regional in- formation. Figure.3 illustrates the components of such a model in which the atmosphere is contained within a 'box' with appropriate radiative inputs and out- puts. Exchange of heat occurs at the land surface (another 'box') and the ocean surface. Within the ocean allowance is made for vertical diffusion and vertical circulation. Such a model is appropriate for simulating changes in global average surface temperature with increasing greenhouse gases or aerosols. When exchanges of carbon dioxide across the interfaces between the atmosphere, the land and the ocean are also included, the model can be em- ployed to simulate the carbon cycle.

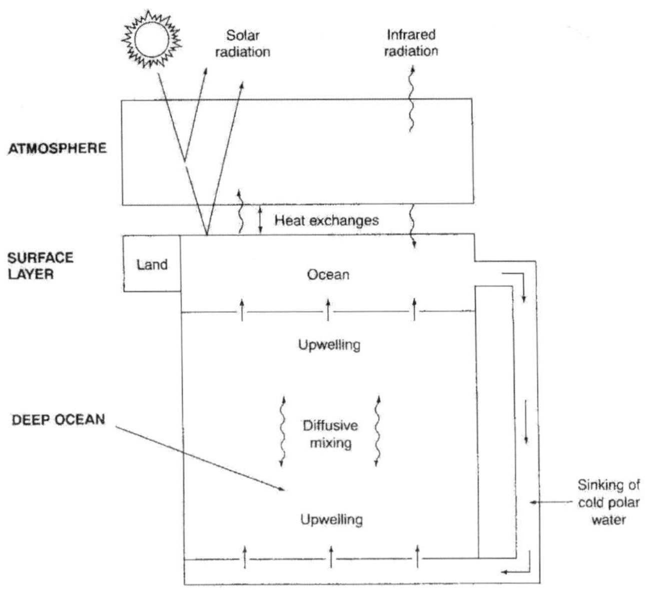

Fig. 3: The components of a simple 'upwelling–diffusion' climate model.

Compared with the temperature changes normally experienced from day to day and throughout the year, changes of between 2 C and 6 C may not seem very large. But, as was pointed out , it is in fact a large amount when considering globally averaged temperature. Compare it with the 5 C or 6 C change in global average temperature that occurs between the middle of an ice age and the warm period in between ice ages. The changes projected for the twenty-first century are from one-third to a whole ice age in terms of the degree of climate change!

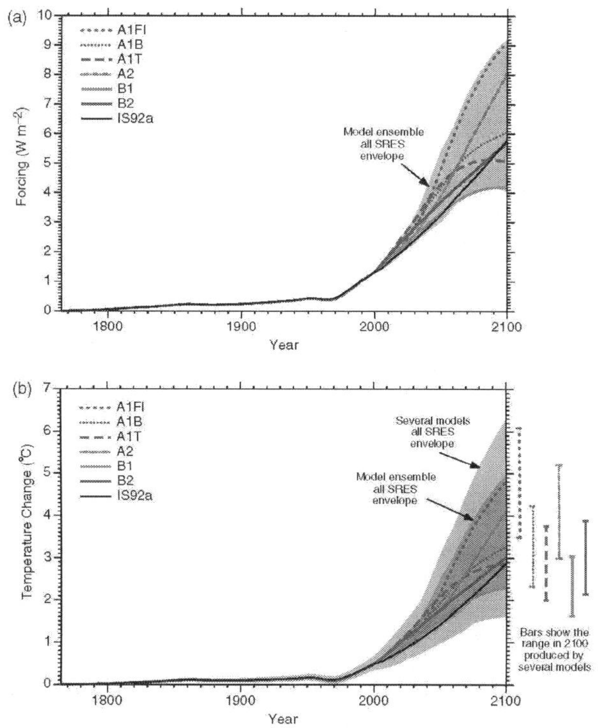

Fig. 4: Simple model results. (a) Anthropogenic globally averaged radiative forcing, based on historical information about greenhouse gas and aerosol concentrations to the year 2000 and the SRES scenarios to the year 2100. The shading shows the envelope of forcing that encompasses the full range of thirty-five SRES scenarios. (b) Historic anthropogenic global mean temperature change and future changes for the SRES scenarios and the IS 92a scenario calculated using a simple climate model tuned to seven AOGCMs (with climate sensitivity in the range 1.7 to 4.2 C). The darker shading represents the envelope of the full set of thirty-five SRES scenarios using the average of the model results (mean climate sensitivity is 2.8 C). The lighter shading is the envelope including all seven model projections (the range of model results for each scenario is also shown by the bars on the right hand side).

The rate of change of global average temperature projected for the twenty-first century is in the range of 0.15 C to 0.6 C per decade. These might seem small rates of change; most people would find it hard to detect a change in temperature of a fraction of a degree. But remembering again that these are global averages, such rates of change become very large. Indeed, they are much larger than any rates of change the climate has experienced for at least the past 10 000 years as inferred from paleoclimate data. As we shall see in the next chapter, the ability of both humans and ecosystems to adapt to climate change depends critically on the rate of change.

The changes in global average temperature shown in Figure 4(b) from the IPCC 2001 Report are substantially greater than those shown in the IPCC 1995 Report. The main reason for the difference is the much smaller aerosol emissions in the SRES scenarios compared with the IS 92 scenarios. This is illustrated by the temperature in 2100 shown in Figure 4(b) for the IS 92a scenario which is similar to that for the SRES B2 scenario even though the carbon dioxide emissions at that date for IS 92a are fifty per cent greater than those for B2.

In many of the modelling studies of climate change, the situation of doubled pre-industrial atmospheric carbon dioxide has often been introduced as a benchmark especially to assist in comparisons between different model projections and their possible impacts. Since the pre- industrial concentration was about 280 ppm, doubled carbon dioxide is about 560 ppm. From the curves in Figure 2 this is likely to occur sometime in the second half of the twenty-first century, depending on the scenario. But, other greenhouse gases are also increasing and con- tributing to the radiative forcing. So as to achieve an overall picture more easily, it is often convenient to convert other greenhouse gases to equiv- alent amounts of carbon dioxide, in other words to amounts of carbon dioxide that would give the same radiative forcing. The information in Table 1 enables the conversion to be carried out. For instance, the increases in the greenhouse gases other than carbon dioxide (including ozone) to date produce about eighty per cent of the radiative forcing due to the increase in carbon dioxide to date. This propor- tion will drop substantially during the next few decades as the growth in carbon dioxide becomes more dominant in nearly all scenarios. Refer- ring to Figure 6.4(a), and noting that doubled carbon dioxide produces a radiative forcing of about 3.7 W m−2, it can be seen that doubling of the equivalent carbon dioxide amount from pre-industrial times will occur between 2040 and 2070 depending on the scenario.

- Now, in Figure 4(a), select one of the scenarios, say A1B, and note that the radiative forcing for this scenario reaches that equivalent to doubled carbon dioxide (i.e. 3.7 W m−2) in about 2040. Then move to:
- Figure 4(b) and note that in 2040 the temperature rise is about 1.7 C.
- This is only just over half the 2.8 C (the value for climate sensitivity used for the results presented in Figure 6.4(b) see figure caption) that would be expected for doubled carbon dioxide under steady conditions. This difference occurs because of the slowing effect of the oceans on the temperature rise. But this means that, as the carbon dioxide concentration continues to increase, at any given time there exists a commitment to further significant temperature rise which has not been realised at that time.

10.4 REGIONAL PATTERNS OF CLIMATE CHANGE

So far, we have been presenting global climate change in terms of likely increases in global average surface temperature that provide a useful overall indicator of the magnitude of climate change. In terms of regional implications, however, a global average conveys rather little information. What is required is spatial detail. It is in the regional or local changes that the effects and impacts of global climate change will be felt.

With respect to regional change, it is important to realise that, be- cause of the way the atmospheric circulation operates and the interactions that govern the behaviour of the whole climate system, climate change over the globe will not be at all uniform. We can, for instance, expect substantial differences between the changes over large land masses and over the ocean; land possesses a much smaller thermal capacity and so can respond more quickly. Listed below are some of the broad features on the continental scale that characterise the projected temperature changes; more detailed patterns are illustrated in Figure 5(a). Refer- ence to Chapter 4 indicates that many of these characteristics are already being found in the observed record of the last few decades.

- Generally greater surface warming of land areas than of the oceans typically by about forty per cent compared with the global average, greater than this in northern high latitudes in winter (associated with reduced sea ice and snow cover) and less than forty per cent in south and southeast Asia in summer and in southern South America in winter.
- Minimum warming around Antarctica and in the northern North
- Atlantic which is associated with deep oceanic mixing in those areas.
- Little warming over the Arctic in summer.
- Little seasonal variation of the warming in low latitudes or over the southern circumpolar ocean.
- A reduction in diurnal temperature range over land in most seasons and most regions; night time lows increase more than daytime highs.

So far we have been presenting results solely for atmospheric tem- perature change. An even more important indicator of climate change is precipitation. With warming at the Earth's surface there is increased evaporation from the oceans and also from many land areas leading on average to increased atmospheric water vapour content and therefore also on average to increased precipitation. Since the water-holding capacity of the atmosphere increases by about 6.5% per degree celsius, the in- creases in precipitation as surface temperature rises can be expected to be substantial. In fact, model projections indicate increases in precipitation broadly related to surface temperature increases of about three per cent per degree celsius. Further, since the largest component of the energy input to the atmospheric circulation comes from the release of latent heat as water vapour condenses, the energy available to the atmosphere's circulation will increase in proportion to the atmospheric water content. A characteristic therefore of anthropogenic climate change due to the increase of greenhouse gases will be a more intense hydrological cycle.

In Figure5(b) are shown the projected changes in the distribution of precipitation. Although on average precipitation increases there are large regional variations and large areas where there are likely to be decreases in average precipitation and also changes in its seasonal distribution. For instance, at high northern latitudes there are large increases in winter and over south Asia in summer. Southern Europe, Central America, southern Africa and Australia are likely to have drier summers.

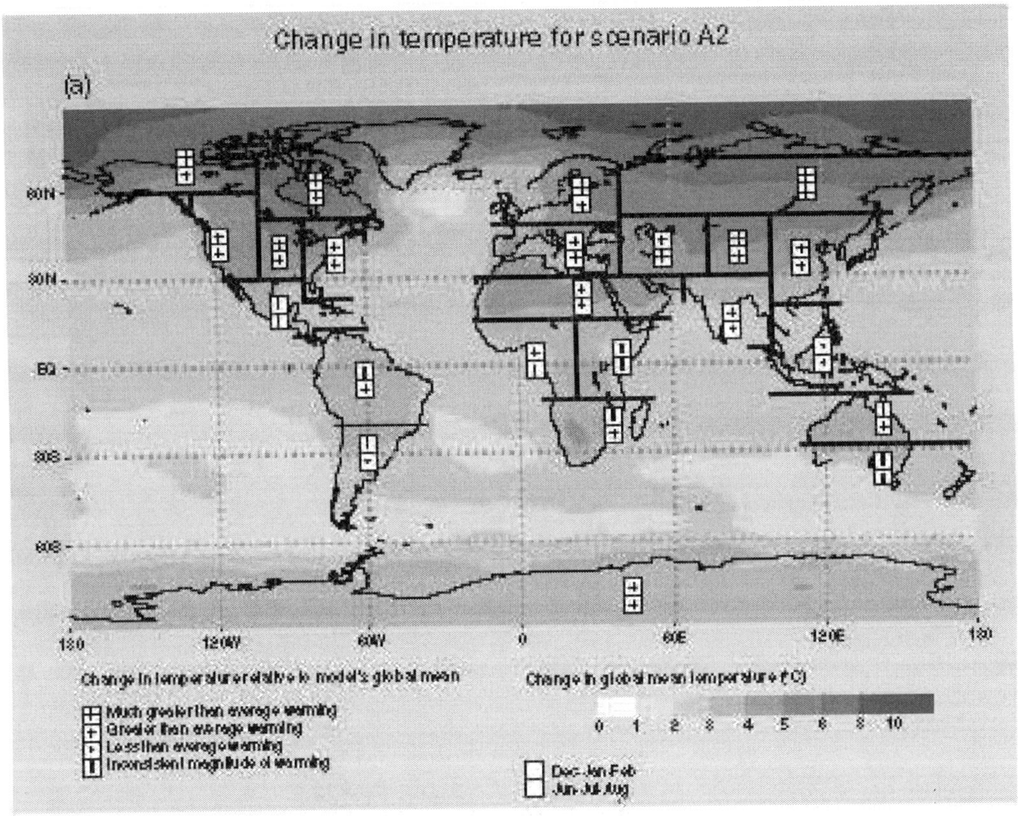

Fig. 5: Projections for the SRES scenario A2 for the period 2071–2100 relative to 1961–1990 from an ensemble of nine different ocean–atmosphere general circulation models. (a) The annual mean change of temperature in C shown by the shading. The boxes show an analysis of inter-model consistency in regional relative warming (i.e. warming relative to each model's global average warming) for winter and summer seasons. Regions are classified as showing either agreement on warming in excess of forty per cent above the global mean annual average (much greater than average warming), agreement on warming greater or less than the global mean annual average (greater or less than average warming) or disagreement amongst models on the magnitude of regional relative warming (inconsistent magnitude of warming). A consistent result from at least seven of the nine models is defined as being necessary for agreement. The global mean average warming of the models used are in the range 1.2 to 4.5 ⊠C for this scenario. (b) The boxes show an analysis of inter-model consistency in regional precipitation change for winter and summer seasons. Regions are classified as showing

agreement on change of greater than +20% (large increase), between +5% and +20% (small increase) and more than −20% (large decrease), or disagreement (inconsistent sign). A consistent result from at least seven of the nine models is defined as being necessary for agreement.

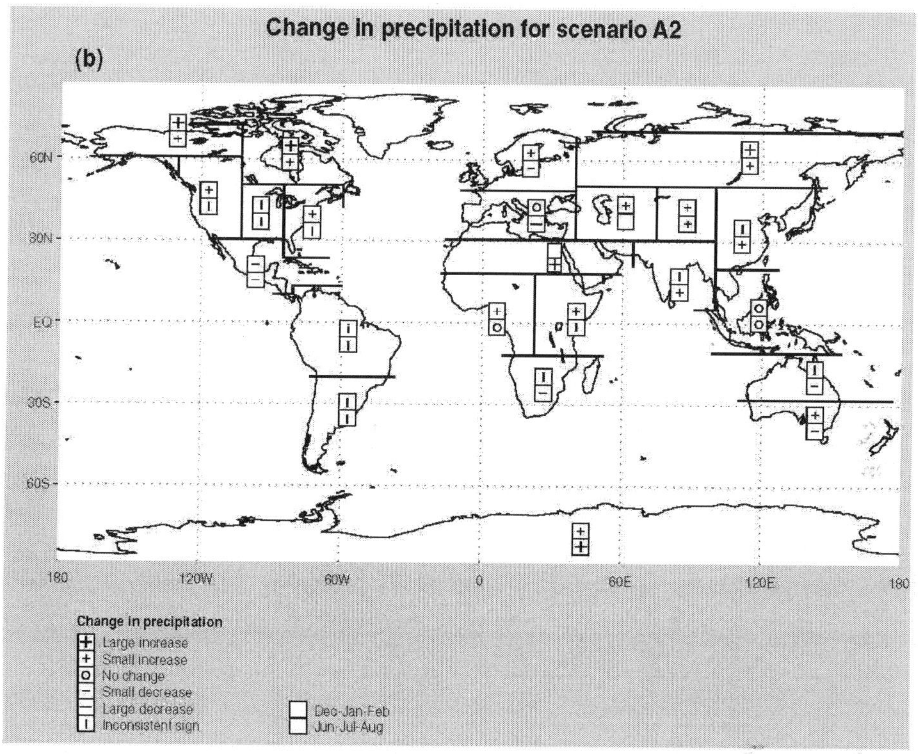

Figure 5 (Cont.).

Much natural climate variability occurs because of changes in, or oscillations between, persistent climatic patterns or regimes. The Pacific, North Atlantic Anomaly (PNA – that is dominated by high pressure over the eastern Pacific and western North America and which tends to lead to very cold winters in the eastern United States), the North Atlantic Oscillation (NAO – that has a strong influence on the character of the winters in north west Europe) and the El Nin˜o events mentioned. Important components of climate change in response to the forcing due to the increase in greenhouse gases can be expected to be in the form of changes in the intensity or the frequency of established climate patterns illustrated by these regimes. There is little consistency at the present time between models regarding projections of many of these patterns. However, recent trends in the tropical Pacific for the surface temperature to become more El Nin˜o-like, with the warming in the eastern tropical Pacific more than that in the western tropical Pacific and with a corresponding eastward shift of precipitation, are projected to continue by many models. There is also evidence that warming associated with increasing greenhouse gas concentrations will cause an intensification of the Asian summer mon- soon and an increase of variability in its precipitation. The influence of increased greenhouse gases on these major climate regimes, especially the El Nin˜o, is an important and urgent area of research.

A complication in the interpretation of patterns of climate change arises because of the differing influence of atmospheric aerosols as com- pared with that of greenhouse gases. Although in the projections based on the SRES scenarios the influence of aerosols is much less than in those based on the IS 92 scenarios published by the IPCC in its 1995 Report, their projected radiative forcing is still significant. When considering global average temperature and its impact on, for instance, sea level rise it is appropriate in the projections to use the values of globally averaged radiative forcing. The negative radiative forcing from sulphate aerosol, for instance, then becomes an offset to the positive forcing from the increase in greenhouse gases. However, because the effects of aerosol forcing are far from uniform over the globe, it is not correct, when considering climate change and its regional characteristics, to consider the effects of increasing aerosol as a simple offset to those of the increase in greenhouse gases. The large variations in regional forcing due to aerosols produce substantial regional variations in the climate response. Detailed regional information from the best cli- mate models needs to be employed to assess the climate change under different assumptions about the increases in both greenhouse gases and aerosols.

10.5 CHANGES IN CLIMATE EXTREMES

Can anything be said about likely changes in the frequency or intensity of climate extremes in the future? It is, after all, not the changes in average climate that are generally noticeable, but the extremes of climate – the droughts, the floods, the storms and the extremes of temperature in very cold or very warm periods – which provide the largest impact on our lives.

The most obvious change we can expect in extremes is a large increase in the number of extremely warm days (Figure. 6) coupled with a decrease in the number of extremely cold days. A number of model projections show a generally decreased daily variability of surface air temperature in winter and increased daily variability in summer in

Northern Hemisphere land areas, suggesting that the situation in Figure 6(c) could apply in these areas. However, the changes that are likely to lead to most impact are those connected with the hydrological cycle. In the last section it was explained that in a warmer world with increased greenhouse gases, average pre- cipitation increases and the hydrological cycle becomes more intense.16 Consider what might occur in regions of increased rainfall. Often in such regions with the more intense hydrological cycle the larger amounts of rainfall will come from increased convective activity: more really heavy showers and more intense thunderstorms. The result of a study with an Australian climate model of the effect on rainfall amounts of doubling the carbon dioxide concentration is shown in Figure 7. The number of days with large rainfall amounts (greater than 25 mm day^{-1}) doubled. The probability of conditions leading to floods would also at least have doubled. Similar results (fewer rainy days, higher maximum daily rain- falls for a given mean rainfall rate) have been obtained from many other climate models. For instance, a recent modelling study (Figure 8) has shown that if atmospheric carbon dioxide concentration is doubled from its pre-industrial value, the probability of extreme seasonal precipitation in winter is likely to increase substantially over large areas of central and northern Europe and likely to decrease over parts of the Mediterranean and north Africa. The increase in parts of central Europe is such that the return period of extreme rainfall events would decrease by about a factor of five (e.g. from fifty years to ten years). Similar results have been obtained in a study of major river basins around the world.

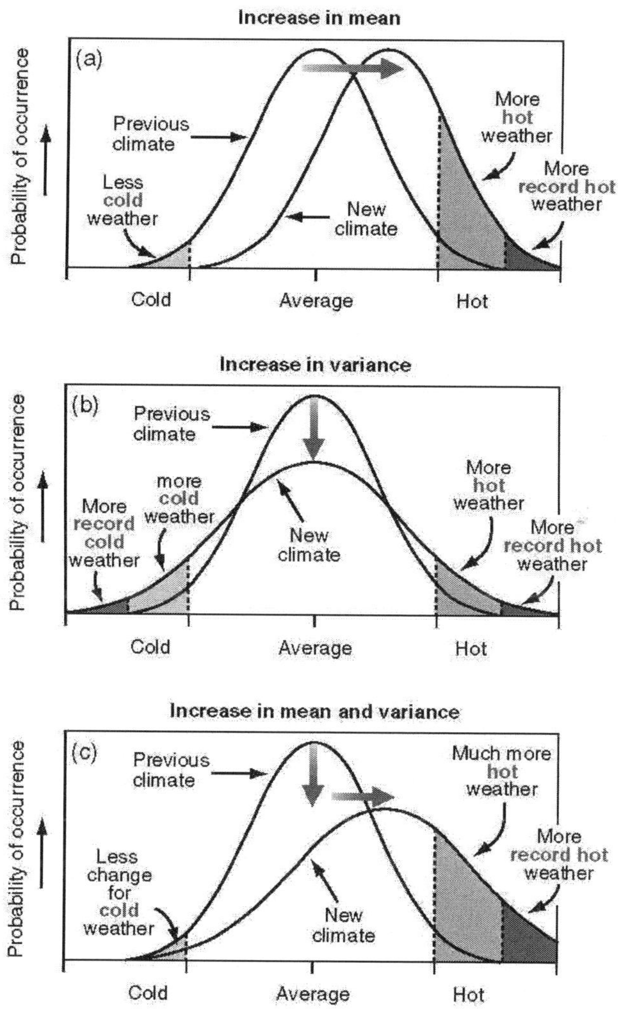

Figure 6 Schematic diagrams showing the effects on extreme temperatures when (a) the mean increases leading to more record hot weather, (b) the variance increases and (c) when both the mean and the variance increase, leading to much more record hot weather.

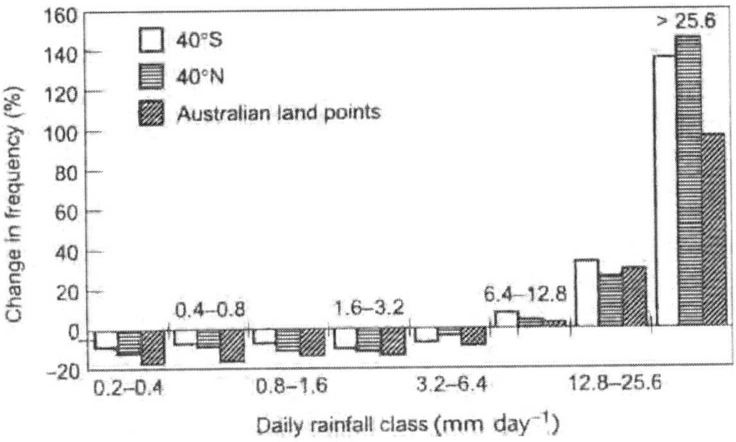

Figure 7 Changes in the frequency of occurrence of different daily rainfall amountswith doubled carbon dioxide as estimated by a CSIRO model in Australia.

Note also from Figure .7 that the number of days with lighter rain- fall events (less than 6 mm/day) is expected to decrease in the globally warmed world. This is because, with the more intense hydrological cy- cle, a greater proportion of the rainfall will fall in the more intense events and, furthermore, in regions of convection the areas of downdraught be- come drier as the areas of updraught become more moist. In many areas with relatively low rainfall, the rainfall will tend to become less – which has implications for the likelihood of drought.

Take, for instance, the likelihood of drought in regions where the average summer rainfall falls by perhaps twenty per cent as is likely to occur, for instance, in southern Europe (Figure 5(b)). The likely result of such a drop in rainfall is not that the number of rainy days will remain the same, with less rain falling each time; it is more likely that there will be substantially fewer rainy days and considerably more chance of prolonged periods of no rainfall at all (see Figure 9). In other words, much more likelihood of drought. Further, the higher temperatures will lead to increased evaporation reducing the amount of moisture available at the surface – thus adding to the drought conditions. The proportional increase in the likelihood of drought is much greater than the proportional decrease in average rainfall.

Thus, in the warmer world of increased greenhouse gases, different places will experience more frequent droughts and floods – we noted that these are the climate extremes which cause the greatest problems.

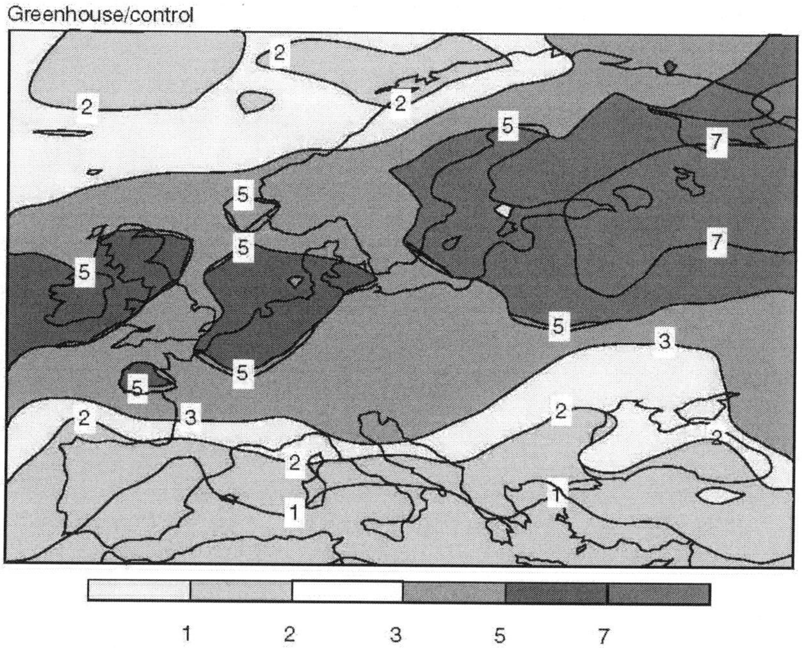

Figure 8 The changing probability of extreme season precipitation in Europe in winter as estimated from an ensemble of nineteen runs with a climate model starting from slightly different initial conditions. The figure shows the ratio of probabilities of extreme precipitation events in the years sixty-one to eighty of eighty-year runs that assumed an increase of carbon dioxide concentration of one per cent per year (hence doubling in about seventy years) compared with control runs with no change in carbon dioxide.

What about other climate extremes, intense storms, for instance? How about hurricanes and typhoons, the violent rotating cyclones that are found over the tropical oceans and which cause such devastation when they hit land? The energy for such storms largely comes from the latent heat of the water which has been evaporated from the warm ocean surface and which condenses in the clouds within the storm, re- leasing energy. It might be expected that warmer sea temperatures would mean more energy release, leading to more frequent and intense storms. However, ocean temperature is not the only parameter controlling the genesis of tropical storms; thc nature of the overall atmospheric flow is also important. Further, although based on limited data, observed vari- ations in the intensity and frequency of tropical cyclones show no clear trends in the last half of the twentieth century. Models can take all the relevant factors into account but, because of the relatively large size of their grid, they are unable to simulate very reliably the detail of rela- tively small disturbances such as tropical cyclones. There is no consistent evidence from model projections of changes in the frequency of tropi- cal cyclones or their areas of formation. However, model projections18 and theoretical studies suggest that, if carbon dioxide concentration is doubled, the peak wind intensities will tend to increase by perhaps five or ten per cent and the mean and peak precipitation intensities by twenty to thirty percent.

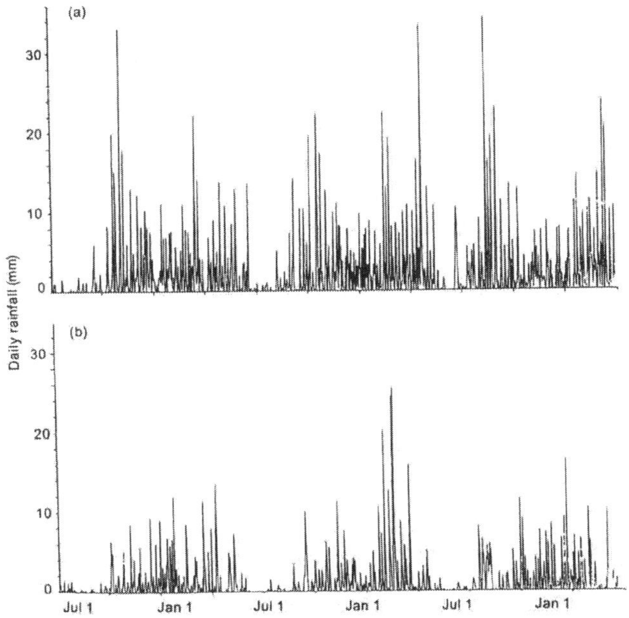

Figure .9 Daily rainfall for Italy for a three-year period simulated by a climate model,

(a)for the current climate situation and (b) for the climate if the equivalent carbon dioxide concentration increased by a factor of four from its pre-industrial value (predicted for instance to occur before 2100 under the A2 scenario).

Regarding storms at mid latitudes, the various factors that control their incidence are complex. Two factors tend to an increased intensity of storms. The first, as with tropical storms, is that higher temperatures, es- pecially of the ocean surface, tend to lead to more energy being available. The second factor is that the larger temperature contrast between land and sea, especially in the Northern Hemisphere, tends to generate steeper temperature gradients, which in turn generate stronger flow and greater likelihood of instability. The region around the Atlantic seaboard of Europe is one area where such increased storminess might be expected, a result that some model projections have shown. However, such a picture may well be too simple; other models suggest changes in storm tracks that result in very different changes in some regions and there is little overall consistency between model projections.

For some other extremes such as very small-scale phenomena (e.g. tornadoes, thunderstorms, hail and lightning) that cannot be simulated in global models, although they may have important impacts, there is cur- rently insufficient information to assess recent trends, and understanding is inadequate to make firm projections.

Table 2 summarises the state of knowledge regarding the likely fu- ture incidence of extreme events. Although general indications of trends can be given, there have been few projections with quantitative estimates of likely changes in the frequency or intensity of extreme events. In many research centres, work is under way on more detailed studies of the in- fluence of increased greenhouse gases on extreme events and changes in climate variability.

10.6 REGIONAL CLIMATE MODELS

Most of the likely changes that we have presented have been on the scale of continents. Can more specific information be provided about change for smaller regions? Example are shown in Figures 10 of the improvement achieved by RCMs in the simulation of extremes and in providing regional detail that in many cases (especially for precipitation) shows substantial disagreement with the averages provided by a GCM.

Table .2 Estimates of confidence in observed and projected changes in extreme weather and climate events

Confidence in observed changes (latter half of the twentieth century)	Changes in phenomenon	Confidence in projected changes (during the twenty-first cenury)
Likely[b]	Higher maximum temperatures and more hot days over nearly all land areas	Very likely
Very likely	Higher minimum temperatures, fewer cold days over nearly all land lareas	Very likely
Very likely	Reduced diurnal temperature range over most land areas	Very likely
Likely, over many areas	Increase of heat indexa over land areas more intese precipitation events	very likely, over most areas
Likely, over many Northern Hemisphere mid to high latitude land areas	Increased summer continental drying and associated risk of droght	Very likely, over most areas
Likey, in a few areas	increase in tropical cyclone peask wind intensites	Likely, over most mid-atiude continental interior (lack of projection consisten in other areas Likey, over some ares
Not observed in the few analyses available	Increase in tropical cycle mean and peak precipitation intensities	Likely, over some areas
Insuffcient data for assessment		

Source: Table SPM-1 from Summary for policymakers. In Houghton, J. T., Ding, Y., Griggs, D. J., Noguer, M., van der Linden, P. J., Dai, X., Maskell, K., Johnson, C. A. (eds.) Climate

Change 2001: The Scientific Basis. Contribution of Working Group I to the Third Assessment Report of the Intergovernmental Panel on Climate Change. Cambridge: Cambridge University Press.

Regional models are providing a powerful tool for the investigation of detail in patterns of climate change. However, it is important to realise that, even if the models were perfect, because much greater natural variability is apparent in local climate than in climate averaged over continental or larger scales, projections on the local and regional scale are bound to be more uncertain than those on larger scales.

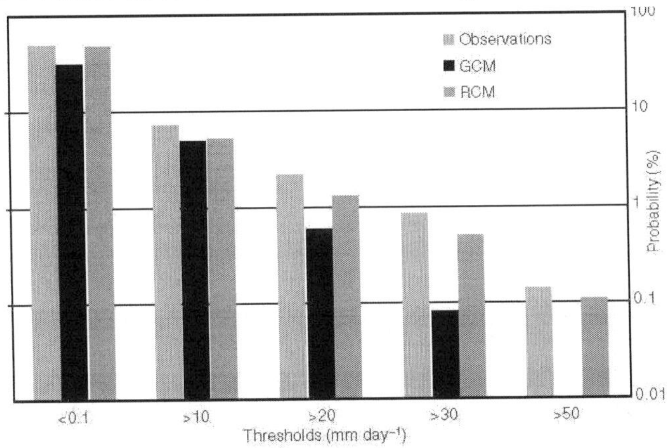

Figure 10 Example of simulations showing the probability of winter days over the Alps with different daily rainfall thresholds, as observed, simulated by a 300-km resolution GCM and by a 50-km resolution RCM. The RCM shows much better agreement with observations especially for higher thresholds.

10.7 LONGER-TERM CLIMATE CHANGE

Most of the projections of future climate that have been published cover the Twenty-first century. For instance, the curves plotted in Figure 2 ex- tend to the year 2100. They illustrate what is likely to occur if fossil fuels continue to provide most of the world's energy needs during that period.

From the beginning of the industrial revolution until 2000 the burning of fossil fuels released approximately 600 Gt of carbon in the form of carbon dioxide into the atmosphere. Under the SRES A1B scenario it is projected that a further 1500 Gt will be released by the year 2100. As I show, the reserves of fossil fuels in total are sufficient to enable their rate of use to continue to grow well beyond the year 2100. If that were to happen the global average temperature would continue to rise and could, in the twenty-second century, reach very high levels, perhaps up to 10 C higher than today. The associated changes in climate would be correspondingly large and could well be irreversible.

A further longer-term effect that may become important during this century is that of positive feedbacks on the carbon cycle due to climate change. The +30% uncertainty in 2100 in the atmospheric concentrations of carbon dioxide shown in Figure 2 was introduced to allow for it. Taking this effect into account would add nearly a further degree Celsius to the projected increase in global average temperature in 2100 at the top end of the range shown in Figure 4. Some of the further implications of this feedback will be considered. Especially when considering the longer term, there is also the possi- bility of surprises – changes in the climate system that are unexpected.

The discovery of the 'ozone hole' is an example of a change in the at- mosphere due to human activities, which was a scientific 'surprise'. By their very nature such 'surprises' cannot, of course, be foreseen. How- ever, there are various parts of the system which are, as yet, not well understood, where such possibilities might be looked for 20; for instance, in the deep ocean circulation or in the stability of the major ice sheets. In the next section we shall look in more detail at the first of these possibilities; the second will be addressed in the section of the next chapter entitled 'How much will sea level rise?'.

10.8 CHANGES IN THE OCEAN THERMOHALINE CIRCULATION

The ocean thermohaline circulation (THC) was introduced, where illustrates the deep ocean currents that transport heat and freshwater between all the world's oceans. Also mentioned in the box was the influence on the THC in past epochs of the input of large amounts of fresh water from ice melt to the region in the north Atlantic between Greenland and Scandinavia where the main source region for the THC is located.

With climate change due to increasing greenhouse gases we have seen that the precipitation will increase substantially especially at high latitudes (Figure 5(b)), leading to additional fresh water input to the oceans. The dense salty water in the north Atlantic source region for the THC will become less salty and therefore less dense. As a result the THC will weaken and less heat will flow northward from tropical regions to the north Atlantic. All coupled ocean–atmosphere GCMs show this occur- ring; an example is shown in Figure 11(a), which indicates a weakening of about twenty per cent by 2100. Although there is disagreement be- tween the models as to the extent of weakening, all model projections of the pattern of temperature change under increasing greenhouse gases show less warming in the region of the north Atlantic (Figure 5(a)) – but none show actual cooling occurring in this region during the twenty- first century. In the longer term, some models show the THC actually cutting off completely after two or three centuries of increasing green- house gases. Figure 11(b) illustrates the effect that cut-off would have on the pattern of surface temperature around the globe. Note the severe cooling that would occur in the north Atlantic and north west Europe and the small compensating warming in the Southern hemisphere. In- tense research is being pursued – both observations and modelling – to elucidate further likely changes in the thermohaline circulation and their possible impact.

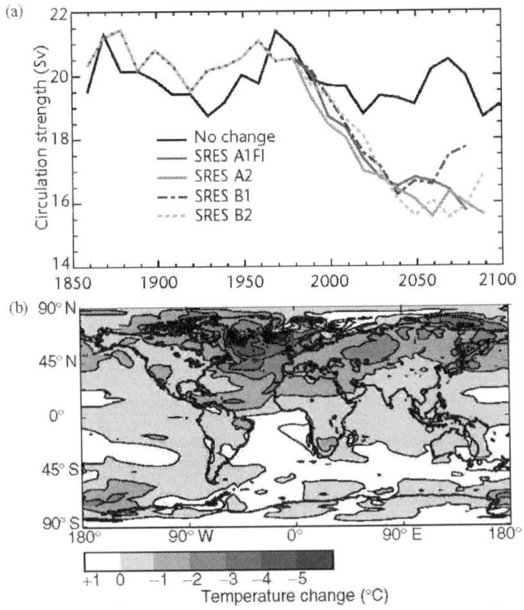

Figure 11 (a) Change in the strength of the thermohaline circulation (THC) in the north Atlantic as simulated by the Hadley Centre climate model for different SRES scenarios. The unit of circulation is the Sverdrup, 106 m^3 s^{-1}. (b) Changes in the surface air temperature, relative to the present day, twenty years after the hypothetical collapse of the thermohaline circulation as simulated by the Hadley Centre climate model. To obtain the temperature distribution for a real situation, the changes due to increased greenhouse gases at the time of collapse would have to be added to these changes resulting from THC.

10.9 OTHER FACTORS THAT MIGHT INFLUENCE CLIMATE CHANGE

So far climate change due to human activities has been considered. Are there other factors, external to the climate system, which might induce change? As showed that it was variations in the incoming solar energy as a result of changes in the Earth's orbit which triggered the ice ages and the major climate changes of the past. These variations are, of course, still going on; what influence are they having now?

Over the past 10 000 years, because of these orbital changes, the solar radiation incident at 60 N in July has decreased by about 35 W m^{-2}, which is quite a large amount. But over one hundred years the change is only at most a few tenths of a watt per square metre, which is much less than the changes due to the increases in greenhouse gases (remember that doubling carbon dioxide alters the thermal radiation, globally averaged, by about 4 W m^{-2}). Looking to the future and the effect of the Earth's orbital variations, over at least the next 50 000 years the solar radiation incident in summer on the polar regions will be unusually constant so that the present interglacial is expected to last for an exceptionally long period. Suggestions therefore that the current increase of greenhouse gases might delay the onset of the next ice age are unfounded.

These orbital changes only alter the distribution of incoming solar energy over the Earth's surface; the total amount of energy reaching the Earth is hardly affected by them. Of more immediate interest are suggestions that the actual energy output of the Sun might change with time.

There have also been suggestions of plausible indirect mechanisms whereby effects on the Sun might influence climate on Earth. Changes in solar ultraviolet radiation will influence atmospheric ozone and hence might affect climate. There is a possibility that the galactic cosmic ray flux, modified by the varying Sun's magnetic field, could influence cloudiness and hence climate. From studies of these possible mechanisms there is as yet insufficient evidence of significant climate effects.

Another influence on climate comes from volcanic eruptions. Their effects, lasting typically a few years, are relatively short-term compared with the much longer-term effects of the increase of greenhouse gases. The recent large volcanic eruption of Mount Pinatubo in the Philippines which occurred in June 1991 has already been mentioned. Estimates of the change in the net amount of radiation (solar and thermal) at the top of the atmosphere resulting from this eruption are of about W m^{-2}. This perturbation lasts for about two or three years while the major part of the dust settles out of the atmosphere; the longer-term change in radiation forcing, due to the minute particles of dust which last in the stratosphere for somewhat longer, is much smaller.

11

MAJOR CAUSES, EFFECTS AND SOLUTIONS OF GLOBAL WARMING

11.1 INTRODUCTION

The scientific and environmental community is on the same page regarding the bitter reality of global warming and the involvement of human factor in it. Global warming is a big hazard and appropriate measures must be taken to tackle this serious problem. This problem is not only causing trouble to the human beings but also to animals and plants. Melting of polar ice caps will lead to floods which can cause mayhem everywhere. Rise of sea levels will devastate agricultural and fishing activities. To embark upon these problems, some remedial steps must be timely taken which include but are not limited to the use of renewable sources of energy and stopping deforestation. Innovative solutions must be brought forward to end this hazard once and forever.

The continuous rise in temperature of the planet is really upsetting. The root cause for this is global warming. Global warming begins when sunlight reaches the Earth. The clouds, atmospheric particles, reflective ground surfaces and surface of oceans then sends back about 30 % of sunlight back into the space, whilst the remaining is absorbed by oceans, air and land. This consequently heats up the surface of the planet and atmosphere, making life feasible. As the Earth warms up, this solar energy is radiated by thermal radiation and infrared rays, propagating directly out to space thereby cooling the Earth. However, some of the outgoing radiation is re-absorbed by carbon dioxide, water vapours, ozone, methane and other gases in the atmosphere and is radiated back to the surface of Earth. These gases are commonly known as greenhouse gases due to their heat-trapping capacity. It must be noted that this re-absorption process is actually good as the Earth's average surface temperature would be very cold if there was no existence of greenhouse gases. The dilemma began when the concentration of greenhouse gases in the atmosphere was artificially increased by humankind at an alarming rate since the past two centuries. As of 2004, over 8 billion tons of carbon dioxide was pumped thermal radiation is further hindered by increased levels of greenhouse gases resulting in a phenomenon known as human enhanced global warming effect. Recent observations regarding global warming have substantiated the theory that it is indeed a human enhanced greenhouse effect that is causing the planet to heat up. The planet has experienced the largest increase in surface temperature over the last 100 years. Between 1906 and 2006, the Earth's average surface temperature augmented between 0.6 to 0.9 degrees Celsius, however out per year. Millions of pounds of methane gas are generated in landfills and agricultural decomposition of biomass and animal manure. Nitrous oxide is released into the atmosphere by various nitrogen-based fertilizers including urea and diammonium phosphate and other soil management utilizations. Once released, these greenhouse gases stay in the atmosphere for decades or even longer. According to Intergovernmental Panel on Climate Change (IPCC), carbon dioxide and methane levels have increased by 35 % and 148 % since the industrial revolution of 1750.

Many researchers, engineers and environmentalists are expressing deep concerns about changes in the overall climate of the planet. Fossil fuels are being continuously used to produce electricity. The burning of these fuels produces gases like carbon dioxide, methane and nitrous oxides which lead to global warming. Deforestation is also leading to warmer temperatures. The hazard of global warming is continuously causing major damage to the Earth's environment. Most people are still unaware of global warming and do not consider it to be a big problem in years to come. What most people do not understand is that global warming is currently happening, and we are already experiencing some of its withering effects. It is and will severely affect ecosystems and disturb ecological balance. Because of the treacherous effects of global warming, some solutions must be devised. The paper introduces global warming, elaborates its causes and hazards and presents some solutions to solve this hot issue. Above all, alternative energy sources (solar, wind, hydro, geothermal, bio mass) need to be seriously pursued. Finding and using renewable sources of energy is one of the methods to combat the ever-increasing global warming effectively.

11.2 GREENHOUSE EFFECT

While other planets in the solar system of the Earth are either roasting hot or bitterly cold, Earth's surface has relatively mild, steady temperatures. Earth enjoys these temperatures because of its atmosphere, which is the thin layer of gases that cover and protect the planet. However, 97 % of climate scientists and researchers agree that humans have changed the Earth's atmosphere in dramatic ways over the past two centuries, resulting in global warming. To understand global warming, it is first necessary to become familiar with the greenhouse effect. As Fig.1 depicts, the natural greenhouse effect normally traps some portion of heat in such a way that our planet is safe from reaching freezing temperatures while human enhanced greenhouse effect leads to global warming. This is due to burning of fossil fuels which increase the amount of greenhouse gases (carbon dioxide, methane and oxides of nitrogen) present in the atmosphere.

Fig. 1: Types of greenhouse effects

The trade of incoming and outgoing radiation that heats up the Earth is often referred to as the greenhouse effect because a greenhouse works in a similar way (Fig.2). Incoming ultraviolet radiation easily passes through the glass walls of a greenhouse and is absorbed by the plants and hard surfaces inside. Weaker infrared radiation, however, has difficulty passing through the glass walls and is trapped inside, therefore, warming the greenhouse. This effect lets tropical plants prosper inside a greenhouse, even during a cold season.

Fig. 2: Plants embodied in a greenhouse

A similar phenomenon takes place in a car which is parked outside on a cold sunny day. Incoming solar radiation warms the interior of the car but outgoing thermal radiation is trapped inside the closed windows of the cars. This entrapment basically warms up the car. This trapping occurs in such a way that the hot air does not rise and does not lose energy though convention. This phenomenon is depicted in Fig. 3.

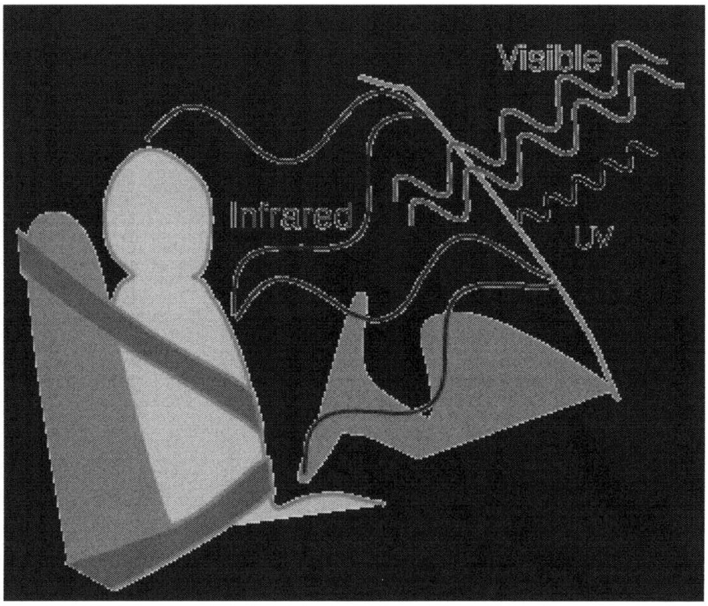

Fig. 3: Greenhouse effect example

In the words of Michael Daley, an Associate Professor of Environmental Science at Lasell College: "Gas molecules that absorb thermal infrared radiation, and are in significant enough quantity, can force the climate system. These types of gas molecules are called greenhouse gases". Carbon dioxide and other greenhouse gases act like a mantle, absorbing infrared radiation and preventing it from escaping into the outer space. The net effect is the regular heating of the Earth's atmosphere and surface.

The greenhouse effect, combined with increasing levels of greenhouse gases and the resulting global warming, is expected to have philosophical implications. If global warming continues unrestrained and nothing effective is done to limit this evil, it will cause significant climate change, a rise in sea levels, extreme weather events and other ruthless natural, environmental and social impacts [2].

11.3 GREENHOUSE GASES: A HAZARD

There are many greenhouse gases which are mainly emitted by human activity. The first and foremost in the list is carbon dioxide. Excessive burning of fossil fuels like coal and oil is the major factor for producing this gas. Moreover, deforestation i.e., removal of trees for acquiring lands also causes large amount of carbon dioxide in the atmosphere. Cement manufacture also contributes carbon dioxide to atmosphere when calcium carbonate is heated generating lime and carbon dioxide. The second culprit gas is methane, commonly known as natural gas. It is produced as a result of agricultural activities such as livestock digestion, paddy rice farming and use of manure. Methane is also produced due to improper management of waste. Nitrous oxides are generated mainly by fertilizers. Moreover, fluorinated gases such as chlorofluorocarbons (CFCs) are chiefly a result of various industrial processes and refrigeration. Fig.4 shows pictorially the distribution of greenhouse gases. These gases are playing their negative part in increasing the havoc of global warming. They are continuously causing an increase in the earth's temperature.

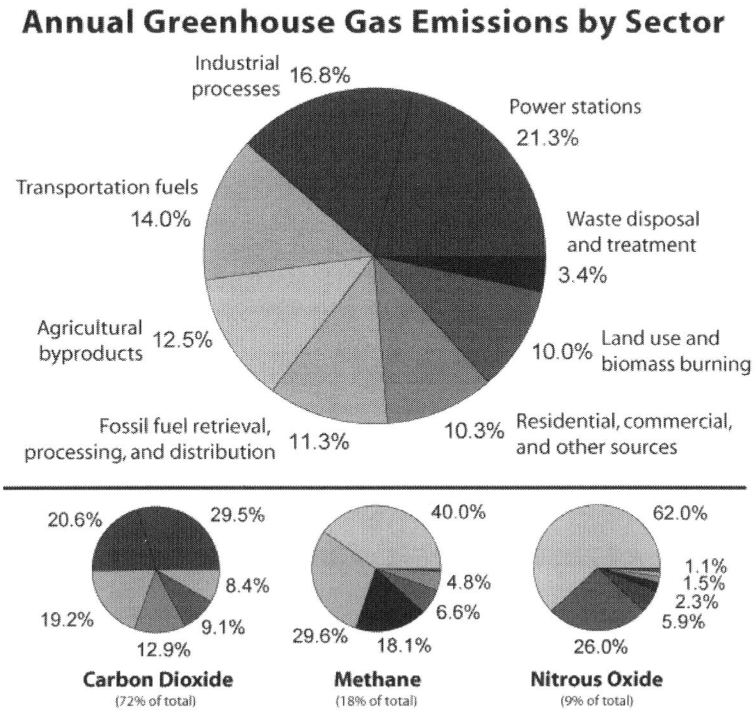

Fig.4: Distribution of greenhouse gases

11.4 CAUSES OF GLOBAL WARMING

The major cause of global warming is the greenhouse gases. They include carbon dioxide, methane, nitrous oxides and in some cases chlorine and bromine containing compounds. The build-up of these gases in the atmosphere changes the radiative equilibrium in the atmosphere. Their overall effect is to warm the Earth's surface and the lower atmosphere because greenhouse gases absorb some of the outgoing radiation of Earth and re-radiate it back towards the surface. The net warming from 1850 to the end of the 20th century was equivalent to nearly 2.5 W/m2 with carbon dioxide contribution about 60 % to this figure, methane about 25 per cent, with nitrous oxides and halocarbons providing the remainder. In 1985, Joe Farman, of the British Antarctic Survey, published an article showing the decrease in ozone levels over Antarctica during the early 1980s. The response was striking: large scale international scientific programmes were mounted to prove that CFCs (used as aerosol propellants in industrial cleaning fluids and in refrigeration tools) were the cause of the problem. Even more important was abrupt international action to curb the emissions of CFCs. The second major cause of global warming is the depletion of ozone layer. This happens mainly due to the presence of chlorine-containing source gases.

When ultraviolet light is present, these gases dissociate releasing chlorine atoms which then catalyses ozone destruction. Aerosols present in the atmosphere are also causing global warming by changing the climate in two different ways. Firstly, they scatter and absorb solar and infrared radiation and secondly, they may alter the microphysical and chemical properties of clouds and perhaps affect their lifetime and extent. The scattering of solar radiation acts to cool the planet, while absorption of solar radiation by aerosols warms the air directly instead of permitting sunlight to be absorbed by the surface of the Earth. The human contribution to the number of aerosols in the atmosphere is of various forms. For instance, dust is a by-product of agriculture. Biomass burning generates a mixture of organic droplets and soot particles. Many industrial processes produce a wide diversity of aerosols depending on what is being burned or generated in the manufacturing process. Moreover, exhaust emissions from various sorts of transport produce a rich mixture of pollutants that are either aerosols from the outset or are transformed by chemical reactions in the atmosphere to form aerosols.

11.5 GLOBAL WARMING: THE EFFECTS

Predicting the consequences of global warming is one of the most difficult tasks faced by the climate researchers. This is due to the fact that natural processes that cause rain, snowfall, hailstorms, rise in sea levels is reliant on many diverse factors. Moreover, it is very hard to predict the size of emissions of greenhouse gases in the future years as this is determined majorly through technological advancements and political decisions. Global warming produces many negative effects some of which are described here. Firstly, extra water vapour which is present in the atmosphere falls again as rain which leads to floods in various regions of the world. When the weather turns warmer, evaporation process from both land and sea rises. This leads to drought in the regions where increased evaporation process is not compensated by increased precipitation. In some areas of the world, this will result in crop failure and famine particularly in areas where the temperatures are already high. The extra water vapour content in the atmosphere will fall again as extra rain hence causing flood. Towns and villages which are dependent on the melting water from snowy mountains may suffer drought and scarcity of water supply. It is because the glaciers all over the world are shrinking at a very rapid rate and melting of ice appears to be faster than previously projected. According to Intergovernmental Panel on Climate Change (IPCC), about one-sixth of the total population of the world lives in the regions which shall be affected by a decrease in melting water. The warmer climate will likely cause more heat waves, more violent rainfall and also amplification in the severity of hailstorms and thunderstorms. Rising of sea levels is the deadliest affect of global warming, the rise in temperature is causing the ice and glaciers to melt rapidly. This will lead to rise of water levels in oceans, rivers and lakes that can pilot devastation in the form of floods.

As evident from Fig. 5, temperature anomalies are projected to increase in coming years. Before, the 20th century, the situation was well under control but the beginning of the current century, the situation started to worsen. This was all due to increase in global warming majorly due to the fact that new industries and power houses started operation and emitted harmful gases which cause the planet to heat up. This data is based on the research carried out by different climate and environmental research agencies.

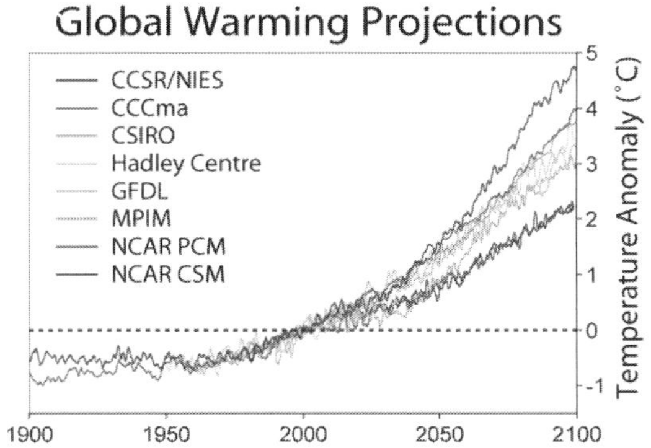

Fig. 5: Global warming projections by various Science and Engineering research agencies

Similarly, Fig.6 elaborates the risks and impacts of global warming in years to come. As can be inferred from figure, we are currently experiencing severity of extreme climate events in the form of thunderstorms, floods and earthquakes. This destruction will take a sharp hike if nothing is done to stop this menace.

Fig. 7 depicts global mean temperature in the recent years according to National Aeronautics and Spac e Administration (NASA). The trend clearly puts up a serious question for us. How will we survive on earth given the rise in temperature to prevail?

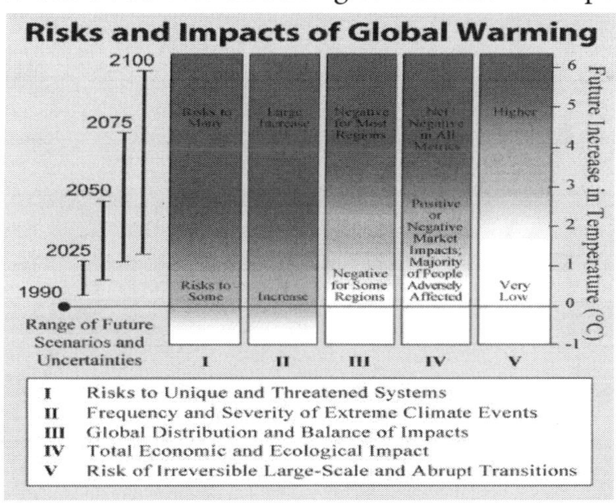

Fig. 6: An assessment of the relative impact and risks connected with global warming. Five categories are assessed. The bars are color-coded to show level of impact/concern for each factor as a function of temperature increase.

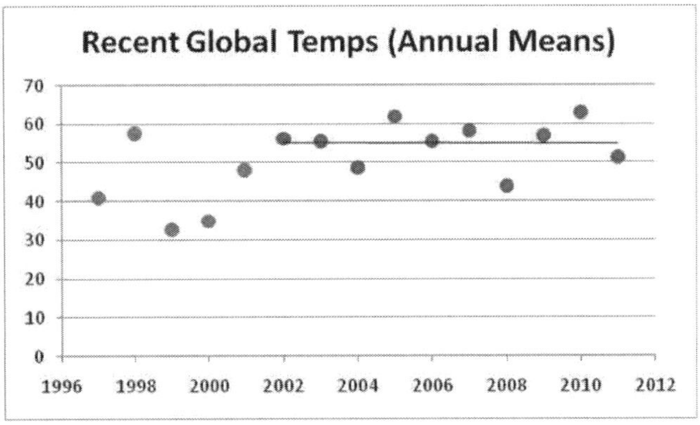

Fig. 7: Recent global mean temperatures according to NASA

11.6 EFFECTS ON LIVING BEINGS

Global warming can severely affect the health of living beings. Excess heat can cause stress which may lead to blood pressure and heart diseases. Crop failures and famines, which are a direct consequence of heating up of earth, can cause a decline in human body resistance to viruses and infections. Global warming may also transfer various diseases to other regions as people will shift from regions of higher temperatures to regions of comparatively lower temperatures. Warmer oceans and other surface waters may lead to severe cholera outbreaks and harmful infections in some types of sea food.

Moreover, it is an established fact that warmer temperatures lead to dehydration which is a major cause of kidney stones. A medical team from The Children's Hospital of Philadelphia examined the health proceedings of more than 60,000 Americans alongside weather records. They discovered that individuals were most likely to be hospitalized with kidney stones three days after a temperature rise. Since 1994, kidney stone incidence has risen from about one in 20 people to one in 11. This trend is likely to increase as the globe gets hotter. According to Luis Ostrosky, M.D. of the Division of Infectious Diseases at The University of Texas Health Science Centre at Houston Medical School and medical director for epidemiology at Memorial Hermann-Texas Medical Centre: "One infection that is definitely making a weird pattern is valley fever". In his words, "This is a fungal infection we used to see only in California, Arizona, New Mexico and a little in Texas, but last year we found it for the first time in Washington State." This potentially deadly condition caused apprehension in California when the number of cases increased drastically during 2010 and 2011. Valley fever infections have been on the rise, probably because of warming climates and drought causing dust storms. Dry soil and wind can carry spores that spread the virus. Hotter and drier climates are projected to increase the amount of dusting carrying this disease. Researchers have already noticed a rise in mosquito-borne disease like dengue fever and malaria due to warmer and longer summers. Perhaps the most prominent mosquito-borne disease, West Nile Virus, has already experienced a sharp increase in annual cases. According to the U.S. Centres for

Disease Control and Prevention, the summer of 2012 was the nastiest West Nile season on record. The likely reason was that summer's scorching heat and drought. Lyme disease is another dangerous disease which is transmitted mainly through bites from certain tick species.

Fig. 8 describes in the form of a block diagram that how alterations in global climate can affect human health. The bitterest fact is that it can cause various diseases and deprive human beings of the food.

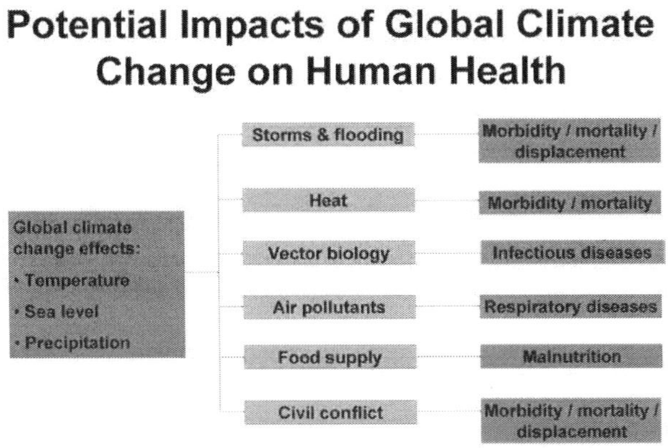

Fig.8: Potential impacts of global climate change on human health

Global warming is also affecting animals. They need to move to cooler places in order to survive. This process has been observed in various places, for instance, in the Alps, in mountainous Queensland in Australia, and in the misty forests of Costa Rica. Fish in the North Sea have been reported to move northwards too. The impacts on species are becoming noteworthy to such an extent that their movements can be used as a sign of a warming world. They are the silent witnesses of the swift changes being inflicted on the Earth. Scientists and researchers predict that global warming is gradually damaging the ecosystems of various species and is playing a very unconstructive role in making them extinct. For instance, Asia's only ape – the orang-utan – is in bottomless trouble. Its last remaining strongholds in the rainforests of Indonesia are being endangered by a range of pressures, including climate change, putting the animal at the menace of extinction within a few decades. With global warming continually increasing the duration and frequency of droughts, bushfires are occurring more often in these heavily logged forests, further fragmenting the orang-utan's living domain. Similarly, in Africa, elephants face a series of threats including shrinking living space, which brings them more regularly into divergence with people. With this reduced living space, elephants will be unable to escape any changes to their natural habitat caused by global warming, including more common and longer dry periods, placing further pressure on their survival.

11.7 ALTERNATIVE ENERGY SOURCES

The hazards caused by global warming are tremendous. Excessive use of fossil fuels such as coal, natural gas and oil play a part in it too. The usage of fossil fuels should be discontinued immediately. The most significant solution to put an end to this disaster is the use of alternative energy sources. They include wind, solar, bio mass, geothermal and hydro. The most noteworthy point in using these sources is their clean nature. They do not produce any sort of pollution or toxic gases that can lead to global warming. They are environmentally friendly and pose no threat to ecological balance. However, their high installation and setup costs may drive energy companies away from them at first but in the long run they are surely beneficial for everyone. Most importantly, fossil fuels will deplete one day and sooner or later, we have to turn to renewable energy sources for energy production. Thus, the eventual solution to end global warming is to use alternative energy sources. Fig. 9 depicts in a pictorial way that earth can be saved from the hazards of global warming if we utilise renewable energy sources.

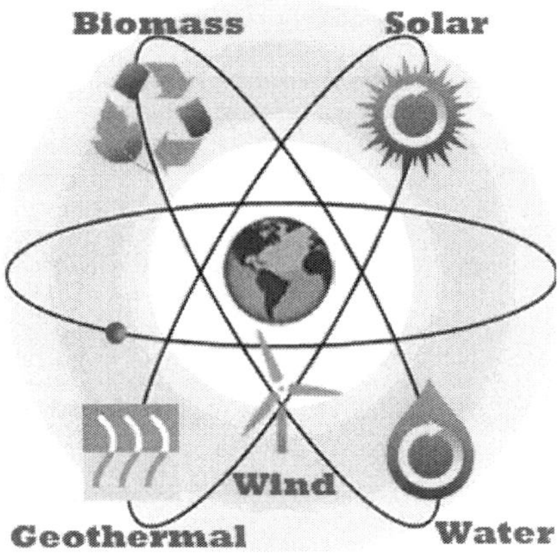

Fig. 9: Save earth from global warming by using renewable energy sources

To counteract the medical hazards of global warming, it is essential to turn to renewable energy sources. Public, in general, should be responsible about their decisions on energy conservation methods. This will ensure a healthy atmosphere and stable climate for our future generations. Governments should devise and pass policies which encourage the energy companies and people, in general, to use renewable energy instead of conventional energy, Nongovernmental organisations (NGOs) should distribute pamphlets to people motivating them to use alternative sources of energy and discourage them from using fossil fuels. They should also explain to them the hazards which the usage of fossil fuels will cause. Many developed countries are already generating huge amounts of power using renewables. These countries should extend their helping hand to developing countries to combat the evil of global warming collectively. Using renewable energy is the most effective way to curtain the emission of gases which play a major role in global warming.

Fig. 10 and Fig. 11 show that the use of renewables is gradually increasing. The figure should be much more than present so that we can tackle the problem of global warming timely and effectively.

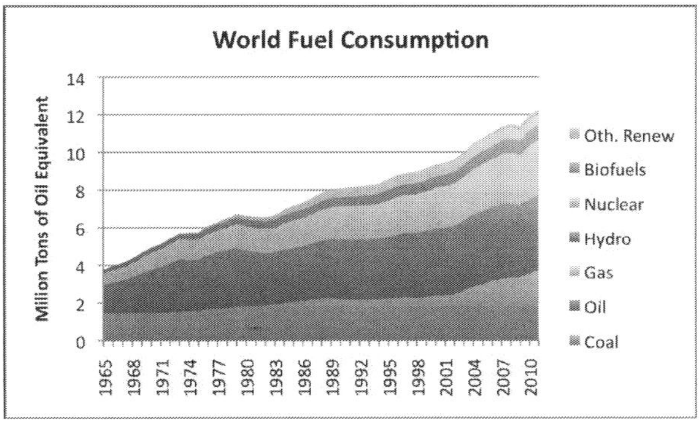

Fig. 10: World fuel consumption in recent years

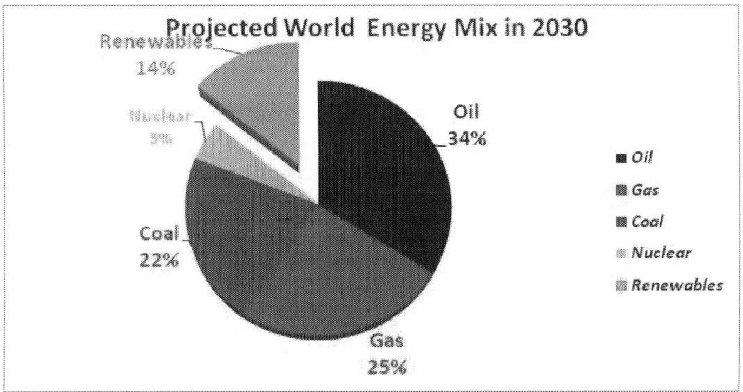

Fig .11: Projected world energy mix in 2030

11.8 OTHER SOLUTIONS

As elaborated earlier, toxic emissions are a major cause of global warming, A likely solution to reduce harmful emissions is to cut the usage of vehicles which produce them. This has not been met with much success as many people refuse to cut down their practice of using cars. No doubt, some people have started to use bicycles and public transport, whereas some other prefer to walk but these numbers are relatively small. It should be noted that fuel economy and emission rates are chief factors to consider regarding the car choice. Hybrid cars have higher efficiency and lower emission rates. Keeping the tires inflated will help improve mileage and air filters should be frequently replaced to cut down harmful emissions. People should share the ride with friends or co-workers to reduce the total number of vehicles on the road. Print and social media can play an effective role in curbing the problem. It should use the philosophy of automobile advertisements to encourage drivers to conserve energy and reduce pollution. Awareness campaigns can be started using placards, posters and logos similar to shown in Figures 12-14. They are a very useful way to demonstrate that global warming is not good for the planet. Recycling is also a good way to reduce global warming. People should use rechargeable batteries instead of disposable ones. Quality products should be bought that have a long life. Shopping should be done from local markets which reduce transportation. Even small individual efforts like lowering the thermostats in winter and using compact fluorescent lamps instead of incandescent lamps can aid to address the issue of global warming. Reforestation schemes must be started to grow a large number of trees. Forest degradation and deforestation must be discouraged at government level. Nuclear power is also a possible solution as this power results in fewer emissions but this method should be used with care as it can lead to severe accidents therefore, the major hurdle is to overcome the security, propagation, waste disposal and high costs of nuclear power if this method has to be made practical.

Fig. 12: shows symbolically how global warming is causing the earth to melt.

Fig. 13: showing a symbolic representation to stop global warming

Fig. 14: depicting that how human beings are destroying the earth for their own benefits

12

CLIMATE CHANGE: CAUSES, EFFECTS AND MITIGATION MEASURES

12.1 INTRODUCTION

Global warming, otherwise known as climate change, has been increasingly recognized as the greatest threat. Of all the holocausts that have afflicted mankind such as plagues, earthquakes, tsunami, smallpox, HIV/AIDS et cetera, none has the greatest threat to wipe out lives on Earth through either continuous flooding or permanent drought than climate change. It has been projected that about 9 billion people will inhabit the earth by 2050, most of which live in developing countries. Consequently, the world faces the most pronounced question over how our planet can sustain and feed this population due to climate change with its implications on food insecurity, access to clean water and sanitation, population migration and the threat of an increased number of both natural and man-made disasters. Scientists have tried to divide the causes into two broad categories, natural and human causes. The natural causes are many including earth's orbital changes, solar variations, volcanic eruptions and ocean currents. The human causes include burning of fossil fuels, land-use and deforestation. The effects of these causes can be seen on rising sea levels, melting of ice caps, heat waves, violent downpours, animal metabolism et cetera. The efforts by researchers, scientists as well as the Inter-Governmental Panel on Climate Change (IPCC), the United Nations Framework Convention on Climate Change (UNFCCC) to identify these factors are noteworthy. The impacts of these various factors,

The Government of India has taken many measures to improve the dire situation of Climate Change. The Ministry of Environment and Forests is the nodal agency for climate change issues in India. It has initiated several climate-friendly measures, particularly in the area of renewable energy. India took several steps and policy initiatives to create awareness about climate change, and help capacity building for adaptation measures. It has initiated a "Green India" programme under which various trees are planted to make the forest land greener and more fertile.

their interplay, synergy and tension on climate change, and the potential mitigation and adaptation strategies are hereunder highlighted. Although climate change has been recognized, studied and debated for decades but the recognition of the present high concentrations of greenhouse gases in the atmosphere growing at an unprecedented rate and magnitudes with its obvious devastating consequences is relatively recent and should call for intensified studies.

The Intergovernmental Panels on Climate Change (IPCC) and the United Nations Framework Convention on Climate Change (UNFCCC) have identified climate change as the greatest threat of the century and reported that the earth's average temperature has risen by 0.740 C. Furthermore, the present atmospheric concentration of carbon dioxide (CO2) is 385 ppm (parts per million) far more than at any time in the last 650,000 years resulting in climate change or global warming. Both natural and human causes of climate change including the earth's orbital changes, solar variations, ocean currents, volcanic eruptions et cetera were reviewed. These various causes have resulted in rising sea levels, melting of ice at the world's poles and on its mountains and violent downpours resulting in over flooding and submerging of coastal areas and consequent migration of animal species; heat waves leading to desertification, depletion of water availability, reduced animal metabolism and agricultural food production, loss of biodiversity, food insecurity, decreased animal health et cetera. Some measures identified to have potentials to bring about climate change mitigation including the use of carbon capture & storage (CCS) technology and trading plants for carbon as carbon tracking devises; breeding of fuzzy-leaved crops and irrigation as techniques for cooling the atmosphere. Strategies for climate change adaptation through a multi-dimensional and multi-sectoral approach have become imperative as a result of inequalities between the developed and the developing nations in terms of limited capital resources and expertise. The successes recorded so far through these efforts are encouraging but they have not been able to drastically reduce the rise in atmospheric temperature, hence the consequences abound, especially in the developing countries like Nigeria. Several scholarly articles were scoured and presented and the paper concludes by positing that atmospheric rise in temperature should be brought below 20C by 2050 in tandem with Copenhagen and Kyoto Summits accord.

12.2 CAUSES OF CLIMATE CHANGE

The natural variability and climate change fluctuations of the climate system have been part of the Earth's history however, there have been changes in concentrations of GHGs in the atmosphere growing at an unprecedented rate and magnitudes in recent years. The causes of climate change can be categorized into two: - natural and manmade causes.

12.2.1 Natural causes:

The earth's climate is grossly influenced and changed through natural causes such as ocean current, volcanic eruptions, the earth's orbital changes and solar variations.

- **Ocean current:** The oceans have been shown to be the major component of the climate system. Ocean currents move vast amounts of heat across the world. Wind pushes horizontally against the sea surface and drive ocean current patterns. Interactions between the ocean and atmosphere can also produce phenomena called El- Nino which occurs every 2 to 6 years. It has been shown that deep ocean circulation of cold water moves from the poles towards the equator and warm water from the equator back towards the poles. Without this movement the poles would be colder and the equator warmer. The oceans play an important role in determining the atmospheric concentration of CO_2. Changes in ocean circulation may affect the climate through the movement of CO2 into or out of the atmosphere.
- Volcanic activities: Volcanic eruptions are known to throw out large volumes of sulphur dioxide (SO2), water vapour, dust and ash into the atmosphere. It is known that large volumes of gases and ash can influence climate patterns for years by increasing planetary reflectivity, causing atmospheric cooling Tiny particles called aerosols are produced by volcanoes. Because these reflect solar energy back into space, they have a cooling effect on the earth's surface.

Explosive Volcanism Source: istockphoto / loops7 changes and solar variations.

- Earth's orbital changes: The earth makes one revolution around the sun once a year, tilted at an angle of 23.50 to the perpendicular plane of its orbital path. Changes in the tilt of the earth can lead to small but climatically important changes in the strength of the seasons, more tilt means warmer summers and colder winters; less tilt means cooler summers and milder winters. Slow changes in the earth's orbit led to small but climatically important changes in strength of the seasons over tens of thousands of years. Climate feedbacks have been shown to amplify these small changes, thereby producing ice ages.

- Solar variation: The sun is known to be the source of energy for the planet's climate system. Although the sun's energy output appears constant from an everyday point of view, small changes over an extended period of time can lead to climate changes. It has been speculated that a portion of the warming in the first half of the 20th Century was due to an increase in the output of solar energy. As the sun is the fundamental source of energy that is instrumental in our climate system, it would be reasonable to assume that changes in the sun's energy output would cause climate to change. But studies by Crowley (2000) and Paehler (2007) have shown that if this were so it would be expected to see warmer temperatures in all layers of the atmosphere. On the contrary, the cooling was observed in the upper atmosphere, a warming at the surface and in the lower parts of the atmosphere. This was shown to be due to greenhouse gases capturing heat in the lower atmosphere.

Layers of Atmosphere Source:.istockphoto/@tbradford

Furthermore, climate models that included solar irradiance changes could not reproduce the last century's observed temperature trend without including a rise in greenhouse gases suggesting that GHGs are the main cause of climate change.

- **Cloud's contribution:** Perkins (2010) reported that global satellite analysis supported by climate models have revealed that cloud cover accentuates warming because as earth's average temperature rises, clouds will accelerate global warming by trapping more heat. Dessler (2010) analyzed satellite data gathered between 2000 and 2010 to estimate the short-term variations in the amount of visible and infrared radiation emitted to space. He made an allowance by subtracting influences such as earth's surface reflectivity and the heat-trapping effect of atmospheric water vapour as well as how clouds affected the planet's radiation balance as a function of temperature over the decade. The result showed that clouds enhance warming by trapping on average, an extra 0.54 watts per square metre for every 10C rise in global average temperature. Dessler (2010) however, posited that uncertainty in the estimate indicates that clouds could actually exert a small cooling effect as temperature rises, although the slight negative feedback wouldn't be nearly enough to cancel out larger, wellconstrained positive climatic feedbacks such as water vapour. In the tropics, for example, Nigeria, the North East Trade wind brings cloudless atmosphere called harmattan with dry air and a cooling effect to the atmosphere. A model that can mimic this will have positive impact on climate change mitigation, provided it does not have negative feedback that could cause permanent drought.

12.2.2 Man-made Causes:

We are accelerating the rate of climate change far past what the influence of the sun or any natural forces could achieve. Processes in our daily lives are expelling chemicals and foreign substances into the air that affect and increase the natural greenhouse gases, which then absorb more energy from the sun. This results in global warming and changes the climate.

- **Greenhouse Gas Emission:** Greenhouse gas emission is said to be the largest man-made climate forcer as it is so prevalent and large-scale. Greenhouse gases occur naturally in the atmosphere and they serve to absorb and radiate the Sun's heat around the globe for the benefit of life. Humans are now producing large amounts of these gasses on such a level that it has upset energy balance. Some of these greenhouse gases are long-lived like carbon dioxide and nitrous oxide, so the damage they cause is obvious. However, short-lived substances like methane still affect the energy balance as they are very potent and there are so many of them. Moreover, human activity continues to release them into the atmosphere.
- **Fossil Fuel Usage:** Fossil fuel burning produces, among other chemicals, carbon dioxide that is the primary greenhouse gas forcer of climate change. In itself it is not a particularly warming gas, but the amount that we are emitting into the atmosphere makes it the most damaging. It also has a long life in the atmosphere, only fully dissipating after 50,000 years during which time it continues to absorb and radiate heat.
- **Black Carbon:** Another by-product of fossil fuels is black carbon, which is produced when fossil fuels, biofuels, and biomass are not properly combusted and emit soot. It enters the atmosphere as fine particles (aerosols) and absorbs large amounts of heat and infrared radiation. The majority of black carbon in

the U.S. is emitted from vehicles, especially diesel-fuelled ones. Wildfires and industrial and residential heating are also black carbon emitters. Black carbon is drifted across the globe on winds, and can be deposited on snow or ice. This darkens snow or ice, thus increasing its ability to absorb heat. This, in turn, forces the ice to melt faster.

- **Industrial Production & Farming:** Emissions from factories contribute directly to the greenhouse effect while runoff waste alters the chemical make-up of water and air. Chemical pesticides, herbicides and fungicides used in industrial farming also wreak havoc to the atmosphere. Livestock farming also releases large amounts of methane into the air, driving up the amount of greenhouse gases already in the atmosphere.
- **Agriculture:** Agriculture accounts for 14% of total greenhouse gas emissions, and with the need for more agriculture to provide for our ever-growing needs, its impact is growing rapidly. The main agricultural contributors to climate forcers include carbon released from farmland soil by inappropriate farming and grazing practices. Changes in land use for agriculture is also a large contributor to climate change. Tillage and water rerouting has accelerated soil erosion, soil organic matter has been lost due to chemicals, and soil itself has been altered and polluted by chemical fertilizers, all of which affect the atmosphere with various emissions.
- **Deforestation:** Vast swatches of primary forests are being wiped out to make way for unsustainable farming, urbanization or for logging and the production of paper goods. Trees absorb carbon dioxide from the atmosphere, so as more trees are cut down, the level of carbon dioxide in the air will increase. In short, deforestation affects climate in a big way.
- **Urbanization:** As our populations increase and our cities reach maximum capacity, we start to expand them in what is called urban sprawl. This decreases the amount of vegetation available to absorb carbon dioxide. In turn, urbanization means more greenhouse gas, pollution and waste output. Other damaging side effects of urbanization include the Heat Island Effect: air and surface temperatures are driven higher above urban areas than above rural areas due to concentrated usage of fossil fuels, electricity etc. As urban areas grow, so will the number of heat islands, which will drive up the overall atmospheric temperature. A side effect of this side effect is that water vapor above urban areas will also be driven up, causing more clouds and rain.
- **Sulfates, Organic Carbons & Aerosols (small particles):** These are actually negative forcers, but warming and cooling aerosols can change some of the processes of clouds. As negative forcers, clouds reflect sunlight back up. As positive forcers, they trap outgoing heat. One cause of climate change that is up for debate as to whether it is natural or man-made is our very own respiration.

Every breath we take releases carbon dioxide into the atmosphere. Whatever it is, we all impact climate change, so we should all make the effort to slow it down too.

12.3 EFFECTS OF CLIMATE CHANGE

The IPCC (2007) report has succinctly identified the most relevant impacts of climate change on human health as "changes in conditions, temperature, rainfall, humidity, and wind likely to alter the intensity and geographical distribution of extreme weather events, raise water levels in coastal regions, alter the distribution of vector insects and mammals, exacerbate health relevant air pollution, intensify the existing burden of malnutrition, and increase human exposure to toxic substances due to the deterioration of natural and manmade environment.

Rising global temperature is accompanied by the changes in weather and climate like changes in rainfall, resulting in more floods, droughts, or intense rain, as well as more frequent and severe heat waves. The planet's oceans and glaciers have also experienced some big changes - oceans are warming and becoming more acidic, ice caps are melting, and sea levels are rising. As these changes become more pronounced in the coming decades, they poses a great challenge to our society and our environment.

12.3.1 Higher Temperatures:

Green House gases release by power plants, automobiles, deforestation and other sources are heating up the planet. In fact, the five hottest years on record have all occurred since 1997 and the 10 hottest since 1990, including the warmest years on record – 2005 and 2010. Increased temperatures are to blame for an increase in heat-related deaths and illness, rising seas, increased storm intensity, and many of the other dangerous consequences of climate change. During the 20th century, the Earth's average temperature has increased by one degree Fahrenheit to its highest level in the past four centuries – believed to be the fastest rise in a thousand years. Research found that if emissions of heat-trapping carbon emissions aren't reduced, average surface temperatures could increase by 3 to 10 degrees Fahrenheit by the end of the century.

12.3.2 Changing Landscapes:

Increasing temperatures and changing patterns of rain and snow have resulted in moving of trees and plants around the world toward Polar Regions and up mountain slopes. These vegetation shifts will affect much of the work the conservation community has accomplished to date, with the potential to permanently change the face of Conservancy preserves, local land trusts, and even our national parks. As plant communities try to adapt the changing climate by moving toward cooler areas, the animals that depend on them will be forced to move in order to survive. Some species and communities such as polar bears and alpine meadows may be left without any remaining viable habitat, putting much of our treasured wildlife at risk.

12.3.3 Wildlife/Ecosystem at Risk

Increased temperatures are changing weather and vegetation patterns across the globe, forcing animal species to migrate to new, cooler areas in order to survive. The rapid change of climate change is likely to exceed the ability of many species to migrate or adjust. Experts predict that one-fourth of Earth's species will be headed for extinction by 2050 if the warming trend continues at its current rate.

Many species are already feeling the heat:

- In 1999, the death of the last Golden Toad in central America marked the first documented species extinction driven by climate change.
- Due to melting in the arctic, polar bears may be gone from the planet in as little as 100 years.
- In the tropics, rising the sea temperatures are causing more coral reefs to "bleach," as the heat kills colourful algate that are necessary to coral health and survival.

12.3.4 Ocean Acidification /Rising Seas level

As the Earth heats up, sea levels rise up because warmer water takes up more area than colder water, a process known as thermal expansion. Melting glaciers compound the problem by dumping even more fresh water into the oceans. Rising seas threaten the populated low-lying areas, islands and dense coastal populations, erode shorelines, damage property and destroy ecosystems such as mangroves and wetlands that protect coasts against storms. Sea levels have risen between four and eight inches in the past 100 years. Current projections suggest that sea levels could continue to rise between 4 inches and 36 inches over the next 100 years. The increased concentration of carbon dioxide (CO2) in the atmosphere has increased the absorption of CO2 in the ocean, which reduces the pH and makes the oceans more acidic. A more acidic ocean would adversely affect the health of many marine species, including plankton, molluscs, and other shellfish. In particular, corals can be very sensitive to rising acidity, as it is difficult for them to create and maintain the skeletal structures needed for their support and protection.

12.3.5 Increased Risk of Drought, Fire and Floods

Higher temperatures increase the amount of moisture that evaporates from land and water, leading to drought in many areas. Lands affected by drought are more vulnerable to flooding once rain falls. As temperatures rise globally, droughts will become more frequent and more severe, with potentially devastating consequences for agriculture, water supply and human health. This phenomenon has already been observed in some parts of Asia and Africa, where droughts have become longer and more intense. Hot temperatures and dry conditions also increase the likelihood of forest fires.

12.3.6 Intensified Storms and Increased Storm Damages

Scientific research indicates that climate change will cause hurricanes and tropical storms to become more intense — lasting longer, unleashing stronger winds, and causing more damage to coastal ecosystems and communities. Higher ocean temperatures are the main culprit, since hurricanes and tropical storms get their energy from warm water. As sea surface temperatures rise, developing storms will contain more energy. At the same time, other factors such as rising sea levels, disappearing wetlands, and increased coastal development threaten to intensify the damage caused by hurricanes and tropical storms.

12.3.7 Illness and Disease

As temperatures increases, so do the risks of heat-related illness and even death for the most vulnerable human populations. Scientists have linked the deadly heat waves to climate change and warn of more to come. In 2003, for example, extreme heat waves caused more than 20,000 deaths in Europe and more than 1,500 deaths in India. Climate change increases the spread of infectious diseases, mainly because warmer temperatures allow disease-carrying insects, animals and microbes to survive in areas where they were once blocked by cold weather. Diseases and pests that were once limited to the tropics — such as mosquitoes that carry malaria — may find hospitable conditions in new areas that were once too cold to support them. The World Health Organization (WHO) estimates that climate change may have caused more than 150,000 deaths in the year 2000 alone, with an increase in deaths likely in the future.

12.3.8 Economic Losses

Research says that if no action is taken to curtail the global carbon emissions, climate change could cost between 5 and 20 percent of the annual global gross domestic product. In comparison, it would take 1 percent of GDP to lessen the most damaging effects of climate change. Climate change may significantly lower lake levels, altering shoreline habitats and costing millions for the relocation of ports and shore infrastructure. Globally, more intense hurricanes and downpours could cause billions of dollars in damage to property and infrastructure. Declining crop yields due to prolonged drought and high temperatures, especially in Africa, could put hundreds of thousands of people at risk for starvation. High sea temperatures also threaten the survival of coral reefs, which generate an estimated $375 billion per year in goods and services.

12.3.9 Agriculture Productivity/Food Security

Solar radiation, temperature, and precipitation are the essential drivers of crop growth; therefore, agriculture has always been highly dependent on climate patterns and variations. Since the industrial revolution, humans have been changing the global climate by emitting huge amounts of greenhouse gases into the atmosphere, resulting in higher global temperatures, affecting hydrological regimes and increasing climatic variability. Climate change is projected to have significant impacts on agricultural conditions, food supply, and food security. Some of these effects are biophysical, some are ecological, and some are economic, including:

- A shift in climate and agricultural zones towards the poles
- Changes in production patterns due to higher temperatures
- A boost in agricultural productivity due to increased carbon dioxide in the atmosphere
- Changing precipitation patterns
- Increased vulnerability of the landless and the poor

12.4 MITIGATION MEASURES

The entire world has applauded the Copenhagen and Kyoto agreements to bring down the atmospheric temperature rise caused by GHGs emission to below 2^{0}C by 2050 due to the obvious consequences of its effects. It has been shown that rainforests are capable of absorbing about 20% of carbon from the atmosphere through the photosynthetic process. Accordingly, scientists are currently employing the great potentials of our ecosystem to bring about climate change mitigation. International efforts to counter climate change international efforts to counter climate change climate change are describes as under:

12.4.1 The Intergovernmental Panel on Climate Change (IPCC)

IPCC was founded in November 1988 by the World Meteorological Organization (WMO) and the United Nations Environment Program (UNEP) jointly as a place to study global warming problems at a governmental level. It is the mechanism that accumulates scientific knowledge on global warming while debates on the international countermeasures have been made in the COPs (Conference of the Parties) of United Nations Framework Convention on Climate Change (UNFCCC). These two mechanisms are complementing each other.

12.4.2 United Nations Framework Convention on Climate Change (UNFCCC)

The UNFCCC entered into force on 21 March 1994. The 195 countries that have ratified the Convention are called Parties to the Convention (Almost universal Membership). The UNFCCC is a "Rio Convention", one of three adopted at the "Rio Earth Summit" in 1992. Others are the UN Convention on Biological Diversity and the Convention to Combat Desertification. Due to intrinsic linkages, the Joint Liaison Group was set up to boost cooperation among the three Conventions, and synergies in their activities on issues of mutual concern. It now also incorporates the Ramsar Convention on Wetlands. The ultimate objective of the Convention is to stabilize greenhouse gas concentrations "at a level that would prevent dangerous anthropogenic (human induced) interference with the climate system." It states that "such a level should be achieved within a time-frame sufficient to allow ecosystems to adapt naturally to climate change, to ensure that food production is not threatened, and to enable economic development to proceed in a sustainable manner." Following its effectuation, the COP1 was held in Berlin, COP2 in Geneva, and the COP3 was held in Kyoto to adopt "Kyoto Protocol", which implemented the objective of the UNFCCC to fight global warming by reducing greenhouse gas concentrations in the atmosphere to 'a level that would prevent dangerous anthropogenic interference with the climate system' (Art. 2).

12.4.3 Kyoto Protocol

The Kyoto Protocol commits its signatories by setting internationally binding emission reduction targets. The Kyoto Protocol was adopted in Kyoto, Japan, on 11 December 1997 and entered into force on 16 February 2005. The detailed rules for the implementation of the Protocol were adopted at COP 7 in Marrakesh, Morocco, in 2001, and are referred to as the "Marrakesh Accords." Its first commitment period started in 2008 and ended in 2012. Recognizing that developed countries are principally responsible for the current high levels of GHG emissions in the atmosphere as a result of more than 150 years of industrial activity, the Protocol places a heavier burden on developed nations under the principle of "common but differentiated responsibilities." When no Protocol exists, the global CO2 emission in 2010 will increase by 24% compared with 1990. When the Protocol is enforced in 2000, the global CO2 emission in 2010 will reduce by 5.2% compared with 1990. Parties to the Kyoto protocol are classified as:

- Annex I: Parties to the UNFCCC listed in Annex I of the Convention. These are the industrialized (developed) countries and "economies in transition" (EITs). EITs are the former centrally-planned (Soviet) economies of Russia and Eastern Europe. The European Union-15 (EU-15) is also an Annex I Party.
- Annex II: Parties to the UNFCCC listed in Annex II of the Convention. Annex II Parties are made up of members of the Organization for Economic Cooperation and Development (OECD). Annex II Parties are required to provide financial and technical support to the EITs and developing countries to assist them in reducing their greenhouse gas emissions (climate change mitigation) and manage the impacts of climate change (climate change adaptation).
- Annex B: Parties listed in Annex B of the Kyoto Protocol are Annex I Parties with first or second round Kyoto greenhouse gas emissions targets.
- Non-Annex I: Parties to the UNFCCC not listed in Annex I of the Convention are mostly low-income developing countries. Developing countries may volunteer to become Annex I countries when they are sufficiently developed.
- Least-developed countries (LDCs): 49 Parties are LDCs, and are given special status under the treaty in view of their limited capacity to adapt to the effects of climate change.

Industrialized countries (Annex I) have to report regularly on their climate change policies and measures, including issues governed by the Kyoto Protocol (for countries which have ratified it). They must also submit an annual inventory of their greenhouse gas emissions, including data for their base year (1990) and all the years since. Developing countries (Non-Annex I Parties) report in more general terms on their actions both to address climate change and to adapt to its impacts - but less regularly than Annex I Parties do, and their reporting is contingent on their getting funding for the preparation of the reports, particularly in the case of the Least Developed Countries. Kyoto Mechanisms are also known as Flexible Mechanisms and they include Emissions Trading, the Clean Development Mechanism and Joint Implementation to lower the cost of achieving emission targets.

- **Emission Trading:** Emissions Trading-mechanism allows parties to the Kyoto Protocol to buy 'Kyoto units' (emission permits for greenhouse gas) from other countries to help meet their domestic emission reduction targets.
- **Joint Implementation:** Any Annex I country can invest in emission reduction projects (referred to as "Joint Implementation Projects") in any other Annex I country as an alternative to reducing emissions domestically.
- **Clean Development Mechanism (CDM):** Countries can meet their domestic emission reduction targets by buying greenhouse gas reduction units from (projects in) non Annex I countries to the Kyoto protocol.
- **Kyoto Units:** The emissions trading can be international or domestic. Under the International Emissions Trading (IET), the countries can trade in the international carbon credit market to cover their shortfall in Assigned amount units. Countries with surplus units can sell them to countries that are exceeding their emission targets under Annex B of the Kyoto Protocol.
- **Certified Emission Reductions (CERs):** Certified Emission Reductions are one of the types of the Kyoto Units. They are issued under the Clean Development Mechanism. The Annex-I countries can use the CERs to comply with their emission limitation targets or by operators of installations covered by the European Union Emission Trading Scheme (EU ETS) in order to comply with their obligations to surrender EU Allowances, CERs or Emission Reduction Units (ERUs) for the CO2 emissions of their installations. The Government and Private entities can hold the CERs on electronic accounts with the UN.

In Doha, Qatar, on 8 December 2012, the "Doha Amendment to the Kyoto Protocol" was adopted. The amendment includes:

- New commitments for Annex I Parties to the Kyoto Protocol who agreed to take on commitments in a second commitment period from 1 January 2013 to 31 December 2020;
- A revised list of greenhouse gases (GHG) to be reported on by Parties in the second commitment period; and
- Amendments to several articles of the Kyoto Protocol which specifically referenced issues pertaining to the first commitment period and which needed to be updated

for the second commitment period.

The Kyoto Protocol is seen as an important first step towards a truly global emission reduction regime that will stabilize GHG emissions, and can provide the architecture for the future international agreement on climate change. In Durban (2011), the Ad Hoc Working Group on the Durban Platform for Enhanced Action (ADP) was established to develop a protocol, another legal instrument or an agreed outcome with legal force under the Convention, applicable to all Parties. The ADP is to complete its work as early as possible, but no later than 2015, in order to adopt this protocol, legal instrument or agreed outcome with legal force at the twenty-first session of the Conference of the Parties and for it to come into effect and be implemented from 2020.

Conference of Parties to the Kyoto Protocol

Conferences	Outcome/Breakthrough
COP 16 16th session of the Conference of the Parties (COP 16) to the UNFCCC and the 6th session of the CMP 6 to the Kyoto Protocol was held in in Cancum, Mexico in 2010.	Called for a large "Green Climate Fund", and a "Climate Technology Centre" and network
	It looked forward to a second commitment period for the Kyoto Protocol.
	The agreement recognizes that climate change represents an urgent and potentially irreversible threat to human societies and the planet, which needs to be urgently addressed by all parties.
COP 17 17th session of the conference of the Parties (COP 17) to the UNFCCC and the 7th session of teh CMP 7 to the Kyoto Protocol was held in Durban, South Africa in 2011.	The conferece led to agreement on a management framework for a future **Green Climate Fund.** The fund is to distribute US$$100bn per year to help poor countries adapt to climate impacts.
	The desing of the new Green Climate Fund for developing countries was completed through a decision on its governance and other practical arrangements. This opens the way for the Fund to become operational in 2012.
	The conferece decided that the second commitment period will start on 1 January 2013 and run until 2017 or 2020. The end date and the emission targets for developed countries taking part will be fixed at the UN climate conferece to be held at the end of 2012 in Qatar.
	The EU it will participate in the second period of the Protocol beginning in 2013.

Conferences	Outcome/Breakthrough
COP 18 18th session of the Conference of the Parties(COP 18) to the UNFCCC and the 8th session of the CMP 8 to the Kyoto Protocol was held at the Qatar National Centre in Doha in 2012.	Doha finalised details of the second Kyoto period and agreed a work plan for negotiations on the new global agreement and on raising ambition under the Durban Platform.
	Operationalised the Technology Mechanism established in Cancun by agrreing the governance arrangements for the Technology Executive Committee and a Climate Technology Centre and Network;
	Launched work programme to elaborate modalities and procedures for the new market mechanism esablished in Durban.
COP 19 19th session of the Conference of the Parties (COP 19) to the UNFCCC and the 9th session of the CMP 9 to the Kyoto Protocol was held in Warsaw, Poland in 2013	The Warsaw conferece agreed a time plan for countries to table their contributions to reducing or limiting greenhouse gas emissions under the new global climate agreement in 2015.
	It also agreed ways to accelerate efforts to deepen emission cuts over the rest of this decade, and to set up a mechanism to address losses and damage caused by climate change in vulnerable developing countries.
	The conferece agreed decisions which enhance the implementation, REDD+, and transparency of reporting on emissions.
COP 20 20th session of the Conference of the Parties (COP 20) to the UNFCCC and the 10th session of the CMp 10 to the Kyoto Protocol was held in Lima, Peru, in 2014	Nations concluded by elaborating the elements of the new agreement, sheduled to be agreed in paris in late 2015, while also agreeing the ground rules on how all countries can submit contrubutions to the new agreement during the first quarter of next year.
	These intended Nationally Determined Contributions (INDCs) will form the foundation for climate action post 2020 when the new agreement is set to come into effect.
	Pledges were made by both developed and developing countries prior to and during the COP that took the capitalization of the new Green Climate Fund (GCF) past an initial $10 billion target.
	Levels of transparency and confidence-builiiding reached new heights as several industrialized countries submitted themselves to questioning about their emissions targets under a new process called a multilateral Assessment.

Conferences	Outcome/Breakthrough
	The Lima Ministerial Declaration on Education and Awareness-raising calls on governments to put climate change into school curricula and climate awareness into national development plans.

12.5 REDUCING EMISSIONS FROM DEFORESTATION AND FOREST DEGRADATION (REDD)

Reducing Emissions from Deforestation and Forest Degradation (REDD) is a set of steps designed to use market/financial incentives in order to reduce the emissions of greenhouse gases from deforestation and forest degradation. Its original objective is to reduce greenhouse gases but it is claimed that it can deliver "co-benefits" such as biodiversity conservation and poverty alleviation. "Reducing emissions from deforestation and forest degradation" implies a distinction between the two activities. Deforestation is the permanent removal of forests and withdrawal of land from forest use. Forest degradation refers to negative changes in the forest area that limit its production capacity. REDD is sometimes presented as an "offset" scheme of the carbon markets and thus, would produce carbon credits.

Main Actors: REDD activities are undertaken by national or local governments, dominant NGOs, the private sector, or any combination of these. It is being pushed strongly by the World Bank and the UN for setting up the bases for the carbon market and the legal and governance frameworks of countries receiving REDD.

Carbon offsets are "emissions-saving projects or programmes" that in theory would "compensate" for the polluters' emissions. The "carbon credits" generated by these projects could then be used by industrialised governments and corporations to meet their targets and/or to be traded within the carbon markets.

REDD ++

"REDD+" goes beyond deforestation and forest degradation, and includes the role of conservation, sustainable management of forests and enhancement of forest carbon stocks. It is predicted that financial flows for greenhouse gas emission reductions from REDD+ could reach up to US$30 billion a year. This significant North-South flow of funds could reward a meaningful reduction of carbon emissions and could also support new, pro-poor development, help conserve biodiversity and secure vital ecosystem services. Further, maintaining forest ecosystems can contribute to increased resilience to climate change. To achieve these multiple benefits, REDD+ will require the full engagement and respect for the rights of Indigenous Peoples and other forest-dependent communities.

US – China Sign Pact on Emission Reductions

China and the United States in November 2014 agreed on a timetable to limit emission of greenhouse gases. In a joint announcement, the U.S. agreed to reduce by 2025 its emission of greenhouse gases by 26 per cent to 28 per cent below its 2005 level.

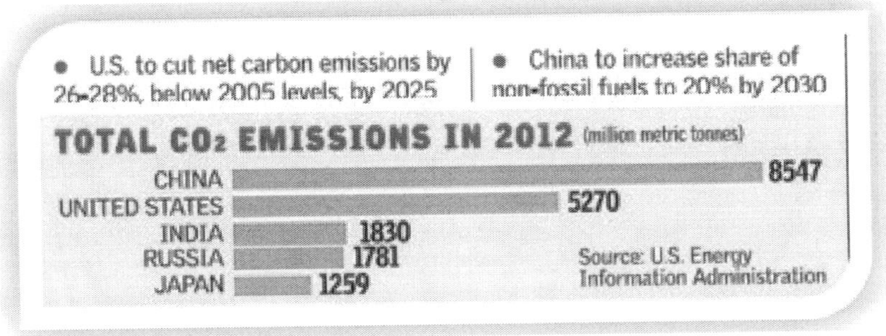

China stated its intent to peak emissions of carbon dioxide in 2030, if not earlier. It also agreed to raise the share of non-fossil fuels to 20 per cent, in its primary energy mix, in the next 16 years. This would entail China shifting towards clean energy generated by nuclear, wind, solar and such zero-emitting resources. The agreement on a timetable for emission cuts to combat climate change has brightened prospects for a climate deal in Paris next year. The joint announcement could impose fresh pressure on India not to become a deal breaker in the run up to the Paris talks. India's per capita emissions are estimated at one-tenth of the United States and one-fourth of China.

HFCs as greenhouse gases:

Hydroflurocarbons (HFCs) are intentionally made fluorinated greenhouse gases used as replacements for ozone-depleting substances. HFCs are used in the same applications where ozone-depleting substances have been used-refrigeration, air-conditioning equipment etc. Since HFCs are not ozone-depleting, they have been kept out of the Montreal Protocol that currently deals with phasing out ozone depleting substances like hydro-chlorofluorocarbon (HCFC) and chlorofluorocarbons (CFC). HFCs, on the other hand, contribute to global warming and come under Kyoto Protocol. In the recent United Nations Climate Change Conference that was held in Bonn, the UNEP expressed its view that the HFCs are a part of Short-Lived Climate Pollutants (SLCPs) but have higher global warming potential when compared to carbon dioxide. Many of the western countries led by US are of the view that HFCs should be dealt under Montreal protocol and the use of HFCs should be reduced. India, however, refused to be part of it. If HFC comes under the Montreal Protocol, it would be binding on emerging economies, including India, to go for phasing it out in an agreed time-bound legal framework. India has consistently maintained that the country will not phase out HFC unless there is availability of safe and economically-viable alternatives. India has demanded for technology transfer without patent restriction, financial assistance, and allowing developing countries, including India, more time for phasing down HFCs.

12.6 INDIA'S EFFORT TO COUNTER CLIMATE CHANGE TO COUNTER CLIMATE CHANGE TO COUNTER CLIMATE CHANGE

India is the world's third largest economy and fifth largest greenhouse gas (GHG) emitter, accounting for about 5% of global emissions. India's emissions increased 65% between 1990 and 2005 and are projected to grow another 70% by 2020. By other measures, India's emissions are low compared to those of other major economies. India accounts for only 2% of cumulative energy-related emissions since 1850. On a per capita basis, India's emissions are 70% below the world average and 93% below those of the United States. India is also at the frontlines of facing the impacts of climate change. Shifting rainfall patterns, recurring floods, stronger cyclones and droughts or soil erosion are exacerbating the challenge of poverty eradication and necessitate the allocation of scarce national resources for preventing loss of human life. Despite resource constraints, India is undertaking ambitious actions to undertake adaptation and mitigation actions, including thorough lowering of the energy intensity of our economic growth, increasing energy efficiency across sectors and making greater use of renewable. India has doubled the Clean Energy Cess on coal, which very few countries have, and the Clean Energy Fund already has over 3 billion US dollars to be used for promoting clean technologies India's National Solar Mission is being scaled up five-fold from 20,000 megawatts to 100,000 megawatts. This will mean an additional investment of 100 billion dollars and savings of about 165 million tonnes of CO2 emissions per year. India is releasing 6 billion US dollars in one go for intensive afforestation which will result in more carbon sinks. India has allocated about 200 million US dollars for the 'National Adaptation Fund', setting-up of Ultra Mega Solar Projects, Ultra-Modern Super Critical Coal Based Thermal Power Technology, and the development of Solar Parks on canals. Another initiative is "100 Smart Cities' with integrated policies for adaptation and mitigation to reduce the vulnerability and exposure of urban areas to climate change and also to improve their energy efficiency for which 1.2 billion US dollars have been allocated. India has put in place stringent norms for cement industry. Our Action Plan for cleaning one of the longest rivers in the world, River Ganga will bring multiple benefits of pollution reduction and climate adaptation. We have also taken initiatives for protecting coastal, Himalayan, and forest areas. India has initiated preparations to develop a National Air Quality Index and have launched a National Air Quality Scheme. Setting-up of Ultra Mega Solar Projects in Tamil Nadu, Rajasthan, Gujarat, Andhra Pradesh and Ladakh to promote renewable energy.

12.7 NATIONAL ACTION PLAN ON CLIMATE CHANGE (NAPCC)

Government of India has launched eight Missions as part of the National Action Plan on Climate Change (NAPCC) in specific areas which include assessment of the impact of climate change and actions needed to address climate change.

12.7. 1. National Solar Mission

The NAPCC aims to promote the development and use of solar energy for power generation and other uses with the ultimate objective of making solar competitive with fossil-based energy options. The plan includes: specific goals for increasing use of solar thermal technologies in urban areas, industry, and commercial establishments; a goal of increasing production of photovoltaic to 1000 MW/year; and a goal of deploying at least 1000 MW of solar thermal power generation. Other objectives include the establishment of a solar research centre, increased international collaboration on technology development, strengthening of domestic manufacturing capacity, and increased government funding and international support.

12.7.2. National Mission for Enhanced Energy Efficiency

Initiatives based on increasing the energy use efficiency were expected to yield savings of 10,000 MW by 2012. Building on the Energy Conservation Act 2001, the plan recommends:

- Mandating specific energy consumption decreases in large energy-consuming industries, with a system for companies to trade energy-savings certificates;
- Energy incentives, including reduced taxes on energy-efficient appliances; and
- Financing for public-private partnerships to reduce energy consumption through demand-side management programs in the municipal, buildings and agricultural sectors.

12.7.3 . National Mission on Sustainable Habitat

Promote energy efficiency as a core component of urban planning, the plan calls for:

- Extending the existing Energy Conservation Building Code;
- A greater emphasis on urban waste management and recycling, including power production from waste;
- Strengthening the enforcement of automotive fuel economy standards and using pricing measures to encourage the purchase of efficient vehicles; and o Incentives for the use of public transportation.

12.7.4. National Water Mission

With water scarcity projected to worsen as a result of climate change, the plan sets a goal of a 20% improvement in water use efficiency through pricing and other measures.

12.7.5. National Mission for Sustaining the Himalayan Ecosystem

The plan aims to conserve biodiversity, forest cover, and other ecological values in the Himalayan region, where glaciers that are a major source of India's water supply are projected to recede as a result of global warming.

12.7.6. National Mission for a "Green India"

Goals include the afforestation of 6 million hectares of degraded forest lands and expanding forest cover from 23% to 33% of India's territory.

12.7.7. National Mission for Sustainable Agriculture

The plan aims to support climate adaptation in agriculture through the development of climate-resilient crops, expansion of weather insurance mechanisms, and agricultural practices.

12.7.8. National Mission on Strategic Knowledge for Climate Change

To gain a better understanding of climate science, impacts and challenges, the plan envisions a new Climate Science Research Fund, improved climate modelling, and increased international collaboration. It also encourages private sector initiatives to develop adaptation and mitigation technologies through venture capital funds.

12.8 NATIONAL ACTION PROGRAMME TO COMBAT DESERTIFICATION

India is a party to the UN Convention to Combat Desertification (UNCCD) and MoEF is the National Coordinating Agency for the implementation of the UNCCD in the country. As an affected party, a 20-year comprehensive National Action Programme (NAP) to combat desertification in the country has been prepared. The objectives are:

- Community based approach to development,
- Activities to improve the quality of life of the local communities,
- Awareness raising,
- Drought management preparedness and mitigation,
- R&d initiatives and interventions which are locally suited,
- Strengthening self-governance leading to empowerment of local communities

It is proposed to initiate activities that include, among others, assessment and mapping of land degradation, drought monitoring and early warning system groups, drought preparedness contingency plans, and on-farm research activities for development of indigenous technology etc. will be taken up.

12.9 INDIA'S STAND ON CLIMATE CHANGE NEGOTIATIONS

Clean air, clean energy and clean power balanced with growth were the priorities for India in its mission to combat climate change. The government had pursued voluntarily set targets with commitment, conviction and followed-up action and had played an active and positive role in tackling the Climate Change. India's stand is based on its domestic obligations of addressing the basic development needs of poverty eradication, food security and nutrition, universal access to education and health, gender equality and women empowerment, water and sanitation, clean energy, employment, sustainable cities and human settlement and its commitment to fight climate change. India is of the view that historical emissions of developed countries as laid down in the Conventions should be the basis for differentiation. The developing countries' need for inclusive growth, sustainable development, poverty eradication, and energy access to all must be recognized as fundamental to the approach to differentiation. India is of the view that Announcement of contributions for Green Climate Fund (GCF) and its actual deposit should be ensured by developed countries. GCF could be used to purchase Intellectual Property Rights (IPRs) of climate friendly technologies. The non-capitalization of the Green Climate Fund was a matter of concern and could affect the 2015 Climate agreement. India grows into a regional power, which can only be accomplished if India is given sufficient development space to grow its economy and eliminate poverty. It is in India's interest to diversify its energy portfolio — a prospect that can be strengthened with the U.S.'s assistance. The way to achieve these objectives is to forge an "India exception" at the global climate talks in Paris; doing so is the only realistic pathway to a global climate deal and will cement the growing ties between the two critical actors in an evolving international order.

13

GLOBALIZATION, CLIMATE CHANGE, AND HUMAN HEALTH

13.1 INTRODUCTION

Rapid globalization has brought new, large-scale influences to bear on patterns of human health. Various global-scale changes — economic, social, demographic, and environmental (particularly climatic) — are linked, for example, to the increased prevalence of obesity, changes in regional food yields, the emergence of infectious diseases, the spread of cigarette smoking, and the persistence of health disparities. Undertaking primary prevention at the source to reduce health risks resulting from these global influences is a formidable challenge. It requires conceptual insights beyond the conventional understanding of causation and prevention, as well as political will, trust, and resources. The complexities of policies to mitigate human-induced climate change are clear. Meanwhile, additional resources and strategies will be needed to reduce the health risks related to global change that have already arisen or are now unavoidable. For populations to live sustainably and with good long-term health, the health sector must work with other sectors in reshaping how human societies plan, build, move, produce, consume, share, and generate energy.

The global scale, interconnectedness, and economic intensity of contemporary human activity are historically unprecedented, as are many of the consequent environmental and social changes. These global changes fundamentally influence patterns of human health, international health care, and public health activities. They constitute a syndrome, not a set of separate changes, that reflects the interrelated pressures, stresses, and tensions arising from an overly large world population, the pervasive and increasingly systemic environmental impact of many economic activities, urbanization, the spread of consumerism, and the widening gap between rich and poor both within and between countries.

In recent decades, international connectivity has increased on many fronts, including the flow of information, movements of people, trading patterns, the flow of capital, regulatory systems, and cultural diffusion. These exponential increases in demographic, economic, commercial, and environmental indexes have been labelled the Great Acceleration. Remarkably, the resultant environmental effects are now altering major components of the Earth system. The current geologic epoch is being called the Anthropocene (successor to the Holocene epoch) in recognition of the global force that Homo sapiens has become, pushing or distorting Earth›s great natural global systems beyond boundaries considered to be safe for continued human social and biologic well-being. The loss of biodiversity, the greatly amplified global circulation of bioactive nitrogen compounds, and human-induced climate change have already reached levels that are apparently unsafe.

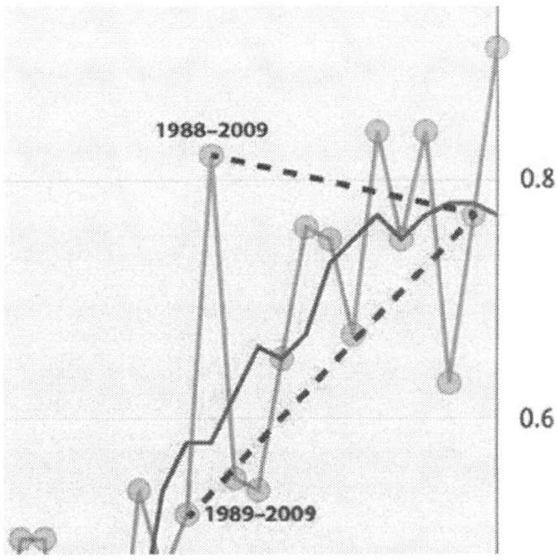

Fig. 1: Global Temperature Trend, 1880–2010

These changes pose fundamental threats to human well-being and health. For example, a positive relationship has been observed between regional trends in climate (rising temperatures and declining rainfall) and childhood stunting in Kenya since 1975, indicating that as projected warming and drying continue to occur along with population growth, food yields and nutritional health will be impaired.8 These human-induced climatic changes often act in concert with environmental, demographic, and social stressors that variously influence regional food yields, nutrition, and health. Furthermore, at the current level of global connectedness and interdependence, the environmental impact of human activity has a wider geographic range, although its influence may be offset somewhat by more effective global alerts and more rapid distribution of food aid. The extreme heat and wildfires in western Russia in the summer of 2010 destroyed one third of that country›s wheat yield, and the subsequent ban on exported grain contributed to a rise in the price of wheat worldwide, exacerbating hunger in Russia (where flour prices increased by 20%) and in low-income urban populations in countries such as Pakistan and Egyp. On the economic front, the recent global financial crisis has underscored the domino-like interdependence of national economies.

13.2 EFFECTS OF GLOBALIZATION ON POPULATION HEALTH

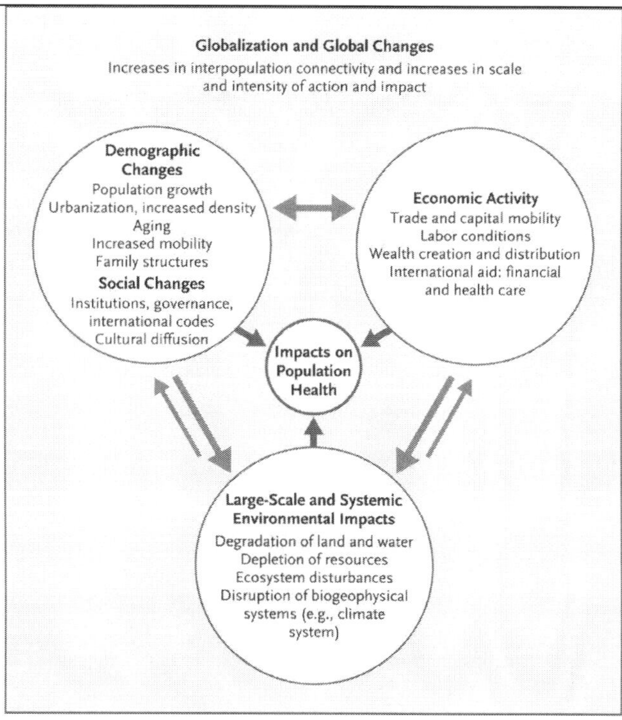

Fig. 2: Influence on Human Health of Changes Related to Globalization.

Global influences on population health such as those described above transcend the more specific, focused frame within which international health issues are addressed. The processes of global change are more systemic, involving disruption or depletion (not merely local pollution). Remediating or adapting to these changes requires an understanding of dynamic systems, their complexity and associated uncertainties, and coordinated policy responses across relevant sectors. The relationships between these pervasive processes of change and human health are shown in Figure 2.

13.3 DEMOGRAPHIC CHANGES

Population growth is often overlooked in the discourse on global change, including its relation to the mitigation (abatement) of climate change, to which the contribution of global emissions is obvious. The projections by the United Nations that today›s population of 7 billion will increase to 9.3 billion by 205012 should reactivate the debate about whether we can succeed in pursuing realistic objectives for a healthy climate without curtailing the actual number of humans pressing on the environment. Furthermore, the negative-feedback loop of excessive population pressure on regional environments (involving soil exhaustion, water depletion, and the loss of various wild animal and plant food species) not only exacerbates various ongoing worldwide environmental and ecologic changes but also entrenches conditions of poverty and disadvantage. In these latter circumstances, fertility rates tend to remain high.

Some additional increase in the world population is inevitable in countries with high fertility rates, given the demographic flywheel momentum of populations weighted toward the young. Meanwhile, moderate gains have been made in facilitating education for girls, although progress in this, as well as in the provision of adequate education about reproduction and reproductive choice, remains slow in many low-income countries. Where unplanned pregnancy rates remain high (e.g., Timor-Leste and Nigeria), so do risks to maternal and child health.

13.4 SOCIAL CHANGES AND ECONOMIC ACTIVITY

Many other aspects of globalization influence population health, including the accelerated emergence of new infectious diseases, the near-ubiquitous rise in the rates of obesity and associated noncommunicable diseases as daily bodily energy budgets (food energy input vs. physical energy output) shift into surplus, the spread of cigarette marketing, the effects of climate change, increases in resistance to antimicrobial agents, and health risks in the workplace due to the deregulation of international labor markets. Looming large in the background as additional determinants of health are the persistent, even increasing, disparities in wealth, education, autonomy, and social inclusion. There are, of course, certain aspects of globalization that are beneficial to health, such as the enhanced flow of information, improvements in internationally coordinated vaccination programs and systems to respond to infectious diseases, and a greater capacity for long-distance responses to disasters.

Adverse global influences on health, such as rising food prices and extended ranges of some infectious diseases, have also impeded attainment of the United Nations Millennium Development Goals. Future global health goals must be better integrated with the fundamental influences of poverty, inequity, illiteracy, climate change, land-use patterns, and food insecurity on health. After the Rio+20 Conference (2012), the Millennium Development Goals are to be replaced by Sustainable Development Goals in 2016, reflecting the principle set forth at the original Rio Declaration on Environment and Development (1992) that concern for humans must be at the center of sustainable development. Nevertheless, concern for human health is not yet near that center. This reflects the continuing misperception of what health means and the dominance of a narrow, clinically based view that seemingly does not take into account the fundamental need, in improving population health, to address the poor fit between environmental and sociocultural conditions and basic human biologic and psychological needs.

13.5 ENVIRONMENTAL AND ECOLOGIC CHANGES

The deep-seated, essentially ecologic risks to population health cannot be countered effectively at the local level alone. Climate change induced by human activities, for example, is due to the globally aggregated excess of greenhouse emissions. Primary prevention of health problems arising from such global environmental and sociodemographic changes therefore requires coordinated international policy, supplemented by more local policy-making and action. For example, the World Trade Organization should give greater priority to averting the adverse health and environmental effects of international free trade. There is also a need for instruments similar to the WHO Framework Convention on Tobacco Control and the WHO Global Outbreak Alert and Response Network, in relation to the emergence of infectious diseases, as well as the United Nations Environmental Programme Montreal Protocol to protect the ozone layer.

The following four examples describe other environmental and ecologic changes on a global scale that will increasingly influence the world's health. First, the probability that new strains of influenza virus will emerge is increasing, particularly in the rural villages of Southeast Asia and East Asia. The risk increases with population growth; the juxtaposition of traditional backyard pig, chicken, and duck farming with intensified commercial poultry production; and environmental changes that affect the flight paths of migrating wild birds.

Second, the decline in available seafood protein (which is important for many low-income coastal populations) is a threat to health and reflects the unprecedented combination of ocean warming, acidification (due to increased uptake of carbon dioxide), deoxygenation, destruction of coastal fish nurseries, and overfishing.

Third, diverse health risks are posed by the deprivation, displacement, and conflict that result from shortages of fresh water. Many populations, such as those in Bangladesh, Vietnam, Egypt, and Iraq, live downstream on great rivers that traverse several countries. In many cases, river flows are threatened by the loss of glacier mass and snowpack due to global warming and by the increased diversion of flow by neighbours upstream.

Finally, the need to maintain food supplies and adequate nutrition for the increasing world population presents a major challenge. Global food production also faces pressures as a result of reduced yield due to land degradation, water shortages, and climate change and the rising demand for animal foods among middle-income populations. Furthermore, agriculture (especially livestock production) accounts for around one fourth of global greenhouse-gas emissions. Thus, there are growing pressures to transform food production (e.g., more mixed cropping and inclusion of acceptable genetically modified crops), distribution, and consumption. Since the environmental, particularly climatic, effects of producing red meat from methane-producing ruminants (e.g., cattle, sheep, and goats) are so great, thought needs to be given to the question of whether production of this protein source will need to be curtailed — while allowing a sufficient increase to ensure safe childhood nutrition in the many poorer populations, which currently consume levels of red meat that are lower than those in the overconsuming rich populations by a factor of 10. The global food security issue is further complicated by the ongoing land grab in eastern Africa and elsewhere by richer countries seeking investment opportunities and self-insurance against future land, food, and biofuel shortages (e.g., Middle Eastern oil-producing states, China, and South Korea).

These four examples also confirm that, in a world of global and systemic changes, these individual changes for the most part do not impinge on population health in isolation; instead, they typically act jointly and often interact. Specific examples are discussed in the next section, which reviews the health risks posed by climate change.

13.5.1 Global Climate Change

Global climate change is part of the larger Anthropocene syndrome of human-induced global environmental changes. These include land degradation, ocean acidification, and disruptions and depletions of the stratospheric ozone concentration, soil fertility, fresh-water resources, biodiversity stocks and ecosystem functioning, and global nitrogen and phosphorus cycles.[4] Greenhouse emissions from fossil fuel–based power generation and transport and from the agriculture and mining sectors increase the heat-retaining capacity of the lower atmosphere, resulting in global warming. In addition, deforestation and ocean saturation have added to greenhouse warming by reducing the capacity of terrestrial and marine environments to absorb extra carbon dioxide (the main greenhouse gas) from the atmosphere. Also contributing to such warming are any ongoing natural variations in climate caused by cosmologic and geologic influences.

Most of the global warming since 1950 (an increase of 0.7°C) has been the result of human activity. Annual global emissions of carbon dioxide have increased over the past decade, as have the rates of sea-level rise, the loss of Arctic Sea ice, and the number of extreme weather events. Without substantial and prompt international action to abate these emissions, average global temperatures (relative to the year 2000) are likely to rise by 1 to 2°C by 2050 and by 3 to 4°C by 2100, including increases of up to 6 to 7°C at high northern latitudes. Additional warming of another 0.7°C is locked in from the extra radiative energy already absorbed by the lower atmosphere and, in turn, by the oceans, though not yet manifested as surface warming. An average rise of 4°C would return Earth's temperature to a level not experienced for 10 million to 20 million years.

Rainfall patterns will also change, with rainfall increasing in some regions and seasons and decreasing in others. Modeling consistently projects an increase in regional aridity, and in the geographic range and severity of droughts, during this century. The frequency, and perhaps intensity, of extreme weather events is also expected to increase in most regions — and may well have already begun to do so.

13.5.2 Effects of Climate Change on Human Health

The complex nature of climate change and its environmental and social manifestations results in diverse risks to human health. A three-way classification of these risks and causal pathways is shown in Table 1.

Table 1. Categories of Climate-Change Risks to Health, According to Causal Pathway.

Table 1. **Categories of Climate-Change Risks to Health, According to Causal Pathway.**

Risk Category	Causal Pathway
Primary	Direct biologic consequences of heat waves, extreme weather events, and temperature-enhanced levels of urban air pollutants
Secondary	Risks mediated by changes in biophysically and ecologically based processes and systems, particularly food yields, water flows, infectious-disease vectors, and (for zoonotic diseases) intermediate-host ecology
Tertiary	More diffuse effects (e.g., mental health problems in failing farm communities, displaced groups, disadvantaged indigenous and minority ethnic groups) Consequences of tension and conflict owing to climate change–related declines in basic resources (water, food, timber, living space)

Our current, rather skewed knowledge of climate–health relationships has come from epidemiologic studies of health risks in relation to differences and extremes in temperature and from quasi-cyclical climatic events such as the El Niño–Southern Oscillation phenomenon. However, most of the health risks will arise from climatic influences on environmental systems and social conditions that affect food yields, water supplies, the stability of infectious disease patterns, and the integrity of natural and human-built protection against natural disasters (including forest cover, windbreaks, mangroves, vulnerable constructed seawalls, and urban water-drainage systems) and from the adverse health consequences of social disruption, displacement of communities, and conflict situations.

Key examples of these types of causal paths are shown in Figure 3. Many of the indirect effects of climate change will be simultaneously influenced by other global changes and sociodemographic pressures that act in conjunction with climate change. Food yields and, hence, nutritional status reflect changes not only in climate but also in water supplies, soil fertility, nitrogen levels, biodiversity (e.g., pollinators and pest predators), and the health and vitality of farm workers.

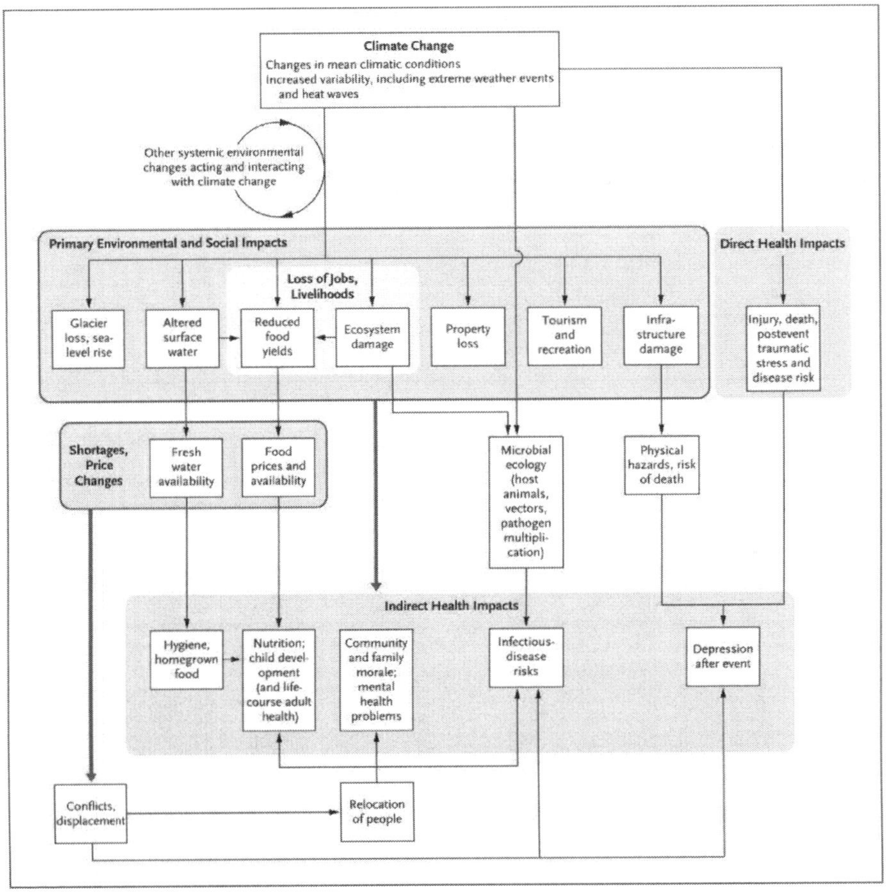

Fig. 3: Processes and Pathways through Which Climate Change Influences Human Health.

It is not surprising that the health effects of climate change will be predominantly adverse. After all, human biology, domesticated food sources, and culture in general have evolved over many millennia within the usual prevailing climate. Furthermore, populations everywhere will be vulnerable to increasingly severe extreme weather events.

Some beneficial health effects are expected to occur, at least in the earlier stages of climate change. In some temperate zones, milder winters may lead to fewer wintertime deaths from myocardial infarction and stroke, and in some low-latitude regions, hotter and drier conditions may reduce mosquito survival and, hence, mosquito-borne infection.

Populations living in diverse social, economic, and physical conditions will be affected differently by climate changes. Low-income and remote populations are more vulnerable to physical hazards, undernutrition, diarrheal and other infectious diseases, and the health consequences of displacement. Populations on low-lying islands and in coastal areas, such as Bangladesh, are vulnerable to increased storm surges and flooding as the sea level rises. In Arctic circumpolar regions, communities may undergo enforced changes in diet as land and marine animal populations migrate or decline and as access to traditional food sources becomes physically more difficult.[41]

The likely future effects of climate change on various health outcomes have been modeled with the use of plausible scenarios of future climate change that have been agreed on internationally. For example, in temperate countries, as summers become hotter and heat waves more severe, modeling indicates that, from around mid-century, additional heat-related deaths will progressively overwhelm the number of deaths averted as a result of milder winters. Such estimates of the extreme effects of weather will improve as the modeling of changes in climatic variability under climate-change conditions improves and as researchers take better account of physiological, behavioral, and technological adaptation by populations over time.

In China, the modeling of medium-scenario warming indicates that the transmission zone of freshwater snail–mediated schistosomiasis will extend northward, putting another 20 million people at risk by 2050. Such model-based estimation of the direction and approximate extent of likely change in health risks is an important resource for decision-making about both climate-change abatement and localized adaptation.

Table 2. Indicators of Early Health Effects of Climate Change

Increases in annual numbers of deaths and hospitalization due to extreme heat, observed in a range of high-income and low-income countries.
Increases in rates of injuries and deaths due to the rising frequency of weather disasters in many regions.
Extensions in the geographic range of several vector-borne infectious diseases of their vectors, including tickborne encephalitis in Sweden, the tick vector of Lyme diseases in eastern Canada, and malaria in the western Kenyan highlands
Although less certain, increases in the tempo of coastal outbreaks of cholera relative to the warming of coastal waters and EI Nino events
Increases in the price of some staple foods, especially in vulnerable, food-insecure regions, leading to nutritional deprivation in low-income households.

Meanwhile, an important research task is to identify ongoing changes in health risks and outcomes that can be reasonably attributed to recent climate change. Given the multivariate causation of most human health outcomes, attribution is rarely simple. Nevertheless, over the past decade, observed changes in some health outcomes, viewed collectively, suggest a climate signal (Table 2).

Health Risks and Benefits of Climate-Change Mitigation and Adaptation

The mitigation of climate change is a crucial first-order task for the world. However, while governments continue to wrestle with this unprecedented, complex political and ethical task, the more immediate challenge for the health sector is to identify the main regional health threats posed by climate change and ensure the development of risk-lessening adaptations.

13.6 ADAPTATION STRATEGIES

Adaptation capacities and strategies will differ greatly among populations. They will be particularly important where the rates of preexisting disease (e.g., childhood diarrhea and malnutrition) are already high and therefore, in absolute terms, would become considerably higher because of the multiplier effects of climate change. Effective adaptive strategies will mostly require collaboration among diverse government sectors, research disciplines, and communities.

During heat waves, deaths and hospitalizations predominate among the elderly, patients with chronic cardiorespiratory disease, and persons living in low-grade housing. Early heat wave–warning systems, community caregiver schemes for vulnerable persons, well-insulated housing, and educational advice from primary health care providers would lower this risk. Meanwhile, longer-term planning is needed to climate-proof urban residential areas.[4]

13.7 HEALTH BENEFITS OF CLIMATE-CHANGE MITIGATION

One favorable aspect of efforts to mitigate climate change is that local health gains will quickly accrue to populations that undertake such efforts. Awareness of this potential health dividend — in addition to the longer-term global health benefits — should strengthen support for such actions.

Health benefits will result from mitigating actions that address modes of transport, housing-design standards, energy generation, and agricultural systems (including livestock production). In many poor populations, improvements in environment-related technologies will help to replace indoor-polluting cooking fuels with low-carbon fuels, and improvements in reproductive literacy will lead to fewer, better-spaced pregnancies; both types of improvement reduce pressures on the climate system. All these actions will directly reduce well-known risk factors for disease and premature death (e.g., air pollution, sedentary living, and dietary excesses). Innovative urban design can have wide-ranging positive effects with regard to energy use, greenhouse-gas emissions, the effects of urban heat islands, patterns of physical activity, social relations, and community cohesion.

13.8 CHALLENGES FOR THE HEALTH SECTOR

The health sector has important roles to play in relation to climate-change abatement and adaptation strategies for lessening unavoidable risks to health (Table 3). Such strategies would include the "greening" of health care institutions and participation in national health impact assessments and in intersectoral planning of sustainable energy systems, transportation, and urban design. National delegations to international policymaking meetings that address global trends and threats (e.g., the annual conferences convened under the United Nations Framework Convention on Climate Change) should include representatives from, or at least substantive briefing by, the formal health sector.

Table 3. Role of the Health Sector in Climate-Change Mitigation (Primary Prevention) and Adaptation (Preparedness, or Secondary Prevention).

Table 3. Role of the health sector in climate-change mitigation (primary prevention) and adaptation (preparedness, or secondary prevention).

Goal and generic action mitigation of climate change	**Suggested strategies**
Carry out health impact assessment of mitigation strategies	Conduct epidemiologic research to estimate and document changes in health outcomes that result directly from mitigation actions
Limit the carbon (and other environmental) footprint related to the health care system	Design buildings, transport services and facilities to achieve energy efficiency in terms of energy sources and use and to minimize waste
Enlist health professional organizations and government health departments	Educate the public about risks to health from climate change and explain that mitigating actions can confer additional local health benefits
Include physicians and other health care workers as citizens	Participate in wider public discussion and moderate personal behaviors
Adaptation to lessen health risks	
Provide adequate health care facilities and services	Improve facilities for handing increased patient volume resulting from extreme weather events and ensure adequate stocks of vaccine
Anticipate necessary surge capacity (e.g., for major heat waves, fires, epidemics)	Coordinate with emergency-services agencies and ambulance facilities and consider morgue capacity
Reinforce and extend public health programs to provide a foundation for dealing with most types of climate related health effects	Develop early-warning systems (e.g., for heat waves, floods, and possible epidemics); programs for infectious-disease surveillance and analysis, vaccination and vector control (e.g., mosquitoes, ticks); support for vulnerable communities; and mental health services (e.g., for postevent trauma and depression)
Educate and train the health workforce	Develop programs that prepare health care workers to contribute to public education and to be on the alert for unexpected diagnoses
Engage in broader collaboration with other sectors	Institute policies for creating green spaces in cities (to promote physical and mental health); develop housing design and insulation to optimize health protection; consider livestock and wild animals as possible risks for infection

*The listed adaptation activities are intended to reduce health risks on the local and regional levels.

14

THE GLOBAL ENVIRONMENTAL CHANGE AND HUMAN SECURITY (GECHS)

14.1 INTRODUCTION: AN OVERVIEW

The GECHS project made important contributions to a field that is likely to become more significant in the future, and both the approach and the findings can inform critical research on responses to environmental change, including adaptation and transformation. New calls for integrated, transdisciplinary research for global sustainability and initiatives such as Future Earth can draw on some of the insights from human security research, including a recognition that the goal of global sustainability is neither neutral nor apolitical and is meaningless unless it identifies and challenges the powers that have long enabled the security of some to be achieved at the expense of others. The urgency of global environmental issues may create a situation where responses are offered and implemented at a rate and speed that overpower the capacity for critical reflection.

The concept of human security has been used for over two decades to frame and analyze problems of social change, and since the late 1990s, it has been used for these purposes with respect to global environmental change. Yet research on the intersections between global environmental change and human security is still an emerging field: a field that is currently being redefined in recognition that responses to environmental changes may also have profound implications for human security. It is a heterogeneous field, with diverse disciplines and theoretical perspectives producing knowledge about an array of environmental issues across different places, sectors, and scales of analysis. The development and diversity of this research area is reflected in, but by no means completely captured by, the Global Environmental Change and Human Security (GECHS) project that operated between 1999 and 2010. As a core project of the International Human Dimensions Programme on Global Environmental Change (IHDP), the GECHS project facilitated, supported, and to some extent institutionalized research on global environmental change and human security. In the aftermath of the GECHS project, a transition of sorts is now underway, with many new researchers pushing an expanding frontier of knowledge about the myriad of challenges associated with promoting, sustaining, and enhancing human security in the context of global environmental change. In this endeavor, the concept of human security helps to frame both research and policy on global environmental change.

This chapter on global environmental change and human security provides retrospective and tentative prospective views of research in this field. We discuss the meaning of human security and the five roles that it plays in environmental research. We then provide a synopsis of the field and present the GECHS project as an example of how this research has contributed to more critical understandings of global environmental change.

In addition, we consider what a human security lens has to offer the next generation of global environmental change research, including research for global sustainability organized around the new international Future Earth initiative. In particular, we discuss how a human security approach complements Earth systems approaches to environmental change through a more pluralistic understanding of scales and different forms of knowledge, and amore politicized understanding of the problem that draws attention to the role of power and agency. We conclude by exploring some potential directions for critical research that advances knowledge on transformations that take into account both human security and sustainability.

14.2 A HUMAN SECURITY LENS

Human security can serve as a valuable lens for understanding and addressing the complex challenges facing individuals and communities in the twenty-first century, including global environmental change. Definitions of human security vary according to discipline, sector, and norms concerning the meaning of security. In terms of disciplines, human geographers, political scientists, psychologists, and sociologists engage with the concept of human security. In terms of sectors, human security is used by governments, intergovernmental institutions, and nongovernmental organizations as a means to frame policies and decisions on issues related to development, environment, health, peace, conflict, and human migration.

Finally, human security itself marks one end of a spectrum of norms about the breadth and depth of security, suggesting that individuals (as opposed to states) are a referent object of security and that their security can be compromised by a diverse set of risks that include, but are not restricted to, the military actions of other countries.

Define human security as a condition in which people and communities have the capacity to respond to threats to their basic needs and rights, so that they can live with dignity. This definition builds on basic needs and human rights approaches to development, as well as the capabilities and freedoms approaches of Amartya Sen. The capabilities approach recognizes that peoples' actions to advance their well-being are most effective when given economic and social opportunities, political freedoms, transparency guarantees, and protective security. The definition of human security focuses on the freedom to choose and act—even if people choose not to choose, and not to act. As a normative concept, human security describes a condition with interdependent subjective and objective components that are achieved when such freedoms and opportunities are recognized, protected, and enhanced.

The concept of human security performs five key roles or functions, although rarely are all of these explicitly recognized. First, it contests the notion that security applies only to states and that the major risk facing states is violent conflict. Indeed, the concept has its origins in this critique of the dominant discourse on national security. In terms of global environmental change, this function of human security is critical. Although there is a growing literature on this topic that reflects concerns about sovereignty, territoriality, resources, and migration, the concept of human security explores how multiple stressors, including conflict, influence individuals and communities and their capacity to respond to threats to their social, environmental, and human rights.

The second function of human security is to highlight that environmental change is a social problem with environmental characteristics, rather than an abstract scientific problem that can be disarticulated from social processes. It shows that many of the drivers of vulnerability to environmental change arise from political, economic, social, and cultural processes that cause inequalities both in exposure to environmental changes and in the capacities to respond to them. A human security approach supports existing challenges to mainstream development discourses and calls for addressing environmental issues within a much larger socio-political context. From this perspective, environmental change is not a neutral, apolitical process but rather one that raises important issues of equity and etics. The third function of human security is to acknowledge the power of agents to navigate and contest systems and structures that influence their everyday lives. Human security recognizes that people matter in the science and politics of environmental change.

It also shows that how individuals and groups engage with power can be decisive in terms of socioecological outcomes. Although there has been growing attention to how people adapt to environmental changes, human security also points to the potential for individuals and groups to challenge systems and structures that contribute to vulnerability in the first place. This includes mitigating the drivers of environmental changes, as well as addressing the social drivers of vulnerability, such as the denial of human rights, inadequate access to resources, or a lack of voice in decisions. The fourth function of human security is to elevate the problem of environmental change above ordinary policy problems. This superordinate function applies to all invocations of security, whether with respect to problems of migration, poverty, or drug trafficking .

Thus, describing environmental change as an issue of human security elevates the significance of the problem, adding a sense of gravitas that can influence political agendas When considered within the context of complex social and ecological interconnections, human security becomes not just a concern for individuals and communities that are most vulnerable to environmental changes, but also a concern for all of humanity. However, as a discursive tactic or a speech act, invocations of security often lead not to reduced military budgets and increased spending on environmental problems but rather to environmental issues being appropriated by security institutions to reconstitute their legitimacy.

Finally, bearing in mind the array of disciplines and institutions that work with the concept of human security, it can be considered a boundary object that facilitates interactions between diverse and otherwise disconnected communities. Human security brings the security, development, and environmental change policy communities together, forging new conversations, learning, and cooperation.

This role is significant because transdisciplinary research is often problem oriented, and such boundary objects serve to pull together diverse voices and perspectives. Even the debates over the meaning of human security and its relevance to environmental change can be considered important steps in reframing research agendas.

In summary, human security entails a focus on individuals and communities and the social and ecological context in which they meet and respond to complex challenges. Because it puts people at the center of the analysis, human security is increasingly being used as a lens through which to understand and respond to the human dimensions of global environmental change.

14.3 ENVIRONMENTAL CHANGE AND HUMAN SECURITY

One of the difficulties in reviewing research on global environmental change and human security is in identifying the boundaries of the field. Depending on how one defines security, the field could include, at one extreme, almost all knowledge about environmental changes or, at the other extreme, only issues that increase the risk of violent conflict. We can, however, point to several characteristics that are generally shared by research in this field. For the most part, the environmental problems of concern are global. Although the differences between local and global issues have become increasingly blurred, a general rule of thumb for what constitutes global might be those problems that cannot be addressed without some sort of international cooperation. The existence of multilateral treaties and agreements signifies that more than a few countries care enough about environmental issues to commit time to consider their solutions. The history of global environmental diplomacy is, therefore, an integral part of the history of this field, and the science and the politics of global environmental change have influenced each other over the past 40 years. Though there are many multilateral treaties on environmental issues dating back to the early 1970s, it was the agreements signed at the United Nations Conference on

Environment and Development in 1992 that most stimulated research on the social dimensions of global environmental change. Of these agreements, the United Nations Framework Convention on Climate Change has arguably been the most galvanizing. However, although the United Nations Conference on Environment and Development was informed by and promoted the idea of sustainable development and the associated idea that environmental problems cannot be separated from development, the splitting up of environmental problems into distinct conventions, with their own secretariats and funding modalities, has affected a similar fragmentation of the research community. This division of labor has meant that the fundamental drivers of environmental change, including inequality and the increasing production and consumption of goods and services, have been marginalized in science and policy.

Another important characteristic common to most research in this field is a concern with the social and human dimensions of global environmental problems with the focus most often on the implications of these changes for society, which is frequently realized through ~~place-based~~ studies or by looking at particular social groups. As such, research on global environmental change and human security is typically informed by social science theories, methods, and concerns. These theories and methods have evolved and expanded over the past decades. Early work on flood hazards established the social production of vulnerability as a central theme of research on the human dimensions of environmental change, i.e., that vulnerability to environmental disasters is largely a product of the way humans occupy and use the natural environment. This theme has since been progressively extended, particularly through social science research on disasters and risk. Since the early 1990s, research has added new insights into how political and economic processes drive environmental degradation and influence people's exposure to it in ways that create winners and losers from environmental change.

These insights are captured in the notion of social vulnerability and they are the key to understanding the effects of global environmental change on human security. A somewhat more arbitrary criterion for identifying research on global environmental change and human security concerns the security dimension. Typically, research in this field has been concerned with the causes and consequences of global environmental changes that are considered to be of such importance to societies that they warrant identification as security problems. Most of the research in this field has explored the risks that global environmental change poses to basic needs and human rights, for example, through food insecurity, water scarcity and pollution, or violent conflict, and here the label security seems unproblematic. But security is a subjective term, and one group's idea of security can translate into insecurities for many others. New research frontiers are less directly focused on traditional material security problems, but rather they explore the risks environmental changes pose to unique cultures and identities (as elements of needs and rights). Furthermore, addressing insecurity need not only entail reducing impacts but should also address the fundamental drivers of environmental changes and social vulnerability. A more recent turn in this field focuses on the ways that responses to environmental change, such as carbon offsets or reducing emissions from deforestation and forest degradation, contribute to insecurities, whether in relation to changes in access to land, resources, and livelihoods or simply through insufficient voice in decision processes. These new areas of research on global environmental change and human security build on knowledge generated from fields such as ecological economics, environmental psychology, environmental politics, and cultural anthropology. It is important to recognize that other science-policy communities also take a broad view of what constitutes human security. Perhaps most notable is the way the concept has been applied in the field of development. The United Nations Development Programme's 1994 *Human Development Report*, for example, defined human security as a "concern with human life and dignity" with seven principal components: economic, food, health, environmental, personal, community, and political security. The United Nations' *Commission on Human Security* also adopted a very broad view, defining human security as protection of "the vital core of all human lives in ways t at enhance human freedoms and human fulfillment," and it discussed education, health, mobility, poverty, and violence as human security issues.

The conceptual and theoretical heterogeneity that attends research on human security is explained by the heterogeneity of peoples, places, and institutions that researcher's study. Yet this breadth does not mean a lack of coherence: In the case of research on environmental change and human security, there is generally a shared focus on the social outcomes of environmental change and the social processes that mediate these outcomes. As we explain below, there are emerging commonalities within research in this field, including a concern to politicize the issue of environmental change, to break free from the hierarchical ontology of scale embedded in much environmental change research, and to produce more situated forms of knowledge.

Another way of gaining an overview of the field on global environmental change and human security is to examine the research carried out by institutions and research groups engaged with this topic. Over the years a few such institutions have emerged, including the GECHS project, the Woodrow Wilson Center's Environmental Change and Security Program, and the United Nations University Institute for Environment and Human Security. Within global change research, the GECHS project has been a key institution in helping to establish, promote, and synthesize research in this field. The following discussion highlights the approach and results of the project, including its contributions to social science research on global environmental change.

14.4 THE GECHS PROJECT

There are many international institutions engaged in research on global environmental change, and these comprise an epistemic network that sustains a certain orthodoxy with respect to the object of study and approach to knowledge. Arguably the most notable cluster of these institutions were those organized under the international global change research programs, including DIVERSITAS, the International Geosphere-Biosphere Program (IGBP), the IHDP, and the World Climate Research Programme (WCRP), as well as joint projects carried out under the Earth System Science Partnership (ESSP). These programs largely framed environmental change within the context of Earth system science, which is defined as "the study of the Earth system, with an emphasis on observing, understanding and predicting global environmental changes involving interactions between land, atmosphere, water, ice, biosphere, societies, technologies and economies". Here the object of study is the Earth system, an epistemology that is focused on understanding and predicting environmental changes through integrated assessments and modeling studies. Within this epistemology, there has been a growing recognition of the need to engage social scientists in integrated research to contribute knowledge on equity/equality, wealth/resource distribution, political systems/governance, political economy, globalization, and social and cultural transitions. Much of this social science research was carried out under the auspices of the IHDP, which aimed to promote, coordinate, and catalyze research on the human dimensions of global environmental change, and work at the interface between science, policy, and practice.

The initial development of the GECHS project was in response to six key issues in global environmental change research. First, there was a recognized need for improved theoretical and conceptual development of the links between environmental change, impoverishment, and insecurity. Second, it was considered that there were too few empirical studies focusing on the elements of environmental change that actually threaten human security, including studies on the role that various economic and social processes play. Third, the extent to which researchers, nongovernmental organizations, and policymakers were jointly engaged in knowledge production was seen to be insufficient. Fourth, understandings of why some communities and organizations were able to adapt to environmental change, whereas others were not was also a key concern. Fifth, issues of inequality and impoverishment were not being adequately included in research on environment and security.

Finally, the development of methods for early warning of environmental change, particularly in regions of high insecurity and vulnerability, was seen to be necessary. In sum, the project sought to develop a new and different perspective on the links between environmental change and security. The project's science plan also recognized the importance of spatial levels of analysis in research on the human dimensions of global environmental change, including the role of cultural and social spaces . The GECHS project brought together an interdisciplinary group of social scientists, including, but not limited to, geographers, political scientists, and sociologists. The first GECHS international project office was established at the University of Victoria, Canada (1999–2002), later moving to Carleton University, Canada (2002–2005), and then the University of Oslo, Norway (2005–2010). The science plan identified five key research foci: (*a*) conceptual and theoretical issues in environment and human security; (*b*) environmental change, resource use, and human security; (*c*) population, environment, and human security; (*d*) modeling regions of environmental stress and human vulnerability; and (*e*) institutions and policy development in environmental security. Crosscutting activities were related to data and methodological issues, as well as to communications, education, and training. These were justified because of the recognized need for a holistic approach to environmental problems and for greater dialogue among different actors and stakeholders. The desires to shift the focus of development from the national level to community and local levels and to move away from traditional, centralized approaches to development planning also formed the context for GECHS research.

The research carried out under the GECHS project resulted in key findings about population, equity, transboundary cooperation, vulnerability, and development. Over time, GECHS research evolved to include the interests and areas of expertise of the scientific steering committee members and "GECHS associates," such that an increasing amount of research focused on vulnerability and adaptation in the context of multiple stressors, the relationship between globalization and global environmental change, and the implications of climate change for human security (64–66). This was carried out through projects, including the Southern Africa Vulnerability Initiative, and through activities, including the International Workshop on Human Security and Climate Change (67, 68). Between 2006 and 2010, the scientific steering committee and GECHS associates published 90 peer reviewed papers in scientific journals and 28 books, as well as over 100 reports and articles.

The GECHS project also produced a series of AVISO policy briefings, which were the basis for a number of policy dialogues. Working at the science-policy interface, the GECHS project engaged in dialogues and discussions about user needs and, in response, produced reports that supported international initiatives, such as the Intergovernmental Panel on Climate Change's *Special Report: Managing the Risks of Extreme Events and Disasters to Advance Climate Change Adaptation* (*SREX*) . GECHS research has been summarized and synthesized in three edited volumes that were direct products of the project. The first of these, titled *Global Environmental Change and Human Security* , presents a collection of research that covers global environmental change and human insecurity, the links to conflict and violence, and relationships between sustainable development and human security. Chapters include critical perspectives on relationships between human security and population, gender, development, health, and conflict and cooperation. Some of the findings that were highlighted in this volume include the following: First, there is general agreement that democratic institutions reduce the risks that environmental changes pose to human security. Although market liberalization can reduce human insecurity, attention is drawn to the distributional effects and their contributions to inequality. Second, oversimplified representations of security concerns linked to global environmental change can lead to counterproductive policy responses.

Thus, there is a need for grounded empirical research, including research that is free from the ideological biases and prejudices that shape many narratives about the poor and vulnerable.

Finally, the concludes that global environmental change is of central importance to all human security assessments and policies, and there should be greater integration of research on human security, environmental change, and sustainable development. This book "establishes the vitality of the social sciences as a central force of inquiry in global environmental studies, engaging the concrete needs, motivations, and actions of people seeking to secure their health and safety, livelihoods, rights, and values in pervasively changing environments." The second book, *Climate Change, Ethics and Human Security*, focuses on framing climate change as an issue of human security, as opposed to an environmental problem. The authors argue that the environmental discourse excludes much more than it explains when it comes to understanding the human dimensions of climate change. A human security framing, in contrast, draws attention to questions related to equity, ethics, and reflexivity. The book includes philosophical perspectives on human security and discusses the links between human security, inequality, and poverty eradication. The findings highlighted in this book, which point to the social, economic, and moral arguments for addressing climate change in a fair and equitable manner, make it clear that inputs from the social sciences and humanities are needed to move beyond an institutionalized discourse on climate change that ignores the driving factors and the normative dimensions of the problems. It calls for a "new science" of climate change that critically engages with the social, institutional, and human context, placing issues of justice, ethics, responsibility, and human security at the forefront of policy debates.

Such a science would recognize that many of the key issues raised by climate change cannot be addressed by scientific or expert responses alone but instead call for new debates and discussions, new research questions, and new types of public deliberations and policies. The third book, a product of the 2009 GECHS synthesis conference, is *A Changing Environment for Human Security: Transformative Approaches to Research, Policy and Action*. This book presents critical analyses, case studies, and reflections on contemporary environmental and social challenges, with a strong emphasis on those related to climate change. It suggests that global environmental change is a deceptively benign label in that it hides the asymmetrical responsibilities for such changes, as well as the unequal outcomes that result from them. It also underplays the capacity for humans to pursue options that address the root causes of multiple, interrelated threats. The book emphasizes that environmental issues are political issues, social issues, economic issues, and development issues, and it shows how a human security approach can provide insights on breakthrough conditions that can enhance sustainability. These conditions include viewing problems from multiple perspectives to link issues and scales, questioning dominant paradigms and framings of problems, focusing on individual and collective empowerment, and integrating knowledge and action . The book recognizes human security as a choice that is tied to values, decisions, and actions, and it draws attention to the need for transformative approaches that question many of the assumptions associated with systems, structures, and paradigms that perpetuate global environmental change and human insecurity.

The GECHS project largely analyzed the causes of human security problems. However, in developing a body of knowledge that shows that the security risks that arise from environmental change are not inevitable and that these risks are within the purview of social systems, the project's message ends up being one of hope: Change for the better is possible, if not easily achieved.

14.5 TOWARD A NEW SCIENCE OF GLOBAL ENVIRONMENTAL CHANGE

The images of Earth from space transmitted by NASA in the late 1960s did much to stimulate concern about changes in the global environment (83). With the growth in the geopolitical imagery of Earth as a system shared by all forms of life came a growth in the science of the global environment based on similar imagery. The resulting Earth system science has tended to frame environmental change as an environmental problem with "human dimensions". Thus, although there have been increasing calls for integrated research on the Earth system and environmental processes, there has been little room within Earth system framings for critical research on the social context and unequal consequences of these changes for different places and groups. This is a significant gap, resulting in a situation where the biophysical processes associated with environmental change are increasingly well understood, the social drivers and human consequences are often overgeneralized (for example, in sweeping claims about impacts on women and the poor) or deterministic (as in accounts of climate change forcing mass migration and warfare). The global environmental change research community is currently transitioning toward a new paradigm that emphasizes the importance of codesigned research and coproduced knowledge for sustainability. This new paradigm forms the basis for Future Earth, a new 10-year international research initiative on global sustainability that builds on all of the global change research programs, and not just the GECHS project (indeed, as described later, the Future Earth's focus on well-being eschews some key concerns of the GECHS project). Key to both the new paradigm and research initiative is the recognition that the social sciences need to be better integrated into Earth system science (87, 88). This involves asking new questions and being open to different perspectives, approaches, methodologies, and validity claims. In the following sections, we discuss how the human security approach to environmental change reviewed above can contribute to a new type of integrated research, one that encompasses a wider understanding of the problem of environmental change, with a less ordered view of scale, a greater focus on the social production of problems, and more situated forms of knowledge .We consider these contributions in relation to an Earth system framing that (*a*) reduces the human dimensions to a narrow notion of well-being, (*b*) has a hierarchical ontology of scale that ascribes too much agency to the environment and too little to people, and (*c*) privileges universalizing and globalizing forms of knowledge over more pluralistic and situated knowledge.

14.6 POLITICIZING GLOBAL ENVIRONMENTAL CHANGE

In considering the impacts, vulnerabilities, resilience, and response capacities of people faced with global environmental change, there is a tendency for Earth system science to focus on human well-being rather than human security. This is consistent with many environmental framings, including in the Millennium Ecosystem Assessment and the UNEP Global Environmental Outlook assessments. The concept of well-being entails a focus on both the subjective and objective determinants of life quality, usually those associated with different dimensions of wealth and health. Yet when operationalized, well-being tends to become that which can be measured and readily included in indexes, rankings, and models that seek to integrate the outcomes of macroprocesses on people's lives. Thus, conventional measures of human development, such as gross domestic product per capita, education, and life expectancy, are easily equated with well-being with little attention to the social processes and politics that threaten the human and environmental rights of individuals and communities.

In short, human well-being tends to become a passive and neutral metric that conforms well to conventional policy objectives. Although there is considerable overlap between the term's human well-being and human security, the latter focuses on a wider range of nonmaterial but no less important dimensions of quality of life, including identity, freedom, opportunity, inclusion, belonging, and culture. It recognizes that these are not merely outcomes but instead are intrinsic to the very processes that contribute to well-being and security. Thus, for example, the threat of violence is a human security concern in and of itself, even if its effects cannot be measured in indicators of well-being; the forced relocation of communities has significant psychosocial impacts, even if these are not always captured in data about wealth or health; and the prospect of no future for places to which people are attached can be devastating for them, even if it has no effect on health and welfare in the present. In addition to being open to the multiple—and not necessarily quantifiable—measures of well being or thc "good life," human security also draws attention to the choices that are available to people, as well as the associated issues of agency, responsibility, and rights. Human security research has investigated the effects of various processes on vulnerability to environmental change, including global markets, patriarchy, human rights abuses), race, class, and violent conflict. Human security approaches, therefore, do not shy away from identifying the role—and in many cases abuse—of power as a cause of environmental change, however unpalatable this might be to decision makers and some researchers. So, although human well-being is a valuable goal, it does not draw critical attention to the processes that undermine it in the same way as human security does. In the context of global environmental change, human well-being is an approach that favours universal measures over meaning, and policy over politics, whereas human security recognizes that elements of wellbeing and the good life can be particular to social groups, difficult to measure, and enjoyed or denied by the operation of power in all its myriad forms.

14.7 ONTOLOGIES OF SCALE AND AGENCY

A human security approach to environmental change emerges from a different ontology of scale and agency than that of Earth system science. Earth system science is concerned with understanding large-scale systems of which humans are a large and undifferentiated component that is both a driver of and profoundly influenced by changes in the global environment. This is made possible by a simple ontology of scale as a hierarchical framework that orders the world into a series of nested and interacting subsystems, each influenced by larger elements in the Earth system. In this scalar framework, people are recognized as the drivers of global environmental change, as most clearly reflected in the notion of a new geologic epoch known as the Anthropocene. In this schema, people and places are at lower levels in the hierarchy and are considered to be highly influenced by changes in the Earth system—a vulnerability based on exposure to global environmental change. The social variables that cause vulnerability (such as markets, the state, or violence) are properties of the surficial social plane and so are not included (and they are in any event not amenable to inclusion) in models. Thus, this ontology of scale is as much an ontology of agency because, once engaged, nature is seen to have the power to shape social life. It is also a hierarchical ordering of what constitutes important knowledge about global environmental changes.

There are two consequences of this ontology of scale and agency that inform Earth system science. First, the power of individuals to make lives for themselves in the context of environmental change is downplayed. It involves questioning the science of the Earth system and its premises that Earth system processes are uniquely decisive to human security, when in fact human agency and collaborative action can influence systems and processes to achieve both security and sustainability. For example, in the literature on climate change and migration, the Earth system discourse casts climate as the principal driver of mass migration: People are "pushed" by environmental changes as if they have no capacity to adapt and no choice but to move. A human security approach, however, also sees migration as the exercise of agency to improve people's lives. Thus, although in some circumstances migration may be largely forced by environmental change, particularly in situations where people are dependent on climate-sensitive resources, for the most part, it is a strategy to improve human security that is sometimes stimulated by environmental changes but rarely wholly explained by them. In this view, migration is not merely an impact to be managed within the context of environmental change, but also an opportunity for people to help them adapt to environmental change in ways that satisfy their needs and values. Thus, among other things, a human security approach helps recast the idea that global environmental change is always geopolitically dangerous; instead, it shows that its social consequences vary and can be managed. As important and alluring as it is, Earth system science's gaze from everywhere and nowhere (i.e., that it is all seeing and objective), and its hierarchical ontology of scale, decultures and depoliticizes environmental change. Global environmental problems are disassociated from any given people and place: Issues of inequitable consumption and vulnerability, justice, and structural power are folded into and subsumed by a focus on the system and its environmental properties. The possibilities that power moves around and that people may transform social and ecological systems for the better are overlooked. Solutions instead tend to favor apolitical, technocentric, and global approaches. As a result, notions of transformation, redistribution, justice, and sufficiency are often given little attention.

In this way, knowledge of the Earth system implicitly serves power by ignoring its existence. Yet complex adaptive systems in which power is multidimensional and multidirectional can be the source of positive innovation and change in directions that are not linear.

In contrast, a human security approach draws attention to humans as agents of change with responsibility for contributing both to environmental problems and their solutions. It recognizes that, through their actions, people contribute to resource degradation and pollution, which harms both them and others, and that they create and sustain the structures that limit access to basic human needs and rights, and silence dissenting voices and actions. In this way, agency is central to the human security approach because it is seen as the creator of environmental change and human insecurity, but also as the solution, for agency can also be transformative: People can respond by mitigating environmental risks and reducing vulnerabilities and insecurities if they so choose.

Thus, in human security research, scale is not preordained, ordered, and hierarchical but is relational and created by people and what they choose to do. This focus on agency and structure distinguishes a human security approach from many paradigms of environmental change research; most of these paradigms consider humans largely as villains and victims without attention to the centrality of power and empowerment in supporting, resisting, or contesting change processes.

14.8 SITUATED KNOWLEDGE OF GLOBAL ENVIRONMENTAL CHANGE

Consistent with its alternative ontologies of scale and agency, a human security approach to global environmental change also carries with it a different philosophy of science. In Earth system science, there is an "instinct" to make certain kinds of knowledge global in scope—a science that is totalizing and universalizing, such that its knowledge is "true, irrespective of scale and place". It is a science of the outside, a panoptic exercise, whereby invulnerable experts model systems that can render people vulnerable by denying their humanity and agency.

Mathematical models are machines used to make approximate truths about the world and to define what counts as knowledge. The things that models do not account for are at best subordinate (or to be factored in if possible), at worst not matters of science, and hence, invisible in the science-policy process. A human security approach is concerned with the root causes of environmental change and its consequences for the needs and rights of individuals and communities. Its preference for locally derived and often qualitative data means that human security research often fails to integrate into Earth system science, for the models do not accommodate lives, values, needs, rights, desires, loves, interests, and the workings of power in all its forms. These cannot be subsumed into mathematical models and are instead unique forms of knowledge. This is the realm of traditional ecological knowledge, culture and place, bodies and embodiment, and of politics and power, all of which influence both the subjective and objective dimensions of human security.

Human security research is, therefore, open to more situated forms of knowledge, which is knowledge that is not of everywhere but of somewhere and that is grounded in peoples' experiences, everyday lives, values, and understandings of the processes that affect them, including changes in the environment. Human security research is sensitive to multiple experiences and to learning about how other people understand and respond to the causes of their insecurity; environmental change is most often one among multiple stressors.

Mike Hulme calls such knowledge "spectral" in the sense that it recognizes and respects differences in beliefs and accommodates the range of perspectives rather than seeks consensus, and "cosmopolitan" in that it dissolves the distinction between global and local. Thus, human security research is emerging as a science of situated knowledge on the basis of multiple partial perspectives that are shared through "webs of connections" so that the learning that emerges is rational, objective, and potentially transformative.

There is more to this emergence of human security as science of situated knowledge about global environmental change than just its sensitivity to partial perspectives, scale, and agency. There is a wealth of research from across the developed and developing world that points to the different ways that environmental change is framed, experienced, and responded to, even if this knowledge is most often produced by researchers, published by commercial publishers, and written in English. There is also the emergence of networks of researchers from across the world, allowing for diverse experiences from many places to emerge.

Finally, and perhaps most importantly, research on global environmental change and human security does not seek consensus, nor assume superior knowledge, and it seeks heterogeneity and eschews hierarchies of knowledge or scale.

14.9 AN EVOLVING RESEARCH AGENDA

In reviewing the literature on global environmental change and human security, we have drawn attention to an approach that can make valuable contributions to a new science on global sustainability. At the same time, we recognize that the field of global environmental change and human security research itself is evolving to take into account a dynamic context, where the pace and variety of responses to global environmental changes are creating new challenges for human security. Although the growing literature on adaptation to climate variability and change discusses ways to reduce vulnerability to multiple stressors, there appears to be an implicit acceptance of climate change, biodiversity loss, ocean acidification, and other global environmental problems as more or less inevitable. Implicit to this is the idea that the drivers of global environmental change are immutable and that individual and collective action to radically transform existing systems and structures is not possible.

This has led to increased attention to transformative adaptation, which is important but does little to question the systems and structures that contribute to global environmental change and social vulnerability in the first place. Research on human security points to a need for greater understanding of transformation processes, including the potential role of transformative politics and democratization.

Human security provides a conceptual and empirical basis for the analysis of transformative approaches to science, policy, and practice . Transformation can be defined as physical and/or qualitative changes in form, structure, or systems of meaning-making, and it often involves questioning assumptions and beliefs, including those that are the foundation of paradigms of economic growth and adaptation without limits. It involves questioning the science of the Earth system and its beliefs that Earth system processes are decisive to human security and that human agency and collaborative action cannot influence systems and processes to achieve human security and sustainability. A truly transformative agenda recognizes that the power to change is not solely vested in the elites who coproduce knowledge about it, but instead involves all whose knowledge, ideas, insecurities, and fears are at stake, such that their knowledge and agency are brought to bear. This involves pluralizing knowledge and power, so that groups at risk, local and indigenous communities, artists, activists, youth groups, and elders, as well as religious and alternative spiritual communities are as influential as the scientists, the owners of capital, and the state. Human security research can contribute to a new science for global sustainability through its alternative ontologies of scale and agency and its focus on how diverse human values and worldviews can be harnessed in support of both the well-being of humans and the wellbeing of the planet. This is, however, a farm ore political endeavor that challenges conventional notions of normal science, and it is one that global environmental change and human security research takes seriously.

Human security offers an alternative lens for looking at global environmental issues, and it captures aspects that have been generally overlooked in Earth system science research. It can be considered a critical project, as it implicitly (through its existence) or explicitly (through more active critique) questions the established order of knowledge about global environmental change. Much of the research on human security reviewed in this article has focused on understanding the conditions and processes that contribute to human insecurity. This includes attention to power, interests, multiple stressors, vulnerability, and response capacities. Moreover, human security research demonstrates that knowledge that is particular to certain contexts and situations can provide key insights into the nature of problems and on how to mobilize human agency to create alternatives to global environmental change.

15

CLIMATE CHANGES AND AGRICULTURE

15.1 CLIMATE CHANGE: A BLESSING OR A CURSE FOR AGRICULTURE?

Global climate change is a change in the long-term weather patterns that characterize the regions of the world. Scientists state unequivocally that the earth is warming. Natural climate variability alone cannot explain this trend. Human activities, especially the burning of coal and oil, have warmed the earth by dramatically increasing the concentrations of heat-trapping gases in the atmosphere. The more of these gasses' humans put into the atmosphere, the more the earth will warm in the decades and centuries ahead. The impacts of warming can already be observed in many places, from rising sea levels to melting snow and ice to changing weather patterns. Climate change is already affecting ecosystems, freshwater supplies, and human health. Although climate change cannot be avoided entirely, the most severe impacts of climate change can be avoided by substantially reducing the number of heat-trapping gases released into the atmosphere. However, the time available for beginning serious action to avoid severe global consequences is growing short. Global Warming or climate change is a topic that increasingly occupies the attention of the world. Is it really happening? If so, how much of it is due to human activities? How far will it be possible to adapt to changes of climate? What action to combat it can or should we take? How much will it cost? Or is it already too late for useful action? Why carbon dioxide is both a blessing and a curse? Is climate change: a blessing or a curse for agriculture? This book sets out to provide answers to all these questions by providing the best and latest information available.

Carbon dioxide might be a greenhouse gas, but it's not necessarily bad for the planet. Without it, there'd be no plant life and no human life as we know it. It's only toxic in high concentrations. And now the most important question is: is the climate changing? It could be answered this question as follows: It seems certain that the world will be even more crowded and more connected. Will the increasing scale of human activities affect the environment? In particular, will the world be warmer? How is its climate likely to change?

Variations in day-to-day weather are occurring all the time; they are very much part of our lives. The climate of a region is its average weather over a period that may be a few months, a season or a few years. Variations in climate are also very familiar to us. We describe summers as wet or dry, winters as mild, cold or stormy. In many parts of the world, no season is the same as the last or indeed the same as any previous season, nor will it be repeated in detail next time round. Most of these variations we take for granted; they add a lot of interest to our lives. Those we particularly notice are the extreme situations and the climate disasters. Most of the worst disasters in the world are, in fact, weather- or climate related.

Not all the climate changes will in the end be adverse. While some parts of the world experience more frequent or more severe droughts, floods or significant sea level rise, in other places crop yields may increase due to the fertilizing effect of carbon dioxide. Other places, perhaps for instance in the sub-arctic, may become more habitable. Even there, though, the likely rate of change will cause problems: large damage to buildings will occur in regions of melting permafrost, and trees in sub-arctic forests like trees elsewhere will not have time to adapt to new climatic regimes. Scientists are confident about the fact of global warming and climate change due to human activities. However, uncertainty remains about just how large the warming will be and what will be the patterns of change in different parts of the world. Although useful indications can be given, scientists cannot yet say in precise detail which regions will be most affected. Intensive research is needed to improve the confidence in scientific predictions.

Climate change is an increasingly urgent problem with potentially far-reaching consequences for life on earth. Humans and wildlife are also exposed to an array of chemical, physical, and biological stressors that arise largely from anthropogenic activity, but also from natural sources. One of the consequences of climate change that has recently attracted attention is its potential to alter the environmental distribution and biological effects of chemical toxicants. There is growing awareness of the importance of anticipating the effects of chemical pollution in the rapidly changing environment, and identifying and mitigating effects in those humans and ecosystems most vulnerable. Today, global climate change is a fact. The climate has changed visibly, tangibly, measurably. An additional increase in average temperatures is not only possible, but very probable, while human intervention in the natural climate system plays an important, if not decisive role). Climate change is a major concern in relation to he minerals sector and sustainable development. It is, potentially, one of the greatest of all threats to the environment, to biodiversity and ultimately to our quality of life. Climate on Earth has changed many times during the existence of our planet, ranging from the ice ages to periods of warmth.

During the last several decades increases in average air temperatures have been reported and associated effects on climate have been debated worldwide in a variety of forums. Due to its importance around the globe, agriculture was one of the first sectors to be studied in terms of potential impacts of climate change. According to studies carried out by the Intergovernmental Panel on Climate Change (IPCC), average air temperatures will increase between 1.4 and 5.8 °C by the end of this century, based upon modeling techniques that incorporated data from ocean and atmospheric behavior (IPCC, 2001). The possible impacts of this study, however, are uncertain since processes such as heat, carbon, and radiation exchange among different ecosystems are still under investigation. Less drastic estimates predict temperature increase rates of 0.088 °C per decade for this century. Other investigators forecast for the near future that rising air temperature could induce more frequent occurrence of extreme drought, flooding or heat waves than in the past. What climate changes are likely? Links between various elements of the climate system, illustrating how changes in any one can interact with others to produce positive or negative feedback effects. These and related phenomena can all interact to accelerate and reinforce climate change, as is illustrated in Figure 1. Climate change impacts are complex in that they can be both direct and indirect. For example, more rain may lead directly to either greater or smaller crop yields, depending on factors such as the type of crop, the soil and the present climate. Indirect effects could include changes in supply and demand as a result of these larger or smaller yields, both regionally and globally, and the resulting changes in commodity prices, the profitability of farming, and the affordability of food and effects on human health. Moreover, impacts can often be made more favorable by changing strategies so as to minimize losses and maximize gains. This is called adaptation. Studies of specific local impacts of climate changes have been conducted by hundreds of research groups, many from organizations concerned with such matters as seasonal crop forecasts, water supply and coastal protection. These groups have found that climate change and sea-level rise of the magnitude and rates suggested would greatly affect many natural systems like forests, rivers and wildlife, as well as human activities and society.

Examples include:

- Changes in natural productivity and biodiversity, with an increased rate of extinctions,
- Decreases in cereal crop yields in most tropical and sub-tropical countries, and in temperate countries for large warmings,
- Increased water shortages in many water-scarce regions due to regional decreases in precipitation, increased evaporation and loss of glaciers and seasonal snow storages,
- Adverse economic effects in many developing countries for even small warmings, and for developed countries for larger warmings,

- Tens of millions of people on small islands and low-lying coastal areas at severe risk of flooding from sea-level rise and storm surges,
- Increased threats to human health,
- Increased inequities between poor and richer countries,
- Increased risk of abrupt and irreversible climate changes. It is intended to answer, in readily understood terms, frequently asked questions about climate change, such as:
- What is the relationship between natural climate variations and human-induced climate change?
- What are the major concerns regarding climate change?
- Why are these arguments about the reality of climate change, and its policy implications?
- How does climate change relate to other problems like population growth, poverty, pollution and land degradation?
- How urgent is the problem? What can we about it, and how much will it cost?

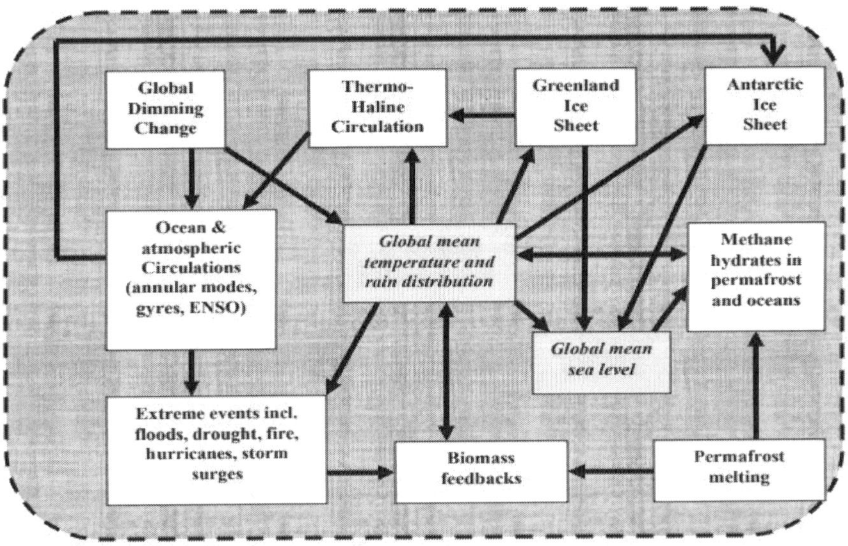

Fig. 1: Links between various elements of the climate system, illustrating how changes in any one can interact with others to produce positive or negative feedback effects

This chapter highlights the interactive effects of climate change, agriculture, and agricultural practices on global food security. The chapter focuses on global crop physiology and productivity, food production and safety, bioenergy and biodiversity and human rights and ethics on land use and soil management options that can enhance the SOC pool, thereby reducing the rate of enrichment of atmospheric CO_2 and improving food production by enhancing soil quality.

15.2 CLIMATE CHANGES AND AGRICULTURE

In fact, climate is a primary determinant of agricultural productivity. In turn, food and fiber production is essential for sustaining and enhancing human welfare. Hence, agriculture has been a major concern in the discussions on climate change. Food supply vulnerability to climate change is an issue in two different ways. First, future food supply may be directly threatened by climate change. Second, food supply capacity may be altered by efforts to reduce GHGE as society tries to mitigate future implications of climate change. Agronomic and economic impacts from climate change depend primarily on two factors: (1) the rate and magnitude of change in climate attributes and the agricultural effects of these changes, and (2) the ability of agricultural production to adapt to changing environmental conditions. Temperature, precipitation, atmospheric carbon dioxide content, the incidence of extreme events and sea level rise are the main climate change related drivers which impact agricultural production. The main agricultural productivity implications of these drivers are indicated in Table 1.

Table 1: Types of impacts on agricultural production and markets

Item subject to impact	Type of Effect				
	Tempera-ture	Precipita-tion	CO_2	Extreme Events	Sea Level
Crops and Forages Plant Size - yield S	X	X	X	X	
Water requirement	X		X		
Soil					
Soil Moisture	X	X		X	
Livestock	X	X			
Rate of gain	X			X	
Feed use	X			X	
Milk production	X			X	
Fertility	X			X	
Carrying Capacity	X	X	X	X	
Irrigation water supply					
Quanity	X	X		X	
Seasonality of supply	X			X	
Non agricultural competition	X	X		X	
Oter					
Navigation	X	X			X
Low lying land inundation				X	X
Weed Competition	X	X	X		
Insects, fungus, diseases	X	X			

15.3 INTRODUCTION

Climate is the single most important determinant of agricultural productivity, primarily through its effects on temperature and water regimes. For example, the physiographic boundaries of principal biomes are determined by mean annual temperature and soil water regimes. Climate change is therefore expected to alter the biophysical environment of growing crops and to influence biomass productivity and agronomic yields. Positive effects may be associated with the fertilization effects of CO2 enrichment, increases in the duration of growing seasons in higher latitudes and montane ecosystems, and possible increase in soil water availability in regions with an increase in annual precipitation. Each 1°C increase in temperature may lead to a 10-day increase in the growing season in northern Europe and Canada.

The CO2 fertilization effect is real. However, the net positive effect may be moderated by other factors, such as the effective rooting depth and nutrient availability. Further, the productivity per unit of available water is expected to rise by 20% to 40%. Negative effects of projected climate change on agriculture may be due to increases in respiration rate as temperature rises with attendant decreases in net primary productivity (NPP); increases in the incidence of pests and diseases; shortening of the growing period in some areas; decrease in water availability as rainfall patterns change; poor vernalization; and increased risks of soil degradation caused by erosion and possible decline in SOC concentration. The yield of rice has been estimated to decrease by 9% for each 1°C increase in temperature. Phillips et al. (1996), using the explicit planetary isentropic coordinate (EPIC) model to examine the sensitivity of corn and soybean yields to climate change, projected a 3% decrease in both corn and soybean yields in response to a 2°C increase in temperature from a baseline precipitation level. However, a 10% precipitation increase balanced the negative effect of a 2°C temperature increase. The effects of climate change on crop yields may be more negative at lower latitudes and generally positive at middle and high-middle latitudes. Further, crop growth is more affected by extremes of weather than by averages. The annual average changes in temperature or precipitation used in most predictive models do not reflect the short-term effects of so-called extreme events — droughts, floods, freezes, or heat waves. But the question must be asked: how remarkable are these extreme events that I have been listing? Do they point to a changing climate due to human activities? Here a note of caution must be sounded. The range of normal natural climate variation is large. Climate extremes are nothing new. Climate records are continually being broken. In fact, a month without a broken record somewhere would itself be something of a record! But what is global warming? Splendid book about the global warming, stated that, because carbon dioxide is a good absorber of heat radiation coming from the Earth's surface, increased carbon dioxide acts like a blanket over the surface, keeping it warmer than it would otherwise be. With the increased temperature the amount of water vapor in the atmosphere also increases, providing more blanketing and causing it to be even warmer. The gas methane is also increasing because of different human activities, for instance mining and agriculture, and adding to the problem. Being kept warmer may sound appealing to those of us who live in cool climates. However, an increase in global temperature will lead to global climate change. If the change were small and occurred slowly enough we would almost certainly be able to adapt to it. However, with rapid expansion taking place in the world's industry the change is unlikely to be either small or slow. Houghton presented that, in the absence of efforts to curb the rise in the emissions of carbon dioxide, the global average temperature will rise by about a third of a degree Celsius or more every ten years –or three or more degrees in a century. This may not sound very much, especially when it is compared with normal temperature variations from day to night or between one day and the next. But it is not the temperature at one place but the temperature averaged over the whole globe. The predicted rate of change of 3 °C a century is probably faster than the global average temperature has changed at any time over the past 10 000 years. And as there is a difference in global average temperature of only about five or six degrees between the coldest part of an ice age and the warm periods in between ice ages, we can see that a few degrees in this global average can represent a big change in climate. It is to this change and especially to the very rapid rate of change that many ecosystems and human communities (especially those in developing countries) will find it difficult to adapt.

15.4 CLIMATE PREDICTION AND AGRICULTURE

It is well established that, agricultural production is highly dependent on weather, climate and water availability, and is adversely affected by weather- and climate-related disasters. Failure of rains and occurrence of natural disasters such as floods and droughts could lead to crop failures, food insecurity, famine, loss of property and life, mass migration, and negative national economic growth. Hence, agricultural communities around the world have always looked for ways and means to cope with the climate variability including the use of various traditional indicators to predict the seasonal climate behavior. In past two decades, significant advances have been made in the science and applications of seasonal climate forecasting. The principal scientific basis of seasonal forecasting is founded on the premise that lower-boundary forcing, which evolves on a slower timescale than that of the weather systems themselves, can give rise to significant predictability of atmospheric developments. These boundary conditions include sea surface temperature (SST), sea-ice cover and temperature, land-surface temperature and albedo, soil moisture and snow cover, although they are not all believed to be generally of equal importance. Climate variations, also called anomalies, are differences in the state of the climate system from normal conditions (averaged over many years, usually a 30-year period) for that time of the year. The strongest evidence for long-term predictability comes largely from the influence of persistent SST anomalies on the atmospheric circulation which, in turn, induces seasonal climate anomalies. The key weather variables for crop prediction are rainfall, temperature and solar radiation, with humidity and wind speed playing also a role. As explained, seasonal climate forecasts are able to provide insight into the future climate evolution on timescales of seasons and longer because slowly-evolving variability in the oceans significantly influences variations in weather statistics. The climate forecast community is now capable of providing an end-to-end multi-scale (in space and time) integrated prediction system that provides skillful, useful predictions of variables with socio-economic interest. Also emphasize the importance of a fully probabilistic approach during all the stages of the forecasting process. Predictions of the climate system evolution in seasonal timescales suffer mainly from two sources of uncertainty: initial condition and structural model uncertainty. To address the first source of uncertainty, forecast models are run many times from slightly different initial conditions, consistent with the error to estimate the effect of this initial condition uncertainty. One way to represent model uncertainty is to incorporate, within the ensemble, independently derived models, resulting in a multi-model ensemble system (Palmer et al. 2004). For agriculture, climate forecasts must be interpreted in terms of production outcomes at the scale of decisions if farmers and other agricultural decisionmakers are to benefit. Interest in linking seasonal climate forecasts from general circulation models (GCMs) with crop models is motivated by: (a) the need for information that is directly relevant to decisions, (b) use for ex ante assessment of potential benefits to enhance credibility and support targeting, and (c) support for fostering and guiding management responses to advance climate information (Hansen, 2005). Several avenues are likely to enhance the quality of forecasts of agricultural impacts of climate variations over the next five to ten years. First, dynamically coupling crop models within climate models will support refined two-way interaction between the atmosphere and agricultural land use. Second, remote sensing and proliferation of spatial environmental databases provide substantial opportunities to expand the use and enhance the quality and resolution of climate based crop forecasts. Finally, climate-based crop forecasts will benefit from climate research in the emerging area of "weather within climate". The IPCC TAR (2001) provided a baseline for the prediction of climatic changes at broad scales by using historical measurements and future predictions made by several global circulation models (GCMs) in addition to what the first and second assessment reports had previously outlined, but included the definitions of more politically oriented scenarios (i.e. the Special Report on Emissions Scenarios (SRES) scenarios; IPCC, 2000) rather than the IS92 emission scenarios described in the first and second assessment reports. Working group I (WGI) reported:

- Average diurnal temperature had increased by 0.6°C in the 20th century, with a significant increase from 1910 to 1945, followed by a slight decrease during the period 1945–1965, and a severe increase from 1976 to 2000;
- Increases in sea level between 0.1 and 0.2 m;
- Geographically differentiated increases and decreases in precipitation of at least 1% per decade;

- Increases in frequency and intensity of heavy rainfall events, increase in cloud cover; and
- Reductions in low temperature extreme events and increases in high temperature extreme events such as *El Niño*-Southern Oscillation (ENSO)

The use of climate model results to assess the impacts of climate change on crop yields and agriculture in general began in the mid-1980s. For the first 10 to fifteen years of this research, the downscaling method used was the simple "delta" method based on results from doubled CO2 experiments. Initially, climate models did not include complete three-dimensional ocean models. Instead very simple "swamp" oceans were used, and trials were conducted for a control period, and then for a future period by doubling the concentrations of CO2 in the models. With the simple ocean models, the simulations of future climate would soon reach equilibrium (i.e., become stable). Fully-coupled atmosphere-ocean global climate models (GCMs) did not come into use until the 1990s and were then capable of producing time-evolving responses to increasing CO2 and other greenhouse gases. Changes in climate from one or more global climate models were appended to point observations of the variables needed in either deterministic or statistical crop yield models. In this earlier period, usually only results from a few doubled CO2 experiments were used, and the uncertainty was presented in the comparison of changes in yields based on the different future climates. Over time as the resolution of global climate models improved, more and more processes were included (three-dimensional oceans, land surface packages, explicit simulation of interactions with aerosols in the atmosphere), and longer simulations were produced. The availability of downscaling techniques further influenced the formation of climate scenarios for use in agricultural impacts. Early research on the effect of statistical and dynamical downscaling on calculations of climate effects on agriculture demonstrated that different changes in yield resulted from whether the large-scale or higher-resolution climate changes were used. As determined that the uncertainty in changes in crop yields was greater across the global models used in PRUDENCE (climate change model in Europe) than across the regional model simulated climate changes.

15.5 CLIMATE CHANGES AND ITS IMPACT ON AGRICULTURE

Driven mainly by population and economic growth, total world food consumption is expected to increase over 50% by 2030 and may double by 2050. Most of the increase in food production in the next decades is expected to occur through further intensification of current cropping systems rather than through opening of new land into agricultural production. Intensification of cropping systems has been a highly successful strategy for increasing food production. The best example is the well-known success of the Green Revolution, where the adoption of modern varieties, irrigation, fertilizers and agrochemicals resulted in dramatic increases in food production. However, this strategy also resulted in unexpected environmental consequences, one of them being the emissions of greenhouse gases (GHGs) into the atmosphere. Therefore, future strategies that promote further intensification of agriculture should aim at the development of sustainable cropping systems that not only consider increasing food production but that also look at minimizing environmental impact. At present, 40% of the Earth's land surface is managed for cropland and pasture. The most important cropping systems globally, in terms of meeting future food demand, are those based on the staple crops, rice, wheat and maize. Rice and maize are each grown on more than 155 million ha. In addition, rice is the staple food of the largest number of people on Earth. The geographic distribution of rice production gives particular significance to Asia where 90% of the world's rice is produced and consumed. Maize is produced mainly in the Americas, followed by Asia and then Africa. Maize is important as a staple crop (mainly in developing countries) but it is also important as animal feed and, increasingly, as biofuel. Wheat is the most widely grown crop, covering more than 215 million ha around the world, with Asia covering close to 50% of the world wheat area. Without additional policies, agricultural N2O and CH4 emissions are projected to increase by 35–60% and ~60%, respectively, to 2030, thus increasing more rapidly than the 14% increase of non-CO2 GHG observed from 1990 to 2005 (Barker et al. 2007). Improved agricultural management enhances resource-use efficiencies, often reducing emissions of more than one GHG. The effectiveness of these practices depends on factors such as climate, soil type and farming system. About 90% of the total mitigation arises from sink enhancement (soil C sequestration) and about 10% from emission reduction. The most prominent mitigation options in agriculture are shown in **Table 2.** In spite of inherent uncertainties in such estimates, it can be concluded that the topic of this review, which addresses the second option (improved cropland management) and the fifth option (improved rice management), comprises a sizable portion of the overall mitigation potential of agriculture.

Promoting agricultural practices that mitigate climate change by reducing GHG emissions is important, but those same practices also have to improve farmer production and income and buffer the production system against the effects of changes in climate. The overall impact predicted by climate change models vary but we are now locked into global warming and inevitable changes to climatic pattern that are likely to exacerbate existing rainfall variability and further increase the frequency of climatic extremes. Where excess rain occurs, extreme rainfall events will increase leading to flooding and soil erosion. In low rainfall, drought prone areas there is general acceptance in the science community of more frequent moisture stress because of failed rainfall patterns and increased evaporation caused by higher temperatures. In Africa specifically, the projected combined impacts of climate change and population growth suggest an alarming increase in water scarcity for many countries, with 22 of the 28 countries considered likely to face water scarcity or water stress by 2025. This in turn will curtail the ability of irrigated agriculture to respond to the expanding food requirements of tomorrow's Africa. This raises the spectre of a worsening food security crisis. In order to cope with the increased climate risk, agricultural systems will have to be more robust and resilient to buffer for extreme weather events such as drought, flooding, etc. It is paramount that new agricultural practices not only prevent further soil degradation but also improve system resilience through increased soil organic matter, improved water-use efficiency as well as nutrient use efficiency, and increased flora and fauna biodiversity. However, the management of agriculture to cope with GHG emissions and the negative effects of climate change on food production lies in the hands of farmers, pastoralists and forest managers whose decisions are determined by multiple goals.

Table 2: Assessing mitigation potentials in agriculture (adapted from Barker et al. 2007)

Mitiagtion option	Million t CO_2-eq/year[a]
Restoration of cultivated organic soils	1260
Improved cropland management (including agronomy, nutrient management, tillage/residue management) and water management (including irrigation and drainage) and set-aside/agroforestry	1110
Improved grazing land management (including grazing intensity, increased productivity, nutrient management, fire management and species introduction)	810
Restoration of degraded lands (using crosion, organic amendments and nutrient amendments)	690
Imporved rice management	210
Improved livestock management (including improved feeding practices, dietary additives, breeding and other structural changes) and improved manure management (imporoved storage and handling and anacrobic digestion)	260

The impacts of climate change on agriculture are expected to be widespread across the globe, although studies suggest that African agriculture is likely to be most affected due to heavy reliance on low-input rainfed agriculture and due to its low adaptive capacity. Broadly speaking, climate change is likely to impact crop productivity directly through changes in the growing environment, but also indirectly through shifts in the geography and prevalence of agricultural pests and diseases, associated impacts on soil fertility and biological function, and associated agricultural biodiversity. While many impact predictions tend towards the negative, increased CO2 will also contribute to enhanced fertilization – although there is significant debate as to the extent to which this may increase plant growth. This section looks at these issues, concentrating entirely on the expected biophysical impacts.

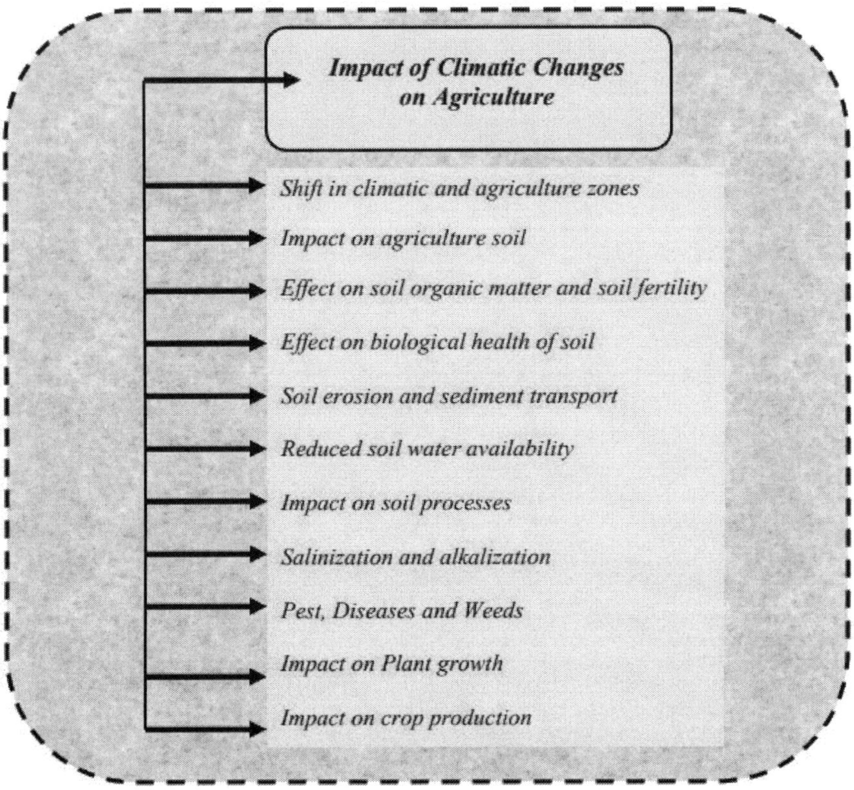

Fig. 2: Points that must be considered while doing the study of impacts of climatic changes on agriculture

Climate change due to anthropogenically-generated greenhouse gases and aerosols has been recognized as a serious threat to the earth's ecosystems and its inhabitants, and the dangers associated with climate change will increase in severity in coming decades in the absence of measures to curb the production of the responsible pollutants (e.g., carbon dioxide, methane, nitrous oxide). Numerous scientific articles and peer-reviewed reports have demonstrated the current and potential future effects of the climate change that result from these increased greenhouse gas emissions. Among the resource sectors that received early attention regarding possible climate change effects has been agriculture and it has continued to receive considerable attention since that early research. Work in this area has become more sophisticated over time and is now connected explicitly to estimates of economics in the agricultural sector and risk of hunger. As is the case with many impact areas, studies of possible adaptation to climate change have come to the fore and become increasingly important, and this is particularly striking for agriculture where studies of adaptation to climate change appeared early in the history of climate change research. The importance of adaptation studies has also put more emphasis on the need for more detailed information regarding future regional climate change.

15.5.1 Current Climate Change:

One of the most striking conclusions of the recent reports issued by IPCC (IPCC, 2007b) is that "most of the observed increase in global temperatures since the mid-20th century is very likely due to the observed increase in greenhouse gases in the atmosphere". The strength of this statement stands in stark contrast to the statement issued in the first IPCC report (IPCC, 1990) that "the observed increase (in temperature) could be largely due to natural variability". Global temperature has increased by about 0.75°C from 1860 to 2005.

Evidence for the attribution of climate change to anthropogenicallygenerated greenhouse gases has steadily increased. Not only has temperature on global and regional scales increased, but other aspects of climate have also changed. These include increases in the amount of water vapor in the atmosphere, increases in global sea level (1.8 mm/year since the early 1960s), decreases in the extent of Arctic sea ice (7.4% per decade in summer), retreat of most glaciers, changes in precipitation amounts (both increases and decreases depending on the region), increases and decreases in ocean salinity, increased ocean acidity, and increased frequency of heavy precipitation events over most land areas (IPCC, 2007b). These tendencies are consistent with what one would expect to see under conditions of increased concentrations of greenhouse gases (Mearns, 2011).

15.5.2 Future Climate:

Future climate (as well as some aspects of current and past climates) is studied mainly through the use of global climate models (GCMs). These are highly complex mathematical models that include representations of physical laws such as Newton's second law of motion, laws of conservation of mass and energy, laws of thermodynamics and the ideal gas law. The models represent important complex processes of the land surface, atmosphere, oceans and sea ice, as well as complex interactions among them (as shown in climate system figure). Over the past 30 years or so, these models have advanced continuously in their complexity, spatial resolution and length of simulations partially due to the rapid, nonlinear increase in computer power. In order to estimate future climate changes, projections of the anticipated pathways of emissions of greenhouse gases and aerosols are generally made. Greenhouse gases in the atmosphere, such as CO2, have very important effects on the energy balance of the earth-atmosphere system. They are relatively transparent to the energy coming from the sun, but they intercept the long wave energy that is emitted from the earth back out toward space. This is known as the greenhouse effect, and it is a natural part of the climate system. However, humans have been steadily increasing the amount of greenhouse gases emitted into the atmosphere. Thus, more and more energy are being trapped in the earth-atmosphere system. Estimating future emissions is a very difficult aspect of the climate change problem. In order to do so, one must also make projections of the socio-economic development of the entire world, including, for example, population growth, evolution of technology, the political and economic future, as well as whether we will manage to decrease (i.e., mitigate) the emissions of these gases. Under the auspices of the IPCC, a report was produced that presented a series of possible future pathways of greenhouse gases (i.e., CO2, CH4, and N2O) and aerosols based on different scenarios about how the world might develop. It should be noted that agriculture (and associated changes in land use) is currently an important contributor to greenhouse gas emissions. It produces about 25% of CO2 emissions (due to deforestation and oxidation of soil organic carbon), 50% of methane emissions (due to livestock and rice cultivation), and 75% of nitrous oxide emissions (primarily fertilizer application). In total, 40 different scenarios of emissions were produced, but these were grouped based on 4 different major story lines. The four-story lines varied based on two major axes of future world development: one axis concerned the degree to which environmental sustainability or protection dominated versus the drive for economic growth, and the other concerned whether the world maintained a primarily global economic and political perspective or if the world became more dominated by regionally-oriented concerns. Embedded within these four different scenarios are assumptions about different rates of population growth, dominance of particular forms of energy (e.g., coal and oil versus renewable resources), and certain aspects of land use. The four scenarios were referred to as A1, B1, A2, and B2. The A1 scenario was divided into three subtypes (A1B, A1FI, A1T), and the resulting six scenarios are collectively referred to as the "representative" scenarios. Table 3) summarizes some major characteristics of these scenarios.

Table 3: Characteristics of the representative SRES scenarios—values are for 2100

Scenario	Description	Population (billion)	Energy (share of coal, %)	CO_2 (GT/C/yr)	SO_2 (areosols) Mt S/yr)
AIFI	Rapid economic growth, rapid introduction of new technologies, but fossil fuel intensive	7.1	29	30.3	40
A2	Economic development regionally oriented-slow development of technology	15.1	53	28.9	60
AIB	Similar to AIFI, but more balanced energy portfolio	7.1	41	3.1	28
B2	Local solutions to economy, intermediate economic development, less rapid tech change compared to "Is"	10.4	22	13.8	48
B1	Service oriented economics, clean and offcient energy technologies	7.0	8	5.2	25
AIT	Similar to AIFI, but strong emphasis on clean technologies	7.0	1	4.3	20

15.6 CLIMATE CHANGE'S IMPACTS ON GLOBAL CROP PRODUCTIVITY

The major concerns for crop productivity as a result of increased levels of greenhouse gases are related to warmer temperatures and altered amounts and patterns of rainfall. Both average temperature and temperature variability are predicted to increase. Average global temperatures are predicted to increase by 0.6–2.5°C over the next 50 years with significant spatial variation. While this will permit cultivation of crops in areas of the world which are currently too cold (e.g. Siberia and northern America) and extend the potential growing season for others, it will also threaten the viability of crops in many of the major areas of production. Simulation models suggest that wheat yields in south-east Australia may decrease by about 29% and direct studies in the Philippines have shown that irrigated rice yields decrease by 10% for each 1°C increase in the minimum night-time temperature although the maximum temperature has no effect. Higher temperatures will shorten the life cycle of most crops, by accelerating development and hastening senescence, thereby decreasing the time available to harvest light and produce biomass. The effects on phenology vary both between species and with environment. Perennial crops may respond more strongly to an increased temperature than annual crops. Other effects such as drought or an increase in ozone concentrations can exacerbate these effects. The decreased time available to harvest light and produce biomass contributes to yield reductions at elevated temperatures. Our knowledge of water use is as poor as our knowledge of water resources perhaps poorer. Information is largely incomplete particularly for agriculture, the largest user and is lacking altogether for some countries. Only limited disaggregated information exists, and even this shows deficiencies of validity and homogeneity and provides extremely poor information on trends. The quality of information systems varies with each country, but there are common difficulties: (1) Statistics on the magnitude of demand and withdrawal are often estimated rather than based on data that are measured or collected from censuses. The level of uncertainty varies, but is particularly high for agriculture. (2) Sectors of use are not defined homogeneously and are not well disaggregated. (3) Adequate historical datasets are rare, and the dates of available statistics are not always explicit. (4) Lack of agreed terminology leads to discrepancies in data compilation and analyses. Agriculture is by far the main user of water. Irrigated agriculture accounts for 70% of water withdrawals, which can rise to more than 80% in some regions (Table 2.4). Although increasing in urbanized economies, industrial (including energy) use accounts for only 20% of total water use and domestic use for about 10%. Water withdrawals for energy generation, hydropower and thermocooling are on the rise, but energy is one of the economic sectors that consumes the least water and it returns most of the water withdrawn back to the water system (about 95%). This is only a partial picture of sectoral usage as there are many unaccounted-for uses. Little is known about water use in informal urban settlements or informal irrigation systems, both of which are generally unaccounted for in official statistics (Connor et al. 2009).

Nowadays, climate changes, their causes and consequences, gained importance in many other areas of interest for sustainable life on Earth. The subject is, however, controversial. Understanding how climate changes will impact mankind in the decades to come is of paramount importance for our survival. Temperature, carbon dioxide (CO_2) and ozone (O_3) directly and indirectly affect the production and quality of fruit and vegetable crops grown in different climates around the world. Temperature variation can directly affect crop photosynthesis, and a rise in global temperatures can be expected to have significant impact on postharvest quality by altering important quality parameters such as synthesis of sugars, organic acids, antioxidant compounds and firmness. Rising levels of CO_2 also contribute to global warming, by entrapping heat in the atmosphere. Prolonged exposure to concentrations could induce higher incidences of tuber malformation and increased levels of sugars in potato and diminished protein and mineral contents, leading to loss of nutritional and sensory quality. Increased levels of O_3 in the atmosphere can lead to detrimental effects on postharvest quality of fruit and vegetable crops. Elevated levels of O_3 can induce visual injury and physiological disorders in different species, as well as significant changes in dry matter, reducing sugars, citric and malic acid, among other important quality parameters.

Table 4: Water resources and withdrawals, 2000 (m^3 per year unless otherwise indicated.

Region	REnewable water resources	Total water withdrawals	Water withdrawals						Withdrawals as precent of renewable resources
			Agriculture		Industry		Domestic (urban)		
			Amount	Percent	Amount	Percent	Amount	Percent	
Africa	3,936	217	186	86	9	4	22	10	5.5
Asia	11,594	2,378	1,936	81	270	11	172	7	20.5
Latin	13,477	252	178	71	26	10	47	19	1.9
America									
Caribbean	93	13	9	69	1	8	3	23	14.0
North	6,253	525	203	39	252	48	70	13	8.4
America									
Oceania	1,703	26	18	73	3	12	5	19	1.5
Europe	6,603	418	132	32	223	53	63	15	6.3
World	43,659	3,829	2,663	70	784	20	382	10	8.8

Note: Water use refers to water that is being put to beneficial use by humans. Detailed water accounting, however, requires more precise definition of terms. Water withdrawal is the gross amount of water extracted from any source in the natural environment for human purposes. Differentiating withdrawals by type of source is useful to understand the pressure put on different parts of the system. Water demand is the volume of water needed for a given activity. If supply is unconstrained, water demand is equal to water withdrawal.

Besides increase in temperature and its associated effects, climate changes are also a consequence of alterations in the composition of gaseous constituents in the atmosphere.

CO_2, also known as the most important greenhouse gas, and O_3 concentrations in the atmosphere are changing during the last decade and are affecting many aspects of fruit and vegetable crops production around the globe. Exposure to elevated temperatures can cause morphological, anatomical, physiological, and, ultimately, biochemical changes in plant tissues and, as a consequence, can affect growth and development of different plant organs. These events can cause drastic reductions in commercial yield. However, by understanding plant tissues physiological responses to high temperatures, mechanisms of heat tolerances and possible strategies to improve yield, it is possible to predict reactions that will take place in the different steps of fruit and vegetable crops production, harvest and postharvest. Temperature increase and the effects of greenhouse gases are among the most important issues associated with climate change. Studies have shown that the production and quality of fresh fruit and vegetable crops can be directly and indirectly affected by high temperatures and exposure to elevated levels of carbon dioxide and ozone. Temperature increase affects photosynthesis directly, causing alterations in sugars, organic acids, and flavonoids contents, firmness and antioxidant activity. Higher temperatures can increase the capacity of air to absorb water vapor and, consequently, generate a higher demand for water. Higher evapotranspiration indices could lower or deplete the water reservoir in soils, creating water stress in plants during dry seasons. For example, water stress is of great concern in fruit production, because trees are not irrigated in many production areas around the world. It is well documented that water stress not only reduces crop productivity but also tends to accelerate fruit ripening. Carbon dioxide concentrations are increasing in the atmosphere during the last decades. The current atmospheric CO_2 concentration is higher than at any time in the past 420,000 years. Further increases due to anthropogenic activities have been predicted. Carbon dioxide concentrations are expected to be 100% higher in 2100 than the one observed at the pre-industrial era. Ozone concentration in the atmosphere is also increasing. Even low-levels of ozone in the vicinities of big cities can cause visible injuries to plant tissues as well as physiological alterations. Carbon dioxide accumulation in the atmosphere has directly effects on postharvest quality causing tuber malformation, occurrence of common scab, and changes in reducing sugars contents on potatoes. High concentrations of atmospheric ozone can potentially cause reduction in the photosynthetic process, growth and biomass accumulation. Ozone-enriched atmospheres increased vitamin C content and decreased emissions of volatile esters on strawberries. Tomatoes exposed to ozone concentrations ranging from 0.005 to 1.0 μ mol mol^{-1} had a transient increase in b-carotene, lutein and lycopene contents. The above mentioned climate changes can potentially cause postharvest quality alterations in fruit and vegetable crops. Although many researchers have addressed climate changes in the past and, in some cases, focused postharvest alterations, the information is not organized and available for postharvest physiologists and food scientists that are interested in better understanding how these changes will affect their area of expertise (Table 5). IPCC (2007) concluded that 'in mid-to high latitude regions, moderate warming benefits crop and pasture yields, but even slight warming decreases yields in seasonally dry and low-latitude regions (medium confidence)'. In IPCC language, moderate warming is in the range of 1–3°C. Smallholder and subsistence farmers, pastoralists and artisanal fisher-folk will suffer complex, localized impacts of climate change (high confidence). Food and forestry trades are projected to increase in response to climate change with increased dependence on food imports for most developing countries (medium to low confidence). The report further concluded that warming beyond 2–3°C was likely to result in yield declines in all areas. This analysis was based on a synthesis of 69 studies, which was a vast improvement on the handful of studies used in the TAR (IPCC, 2001). But even since the IPCC FAR (2007) there has been a much larger number of studies which examine the impacts of climate change on crop production and yields, including global multi-crop studies, down to regional and national studies on individual crops. This chapter summarizes the IPCC findings, and provides a more detailed analysis of impact studies arising from 2006 to 2009. There are fairly consistent pictures drawn by different studies that show the potential effects of changing climates. These all show steeply increasing trends in adverse impacts, particularly in food insecure regions among the tropics, which are likely to increase the extent to which these regions are food insecure, especially taking into account that most of these regions present the least adaptive capacity. Grain yields are expected to fall in developing countries; however, the opposite is likely to happen in developed countries. Geographies of changes may influence yield responses: in high latitudes (where most of the developed countries are located), increased temperatures could increase the duration of growing seasons, thus benefiting farmers. However, in developing countries, which are mostly located in the tropics, this effect would not be observed. Investment capacity within the different agricultural sectors needs to be considered if yield losses are to be offset. Moreover, yield reductions will certainly result in increases in prices of agricultural goods, and this impact will be greater for food insecure regions.

15.7 IMPACTS OF CLIMATE CHANGES ON CROP PHYSIOLOGY

Agriculture accounts for 70% of freshwater withdrawals from rivers, lakes and aquifers up to more than 90% in some developing countries. Furthermore, unlike in industrial and domestic uses, where most of the water returns to rivers after use, in agriculture a large part of water is consumed by evapotranspiration. Many irrigation systems, however, return a large amount of water to the system after use. Biomass cannot be produced without water. The source of all food is photosynthesis. Biomass is processed through the food chain, which describes the flow of energy and feeding relationship between species: from primary producers (plants) to herbivores to carnivores. It could be estimated how much water is needed to sustain our diets by calculating the water lost in evapotranspiration based on crop physiology. Depending on local climate, varieties and agronomical practices, it takes 400-2,000 litres of evapotranspiration daily to produce 1 kilogram (kg) of wheat, and 1,000 – 20,000 litres per kilogram of meat, depending on the type of animal, feed and management practices. Based on these values, researchers have estimated daily water requirements to support diets, ranging from 2,000 to 5,000 litres of water per person per day. FAO uses 2,800 kilocalories (kcal) per person at the national level as a threshold for food security. As a rule of thumb, it can therefore be estimated that 1 litre of water is needed to produce 1 kcal of food. Because of the low energy efficiency of the food chain, protein-rich diets require substantially more water than vegetarian diets (Table.6).

Table 5: Physiological and quality parameters of fruit and vegetable crops affected by exposure to increased CO_2 levels.

Physiological or quality parameter	Effect of high CO_2	Product
Photosynthesis	↑	Potato; spinach
Respiration	↓	Asparagus; broccli; mungean sprouts; blueberries; tomatoes; pears
	↑	Potatoes; lettuce, eggplants; lemons; cucumbers; mango
	=	Apple
Ripening	↓	Tomato
Stomtal conductance	↓	Spinach
Firmness	=	Tomato
	↑	Strawberry; raspberry
Color intensity	↑	Grape
Dry matter	↑	Potato
Starch	↑	Potato
Alcohol	↑	Grape; mango; pear
Titaratable Acidity	=	Grape
Citric Acid	↓	Potato; tomato
Malic Acid	↓	Potato; tomato
Ascorbic acid	↑	Potato; strawberry; orange; tomato

Over the past 800,000 years, atmospheric [CO_2] changed between 180 ppm (glacial periods) and 280 ppm (interglacial periods) as Earth moved between ice ages. From pre-industrial levels of 280 ppm, [CO_2] has increased steadily to 384 ppm in 2009, and mean temperature has increased by 0.76 °C over the same time period. Projections to the end of this century suggest that atmospheric [CO_2] will top 700 ppm or more, whereas global temperature will increase by 1.8–4.0 °C, depending on the greenhouse emission scenario. There is growing evidence suggesting that many crops, notably C_3 crops, may respond positively to increased atmospheric CO_2 in the absence of other stressful conditions, but the beneficial direct impact of elevated CO_2 can be offset by other effects of climate change, such as elevated temperatures, higher tropospheric ozone concentrations and altered patterns of precipitation.

It is now universally accepted that increased atmospheric concentrations of 'greenhouse gases' are the main cause of the ongoing climate change (Forster et al. 2007) and that these changes are expected to have important effects on different economic sectors (e.g.

Physiological or quality parameter	Effect of high CO_2	Product
Photosynthesis	↑	Potato; spinach
Respiration	↓	Asparagus; broccli; mungean sprouts; blueberries; tomatoes; pears
	↑	Potatoes; lettuce, eggplants; lemons; cucumbers; mango
	=	Apple
Ripening	↓	Tomato
Stomtal conductance	↓	Spinach
Firmness	=	Tomato
	↑	Strawberry; raspberry
Color intensity	↑	Grape
Dry matter	↑	Potato
Starch	↑	Potato
Alcohol	↑	Grape; mango; pear
Titaratable Acidity	=	Grape
Citric Acid	↓	Potato; tomato
Malic Acid	↓	Potato; tomato
Ascorbic acid	↑	Potato; strawberry; orange; tomato
Reducing sugars	↑	Potato
		Tomato
Total phenolics	↑	Grape; strawbery
Flavonoids	↑	Grape
Anthocyanins	↑	Grap; strawberry
Glycoalkaloids	↓	Potato
pH	=	Grape
Nitrate	↓	Potato; celery; leaf lettuce; Chinese cabbage
Volatile compuds	↓	Mango
Antioxidant capacity	↓	Scallion; strawberry

agriculture, forestry, energy consumptions, tourism, etc.) Since agricultural practices are climate dependent and yields vary from year to year depending on climate variability, the agricultural sector is particularly exposed to changes in climate. In Europe, the present climatic trend indicates that in the northern areas, climate change may primarily have positive effects through increases in productivity and in the range of species grown, while in southern areas (i.e. the Mediterranean basin) the disadvantages will predominate with lower harvestable yields, higher yield variability and a reduction in suitable areas for traditional crops.

Table.6: Value produced from a unit of water for selected commodities

Product	Water			
	Kilograms per cubic metre	Dollars per cubic metre	Protein grams per cubic metre	Kilocalories per cubic mtre
Cereal				
Wheat ($ 0.2 per kilogram)	0.20-1.2	0.04-0.24	50-150	660-4,000
Rice ($ 0.31 per kilogram)	0.15-1.6	0.05-0.18	12-50	500-2,000
Mazie ($ 0.11 per kilogram)	0.30-2.0	0.03-0.22	30-200	1,000-7,000
Legumes				
Lentils ($ 0.3 per kilogram)	0.3-1.0	0.09-0.30	90-150	1,60-3,500
Faba beans ($ 0.3 per kilogram)	0.3-0.8	0.09-0.24	100-150	1,260-3,360
Groundnut ($ 0.8 per kilogram)	0.1-0.4	0.8-0.32	30-120	800-3,200
Vegetables				
Potatoes ($ 0.1 per kilogram)	3-7	0.30-0.7	50-120	3,000-7,000
Tomatoes ($ 0.15 per kilogram)	5-20	0.75-0.30	50-200	1,000-4,000
Onions ($ 0.1 per kilogram	3-10	0.30-1.0	20-67	1,200-4,000
Fruits				
Apples ($ 0.8 per kilogram)	1.0-5.0	0.8-4.0	Negligible	520-2,600
Olives ($ 1.0 per kilogram)	1.0-3.0	1.0-3.0	10-30	1,150-3,450
Dates ($ 2.0 per kilogram)	0.4-0.8	0.8-1.6	8-16	1,120-2,240
Others				
Beef ($ 3.0 per kilogram)	0.03-0.1	0.09-0.3	10-30	60-210
Fish (auacultue[a])	0.05-1.0	0.07-1.35	17-340	85-1,750

For climate change impact assessment, crop growth models have been widely used to evaluate crop responses (development, growth and yield) by combining future climate conditions, obtained from General or Regional Circulation Models (GCMs and RCMs, respectively), with the simulation of CO_2 physiological effects, derived from crop experiments. Many of these impact studies were aimed at assessing crop development shifts and yield variations under changes in mean climate conditions. These analyses showed that increasing temperatures generally shortened the growing period of commercial crops, resulting in a shorter time for biomass accumulation. On the other hand, changes in yields were not homogeneous and dependent on crop phenology (e.g. summer and winter crops), crop type (e.g. C_3 and C_4 plants) or environmental conditions (water and nutrient availability).

Other studies stressed that changes in climate variability, as can be expected in a warmer climate, may have a more profound effect on yield than changes in mean climate. As such, policy analysis should not rely on scenarios of future climate involving only changes in means. Furthermore, the changes in the frequency of extreme climatic events during the more sensitive growth stages have been recognized as a major yield-determining factor for some regions in the future. Temperatures outside the range of those typically expected during the growing season may have severe consequences on crops, and when occurring during key development stages they may have a dramatic impact on final production, even in case of generally favorable weather conditions for the rest of the growing season. Many studies highlighted the potential of heat stresses during the anthesis stage as a yield reducing factor, while others pointed out that the joint probability of heat stress-anthesis is likely to increase in future scenarios. Accordingly, both changes in mean climate and climate variability (including extreme events) should be considered for a reliable climate change impact assessment in agriculture. An example is the summer heat wave of 2003, taken as an indicator of the future climate change, which reduced cereal production in Europe by 23 MT with respect to 2002. The reason for this reduction was attributed to the shorter growing season combined with a higher frequency of extreme events, both in terms of maximum temperatures and longer dry spells. In contrast, climate change impact assessments carried out so far have not included direct simulations of heat stress impact on crop yield resulting in a probable underestimation of yield losses.

The temperature response of crop growth and yield must be considered to predict the [CO_2] effects. The threshold developmental responses of crops to temperature are often well defined, changing direction over a narrow temperature. High temperatures reduce the net carbon gain in C_3 species by increasing photorespiration; by reducing photorespiration, [CO_2] enrichment is expected to increase photosynthesis more at high than at low temperatures, and thus at least partially offsetting the temperature effects of supraoptimal temperatures on yield. Therefore, yield increases at high [CO_2] should occur most frequently in regions where temperatures approximate the optimum for crop growth. Conversely, in regions where high temperatures already are severely limiting, further increases in temperature will depress crop yield regardless of changes in [CO_2]. In fact, results of mathematical modeling suggest that, in mid- to high-latitude regions, moderate to medium local increases in temperature (1–3 °C), along with associated CO2 increase and rainfall changes, can have beneficial impacts on crop yields, but in low-latitude regions even moderate temperature increases (1–2 °C) are likely to have negative impacts on yield of major cereals. Thus, climate change may impair food production, particularly in developing countries, most of which are located in tropical regions with warmer baseline climates.

In addition to crop growth and yield, crop quality is also expected to be affected by global climatic changes. Crop quality is thought to be a multi-faceted and complex subject involving growth, assimilate partitioning and storage, and pre- and post-harvest, including nutritional, technological and environmental facets. Elemental (e.g., zinc, iodine) and macromolecular (e.g., protein) composition in plant tissues are expected to change in a future high CO_2 world. In this context, crop physiologists will need to take more account of the interests of breeders and processors by studying, quantifying and modeling the differences not only in increasing yields but also in food quality among crop varieties and species in climate change scenarios. The efforts to understand the impact of elevated [CO_2], temperature and other ongoing climatic changes on food crops are crucial to estimate food production in the future. The present review, which is by no means exhaustive, is mainly focused on the current understanding of the consequences of climatic changes (mainly CO_2 enrichment and temperature) on crop physiology and chemistry.

Crops sense and respond directly to rising [CO_2] through photosynthesis and stomatal conductance, and this is the basis for the CO_2 fertilization effect on crop yield. These responses are highly dependent on temperature. Therefore, understanding how crop species will respond to these environmental changes is crucial for maximizing the potential benefits of elevated CO_2, for which agronomic practice needs to adapt as both temperature and CO_2 rise. In fact, both chemical and microbiological risks are foreseen to impair food and feed safety as a consequence of climate change: in particular, mycotoxins, pesticide residues, trace metals and other chemicals could affect food and feed safety. There is, therefore, an urgent need for scientific research that can improve our understanding of the interactions of rising atmospheric [CO_2] with other environmental variables, such as temperature, water supply and ozone concentration, as well as with biotic factors such as pests and diseases, under real field conditions. In doing so, it is necessary not only to quantify the effects of climatic changes on crop production but also on food quality. It is also necessary to assess responses of crops other than the key cereal grains, and in climate regions other than temperate ones, notably those of importance to developing countries in the tropics and subtropics. Furthermore, since distinct varieties seem to respond differently to elevated CO_2 and temperature in terms of harvestable yield, future research should be also directed towards selecting promising genotypes for a changing global climate.

16

IMPACTS OF CLIMATE CHANGES ON FOOD PRODUCTION AND SAFETY

16.1 INTRODUCTION

One-sixth of the world's human population has insufficient food to sustain life, and food supply will need to double by 2050 to meet this demand. Agricultural genetics is one of the components of the solution to meet this challenge. The most serious challenges economies and societies will face over the next decades include providing food and the water needed for food production, to a world that will see its population increase by a third in the face of mounting environmental stresses, worsened by the consequences of global climate change. The challenge of increasing food production in the face of climate change will be greatest for the production of the staple grain crops that form the basis of diets the world over. Wheat, maize and rice are the three major staples, covering together 40% of the global crop land of 1.4 billion ha. Together they provide 37% of all protein, and 44% of all calories for human consumption (Table1). Each crop provides more than 50% of the daily caloric uptake in regions with high consumption, for example North Africa and Central Asia for wheat, sub-Saharan African countries and mesoamerican countries for maize, and South and Eastern Asian countries for rice, and especially among the poorest people in these regions. Wheat is, with 220 million ha, the most widely grown crop followed by maize with 158 million ha and rice with 155 million ha. Average yield of maize, rice and wheat is 5, 3.9 and 3 t/ha, respectively. Although around 135 countries produce more than 10,000 t of maize compared with 100 countries that produce more than 10,000 t of wheat, wheat shows the widest geographical distribution because it is grown from Ecuador to 67°N in Scandinavia to 45°S in Argentina, Chile and New Zealand. Maize is grown from 55°N in Western Europe to 45°S in New Zealand. Rice is grown in a narrower geographic belt between 40°N in Japan and 30°S in Brazil, but is grown over a very wide range of hydrological environments within this area.

Table1: Percentage of calories and protein in the human diet obtained from wheat, maize and rice globally and in the developing world.

Grain crop	Region	Calories (%)	Protein (%)
Maize	World	5	4
	Developing counties	6	5
Wheat	World	19	20
	Developing counties	17	19
Rice	World	20	13
	Developing counties	25	18
Total from wheat, rice and maize	World	44	37
	Developing counties	48	42

There are many factors that affect food safety such as global trade, socioeconomic and technological development, urbanization and agricultural land use. Climate change and variability are among the multiple factors that can provoke changes in the nature and occurrence of food safety hazards. These hazards can arise at various stages of the food chain, from primary production to consumption, and climate change may have direct and indirect impacts on their occurrence. There are many pathways through which climate related factors may impact food safety including: changes in temperature and precipitation patterns, increased frequency and intensity of extreme weather events, ocean warming and acidification, and changes in the transport pathways of complex contaminants (Tirado et al. 2010). Climate change may affect agriculture and food security by altering the spatial and temporal distribution of rainfall, and the availability of water, land, capital, biodiversity and terrestrial resources. It may heighten uncertainties throughout the food chain, from farm to fork and yield to trade dynamics, and ultimately impact on the global economy, food security and the ability to feed nine billion people by 2050. Modelling by IIASA shows that future socioeconomic development and climate change may impact on regional and global irrigation requirements and thus on agricultural water withdrawals. Net irrigation requirements may increase by 45% by 2080. Even with improvements in irrigation efficiency, gross water withdrawals may increase by 20%. Global irrigation requirements with climate change will increase by 20% above the reference base case scenario (without climate change). The simulation shows that the global impacts of climate change on irrigation water requirements could be as large as the projected increase in irrigation due to socioeconomic development. Temperature increases and changes in rainfall patterns have an impact on the persistence and patterns of occurrence of bacteria, viruses, parasites and fungi and the patterns of their corresponding food borne diseases. Such changes also have an impact on microbial ecology and growth, plant and animal physiology and host susceptibility which may result in the emergence, redistribution and changes in the incidence and intensity of plant and animal diseases and pest infestations, all of which could impact food borne diseases and zoonoses (FAO, 2008a). In this context climate change and variability also pose a challenge to pest and diseases control measures such as Good Agriculture and Good Veterinary Practices with potential implications for the presence of chemical residues in the food chain. Extreme weather events such as floods and droughts may lead to contamination of soil, agricultural lands, water and food and animal feed with pathogens, chemicals and other hazardous substances, originating from sewage, agriculture and industrial settings. Emergency situations after natural disasters are of special concern for water and food sanitation. Ocean warming, and climate change related acidification and changes in ocean salinity and precipitation also affect the biochemical properties of water, along with water microflora, fisheries distribution, fish metabolic rates, and persistence and patterns of occurrence of pathogenic vibrios, harmful algal blooms and chemical contaminants in fish and shellfish.

The impacts of climate change on food production, prices and numbers at risk of hunger depend on a number of factors. These include regional climate change, biological effects of increasing atmospheric carbon dioxide, changes in floods, droughts and other extreme events, existing agricultural systems, adaptive capacity, changes in population, economic growth and technological innovation. A number of studies cited in the IPCC (2007) WGII report used the SRES family of scenarios of greenhouse gas emissions and socio-economic change. New conclusions reported by IPCC (2007) include that:

- Increases in frequency of extremes may lower crop yields beyond the impacts of average climate change,
- Impacts of climate change on irrigation water requirements may be large (as also are changes in irrigation water supply),
- Stabilization of CO_2 concentrations through reduced emissions reduces damage to crop production in the long term, despite loss of the physiological benefits of higher CO_2 concentrations,
- The effects of lowers regional and global impacts, and
- The magnitude of climate impacts will be small compared to that from different socioeconomic development paths; sun-saharan africa is likely to surpass asia as the most food-insecure region.

16.2 EFFECTS ON POSTHARVEST QUALITY OF FRUIT AND VEGETABLE CROPS

Climate on Earth has changed many times during the existence of our planet, ranging from the ice ages to periods of warmth. During the last several decades increases in average air temperatures have been reported and associated effects on climate have been debated worldwide in a variety of forums. Due to its importance around the globe, agriculture was one of the first sectors to be studied in terms of potential impacts of climate change. Many alternatives have been proposed to growers aimed at minimizing losses in yield. However, few studies have addressed changes in postharvest quality of fruits and vegetable crops associated with these alterations. Nowadays, climate changes, their causes and consequences, gained importance in many other areas of interest for sustainable life on Earth. According to studies carried out by the IPCC, average air temperatures will increase between 1.4 and 5.8 °C by the end of this century, based upon modeling techniques that incorporated data from ocean and atmospheric behavior. The possible impacts of this study, however, are uncertain since processes such as heat, carbon, and radiation exchange among different ecosystems are still under investigation. Less drastic estimates predict temperature increase rates of 0.088 °C per decade for this century. Other investigators forecast for the near future that rising air temperature could induce more frequent occurrence of extreme drought, flooding or heat waves than in the past (Moretti et al. 2010). Higher temperatures can increase the capacity of air to absorb water vapor and, consequently, generate a higher demand for water.

Higher evapotranspiration indices could lower or deplete the water reservoir in soils, creating water stress in plants during dry seasons. For example, water stress is of great concern in fruit production, because trees are not irrigated in many production areas around the world. It is well documented that water stress not only reduces crop productivity but also tends to accelerate fruit ripening (Henson, 2008). Exposure to elevated temperatures can cause morphological, anatomical, physiological, and, ultimately, biochemical changes in plant tissues and, as a consequence, can affect growth and development of different plant organs. These events can cause drastic reductions in commercial yield. However, by understanding plant tissues physiological responses to high temperatures, mechanisms of heat tolerances and possible strategies to improve yield, it is possible to predict reactions that will take place in the different steps of fruit and vegetable crops production, harvest and postharvest. Besides increase in temperature and its associated effects, climate changes are also a consequence of alterations in the composition of gaseous constituents in the atmosphere. Carbon dioxide (CO2), also known as the most important greenhouse gas, and ozone (O3) concentrations in the atmosphere are changing during the last decade and are affecting many aspects of fruit and vegetable crops production around the globe. Carbon dioxide concentrations are increasing in the atmosphere during the last decades. The current atmospheric CO2 concentration is higher than at any time in the past 420,000 years. Further increases due to anthropogenic activities have been predicted. Carbon dioxide concentrations are expected to be 100% higher in 2100 than the one observed at the pre-industrial era. Ozone concentration in the atmosphere is also increasing. Even low-levels of ozone in the vicinities of big cities can cause visible injuries to plant tissues as well as physiological alterations. It well documented that, climate changes can potentially cause postharvest quality alterations in fruit and vegetable crops. Although many researchers have addressed climate changes in the past and, in some cases, focused postharvest alterations, the information is not organized and available for postharvest physiologists and food scientists that are interested in better understanding how these changes will affect their area of expertise. Reviewed changes in ambient temperature and levels of carbon dioxide and ozone can potentially impact the postharvest quality of fruit and vegetable crops. After crops are harvested, respiration is the major process to be controlled. Postharvest physiologists and food scientists do not have many options to interfere with the respiratory process of harvested commodities, since they are largely dependent on the product specific characteristics. In order to minimize undesirable changes in quality parameters during the postharvest period, growers and entrepreneurs can adopt a series of techniques to extend the shelf-life of perishable plant products.

Postharvest technology comprises different methods of harvesting, packaging, rapid cooling, storage under refrigeration as well as modified (MA) and controlled (CA) atmospheres and transportation under controlled conditions, among other important technologies. This set of strategies is of paramount importance to help growers all over the world to withstand the challenges that climate changes will impose throughout the next decades (Moretti et al. 2010). Temperature increases and the effects of greenhouse gases are among the most important issues associated with climate change. Studies have shown that the production and quality of fresh fruit and vegetable crops can be directly and indirectly affected by high temperatures and exposure to elevated levels of carbon dioxide and ozone. Temperature increase affects photosynthesis directly, causing alterations in sugars, organic acids, and flavonoids contents, firmness and antioxidant activity. Carbon dioxide accumulation in the atmosphere has directly effects on postharvest quality causing tuber malformation, occurrence of common scab, and changes in reducing sugars contents on potatoes. High concentrations of atmospheric ozone can potentially cause reduction in the photosynthetic process, growth and biomass accumulation. Ozone-enriched atmospheres increased vitamin C content and decreased emissions of volatile esters on strawberries. Tomatoes exposed to ozone concentrations ranging from 0.005 to 1.0 $\mu mol\ mol^{-1}$ had a transient increase in β-carotene, lutein and lycopene contents.

It could be summarized the most important atmospheric factors, which affecting on post-harvest fruit and vegetable crops as follows:

i. **Effects of temperature:** Understanding how climate changes will impact mankind in the decades to come is of paramount importance for our survival. Temperature, carbon dioxide and ozone directly and indirectly affect the production and quality of fruit and vegetable crops grown in different climates around the world. Temperature variation can directly affect crop photosynthesis, and a rise in global temperatures can be expected to have significant impact on postharvest quality by altering important quality parameters such as synthesis of sugars, organic acids, antioxidant compounds and firmness. Extensive work has been carried out for more than three decades focusing quality properties of fruit and vegetable crops exposed to high temperatures during growth and development.

- It is well known that; fruit and vegetable growth and development are influenced by different environmental factors.

- During their development, high temperatures can affect photosynthesis, respiration, aqueous relations and membrane stability as well as levels of plant hormones, primary and secondary metabolites.
- Seed germination can be reduced or even inhibited by high temperatures, depending on the species and stress level.
- Most of the temperature effects on plants are mediated by their effects on plant biochemistry.
- For plants that are subjected to water deficit, temperature is a physical facilitator for balancing sensible and latent heat exchange at the shoot, which is modulated by relative humidity and by wind. Most of the physiological processes go on normally in temperatures ranging from 0 °C to 40 °C
- A general temperature effect in plants involves the ratio between photosynthesis and respiration.
- For a high yield, not only photosynthesis should be high but also the ratio photosynthesis/respiration should be much higher than one.
- At temperatures around 15 °C, the above-mentioned ratio is usually higher than ten, explaining why many plants tend to grow better in temperate regions than in tropical ones.
- Higher than normal temperatures affect the photosynthetic process through the modulation of enzyme activity as well as the electron transport chain (Sage and Kubien, 2007). Additionally, in an indirect manner, higher temperatures can affect the photosynthetic process increasing leaf temperatures and, thus, defining the magnitude of the leaf-to-air vapor pressure difference (D), a key factor influencing stomatal.
- Photosynthetic activity is proportional to temperature variations. High temperatures can increase the rate of biochemical reactions catalyzed by different enzymes. However, above a certain temperature threshold, many enzymes lose their function, potentially changing plant tissue tolerance to heat stresses.
- Temperature is of paramount importance in the establishment of a harvest index. The higher the temperature during the growing season, the sooner the crop will mature.
- Flavor is affected by high temperatures. Apple fruits exposed to direct sunlight had a higher sugar content compared to those fruits grown on shaded sides. Grapes also had higher sugar content and lower levels of tartaric acid when grown under high temperatures.

- Fruit firmness is also affected by high temperature conditions during growth. 'Fuerte' avocados exposed to direct sunlight (35 °C) were 2.5 times firmer than those positioned on the shaded side (20 °C) of the tree. Changes in cell wall composition, cell number, and cell turgor properties were postulated as being associated with the observed phenomenon.
- Dry matter content is used as a harvest indicator for avocados due to its direct correlation with oil content, a key quality component.
- Thus, fruit and vegetable growers, packers and shippers must pay close attention to ambient temperatures during growth and development as well as maturity indices to assure harvest at the appropriate time.
- Mineral accumulation was also reported to be affected by high temperatures and/or direct sunlight.
- Antioxidants in fruit and vegetable crops can also be altered by exposure to high temperatures during the growing season. The investigators also observed that high temperature conditions significantly increased the levels of flavonoids and, consequently, antioxidant capacity.
- It was verified that higher temperatures tended to reduce vitamin content in fruit and vegetable crops.
- Exposure of fruit and vegetable crops to high temperatures can result in physiological disorders and other associated internal and external symptoms (Table 2.8).
- Exposure of tomato fruits to temperatures above 30 °C suppresses many of the parameters of normal fruit ripening including color development, softening, respiration rate and ethylene production.
- It is also well known that exposure of fruit to temperature extremes approaching 40 °C can induce metabolic disorders and facilitate fungal and bacterial invasion. Although symptoms of heat injury and disease incidence are easily observed at the end of storage, the incipient incidence of these disorders is often not recognized in time to effect corrective treatment.
- Practical effects of climate change have already been experienced in some parts of the globe.

ii. **Effects of carbon dioxide exposure:** The Earth's atmosphere consists basically of nitrogen (78.1%) and oxygen (20.9%), with argon (0.93%) and carbon dioxide (0.031%) comprising next most abundant gases. Nitrogen and oxygen are not considered to play a significant role in global warming because both gases are virtually transparent to terrestrial radiation. Rising levels of carbon dioxide also contribute to global warming, by entrapping heat in the atmosphere.

Prolonged exposure to CO_2 concentrations could induce higher incidences of tuber malformation and increased levels of sugars in potato and diminished protein and mineral contents, leading to loss of nutritional and sensory quality.

- The greenhouse effect is primarily a combination of the effects of water vapor, CO_2 and minute amounts of other gases (methane, nitrous oxide, and ozone) that absorb the radiation leaving the Earth's surface.
- The warming effect is explained by the fact that CO_2 and other gases absorb the Earth's infrared radiation, trapping heat. Since a significant part of all the energy emanated from Earth occurs in the form of infrared radiation, increased CO_2 concentrations mean that more energy will be retained in the atmosphere, contributing to global warming (Lloyd and Farquhar, 2008).
- Carbon dioxide concentrations in the atmosphere have increased approximately 35% from pre-industrial times to 2005 (IPCC, 2007a). Besides industrial activities, agriculture also contributes to the emission of greenhouse gases.
- Many papers published during the last decade have clearly associated global warming with the increase in carbon dioxide concentration in the atmosphere. Changes in CO_2 concentration in the atmosphere can alter plant tissues in terms of growth and physiological behavior.
- Many of these effects have been studied in detail for some vegetable crops. These studies concluded, in summary, that increased atmospheric CO_2 alters net photosynthesis, biomass production, sugars and organic acids contents, stomatal conductance, firmness, seed yield, light, water, and nutrient use efficiency and plant water potential.

iii. **(3) Effects of ozone exposure:** Ozone in the troposphere is the result of a series of photochemical reactions involving carbon monoxide (CO), methane (CH_4) and other hydrocarbons in the presence of nitrogen species ($NO + NO_2$). It forms during periods of high temperature and solar irradiation, normally during summer seasons. It is also formed, naturally during other seasons, reaching the peak of natural production in the spring. However, higher concentrations of atmospheric ozone are found during summer due to increase in nitrogen species and emission of volatile organic compounds. Increased levels of ozone in the atmosphere can lead to detrimental effects on postharvest quality of fruit and vegetable crops. Elevated levels of ozone can induce visual injury and physiological disorders in different species, as well as significant changes in dry matter, reducing sugars, citric and malic acid, among other important quality parameters.

- Concentrations are at maximum values in the late afternoon and at minimum values in the early morning hours, notably in industrialized cities and vicinities. The opposite phenomenon occurs at high latitude sites. Another potential source for increased levels of ozone in a certain region is via the movement by local winds or downdrafts from the stratosphere.

- The effects of ozone on vegetation have been studied both under laboratory and field experiments. Stomatal conductance and ambient concentrations are the most important factors associated with ozone uptake by plants.
- Ozone enters plant tissues through the stomates, causing direct cellular damage, especially in the palisade cells. The damage is probably due to changes in membrane permeability and may or may not result in visible injury, reduced growth and, ultimately, reduced yield.
- Visible injury symptoms of exposure to low ozone concentrations include changes in pigmentation, also known as bronzing, leaf chlorosis, and premature senescence.
- Since leafy vegetable crops are often grown in the vicinity of large metropolitan areas, it can be expected that increasing concentrations of ozone will result in increased yellowing of leaves. Leaf tissue stressed in this manner could affect the photosynthetic rate, production of biomass and, ultimately, postharvest quality in terms of overall appearance, color and flavor compounds (Table 3).
- Additionally, it is observed that ozone exposure causes reduction in photosynthesis and increased turnover of antioxidant systems.
- The review of the pertinent literature related to plant responses to ozone exposure reveals that there is considerable variation in species response. Greatest impacts in fruit and vegetable crops may occur from changes in carbon transport.
- Underground storage organs (e.g., roots, tubers, bulbs) normally accumulate carbon in the form of starch and sugars, both of which are important quality parameters for both fresh and processed crops. If carbon transport to these structures is restricted, there is great potential to lower quality in such important crops as potatoes, sweet potatoes, carrots, onions and garlic.
- Exposure of other crops to elevated concentrations of atmospheric ozone can induce external and internal disorders, which can occur simultaneously or independently. These physiological disorders can lower the postharvest quality of fruit and vegetable crops destined for both fresh market and processing by causing such symptoms as yellowing

(chlorosis) in leafy vegetables, alterations in starch and sugars contents of fruits and in underground organs.

- Decreased biomass production directly affects the size, appearance and other important visual quality parameters. Furthermore, impaired stomatal conductance due to ozone exposure can reduce root growth, affecting crops such as carrots, sweet potatoes and beet roots.

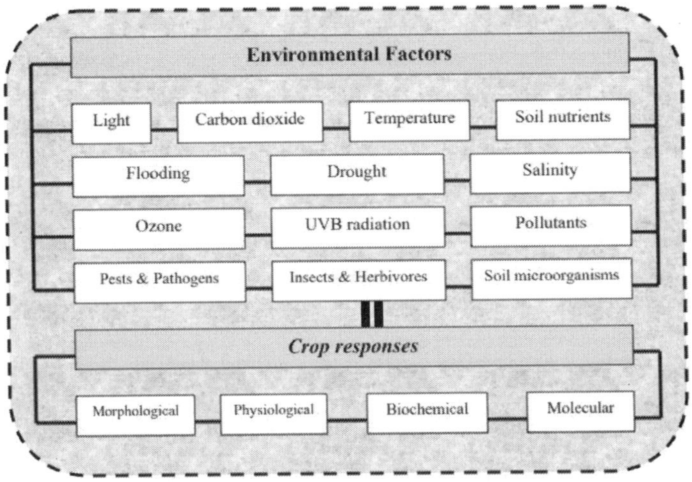

Fig. 1: : Crop responses to CO2, temperature and other environmental factors

Table 2.: Responses of some crop species to the stimulatory effects of elevated CO2 at high temperature

Crop	Environment	
Strong respnse		
Bean	Controlled-environment chamber (CEC)	
Peanut	Greenhouse (GH)	
Rice	Free-ait CO_2 enrichment (SPAR)	
Soybean	Growth chamber (GC)	
Sweet potato	Greenhouse	
Wheat	Open-top chamber	
Weak Response		
Cotton	EGC-soil-plant-atmosphere-reserch unit	
Cowpea	Growth chamber	
Bean	Sunlit-controlled-environment chamber	
Peanut	EGC-soil-plant-atmosphere-reserch unit	
Rice	Greenhouse	
Soybean	Sunlit Controlled-environment chamber	
Wheat	Chamber-Greenhouse	

C, dhamber, OTC, open-top chamber; CEC, controlled-environment chamber; SPAR, soil-plant atmosphere-research unit; GC, growth chamber; FACE, free-air CO_2 enrichment; GH, greenhouse

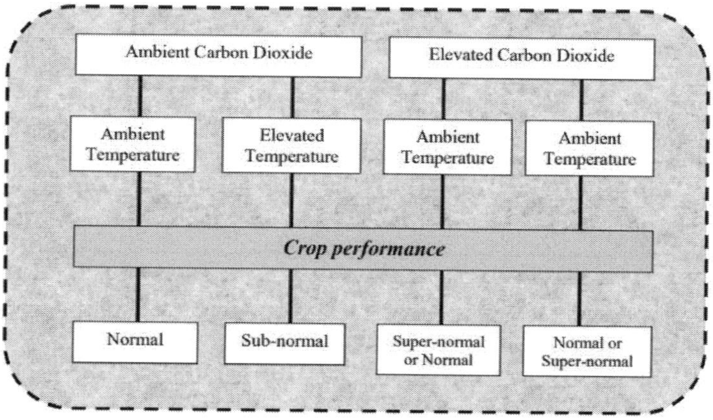

Fig. 2: Crop responses to the combined effects of CO_2 and temperature

17

MANAGING WEATHER AND CLIMATE RISKS IN AGRICULTURE

17.1 AN INTRODUCTION

In general, year to year deviations in the weather and occurrence of climatic anomalies/ extremes respect of the four seasons are old wave, fog, snow storms and avalanches, hailstorms, thunderstorms and dust storms, heat waves, tropical cyclones and tidal waves, floods, heavy rain and landslides, and droughts. Extreme weather events, and climatic anomalies, have major impacts on agriculture. Of the total annual crop losses in world agriculture, many are due to direct weather and climatic effects such as drought, flash floods, untimely rains, frost hail, and storms. High preparedness, prior knowledge of the timing and management of weather events and climatic anomalies and effective recovery plans will do much to reduce their impact on production levels, on land resources and on other assets such as structures and infrastructure and natural ecosystem that are integral to agricultural operations. Aspects of crop and livestock production, as well as agriculture's natural resource base, that are influenced be weather and climatic conditions include air and water pollution; soil erosion from wind or water; the incidence and effects of drought; crop growth; animal production; the incidence and extent of pests and diseases; the incidence, frequency, and extent of frost; the dangers of forest and bush fires; losses during storage and transport; and the safety and effectiveness of all on -farm operations. Although many ideas are available about ways to adapt to climate variability and change, few of these options have been assessed for their effectiveness under projected future climate conditions and for their potential interactions across sectors and with other stressors. Little attention has been given to the processes that decision makers might use to make appropriate adaptation decisions. In brief, it could be suggested that the adaptation process is fundamentally a risk management strategy. Managing risk in the context of adapting to climate change involves using the best available social and physical science to understand the likelihood of climate impacts and their associated consequences, then selecting and implementing the response options that seem most effective. Because knowledge about future impacts and the effectiveness of response options will evolve, policy decisions to manage the risk of climate change impacts can be improved if they are done in an iterative fashion by continually monitoring the progress and consequences of actions and modifying management practices based on learning and recognition of changing conditions. Using a risk-management approach, adaptation options for managing the risks associated with climate impacts can then be identified, evaluated, and implemented (**Figures 1,2 and 3**).

Adaptation to climate change requires attention now because impacts are already being felt across the world and further impacts are unavoidable, regardless of how immediately and stringently greenhouse gas (GHG) emissions are limited. Adaptation to climate variability is nothing new to humanity, but it now seems very likely that climate conditions by the later part of the 21st century will move outside the range of past human experiences. Therefore, historical records and past experience are becoming incomplete guides for the future, and adaptation to climate change needs to become a high national priority. Either we adapt by mobilizing to reduce sensitivities to climate change and to increase coping capacities now, or we will adapt by accepting and living with impacts that are likely in many cases to disrupt our lives and livelihoods. The questions are how, where, and when to adapt—and whether in some cases, if climate change is relatively severe, we may face limits on our ability to avoid painful impacts by adapting. Society and nature have always adjusted to climate variability and weather extremes, but climate change is moving climate conditions outside the range of past human experiences (IPCC, 2007b). While previous experience in coping with climate variability or extremes can provide some valuable lessons for adapting to climate change, there are important differences between coping with variability and planning for climate change. Climate change has the potential to bring about abrupt changes that push the climate system across thresholds, creating novel conditions. Likewise, thresholds in ecosystems and human systems could be crossed, potentially overwhelming their adaptive capacity.

1. Identify current and future climate changes relevant to the system,
2. Assess the vulnerabilities and risk to the system,
3. Develop an adaptation strategy using risk-based prioritization schemes,
4. Identify opportunities for co-benefits and synergies across sectors,
5. Implement adaptation options, and
6. Monitor and reevaluate implemented adaptation options (adapted from NRC, 2010).

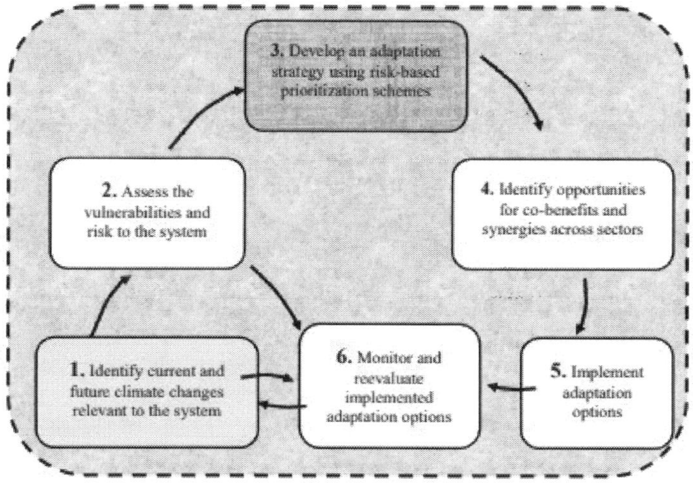

Fig. 1: The planning process is envisioned to incorporate the following steps:

Adaptation is intended to reduce climate change vulnerabilities and impacts. That means any consideration of adaptation planning must begin with consideration of risks associated with climate change vulnerabilities and impacts, to the extent that these can be anticipated. More specifically, adaptation includes

- The strategies, policies, and measures implemented to avoid, prepare for, and effectively respond to the adverse impacts of climate change on natural and human systems (to the extent that they can be anticipated), and
- The social, cultural, economic, geographic, ecological, and other factors that determine the vulnerability of places, systems, and populations. Climate-related changes can create new or interact with existing vulnerabilities to cause impacts, including changes in:
- Temperature, both averages and extremes;
- Precipitation, both averages and extremes;
- The intensity, frequency, duration, and/or location of extreme weather events;
- Sea level; and
- Atmospheric carbon dioxide (CO_2) concentrations

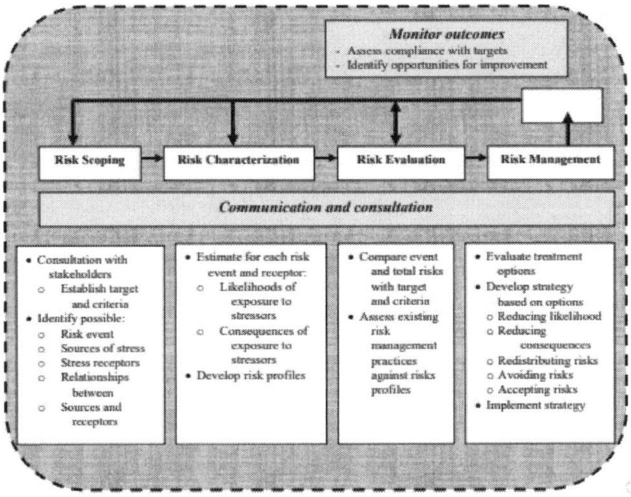

Fig. 2: Generic methodology for characterizing and managing risks

Vulnerability is often defined as the capacity to be harmed. It is a function of the character, magnitude, and rate of climate variation to which a system is exposed, its sensitivity, and its adaptive capacity. Vulnerabilities can be reduced by limiting the magnitude of climate change through actions to limit greenhouse gas (GHG) emissions, reducing sensitivity (the underlying social, cultural, economic, geographic, ecological, and other factors that interact with exposures to determine the magnitude and extent of impacts), or improving coping capacity (the ability to avoid, prepare for, and respond to an impact so that it is not seriously disruptive). Actions to reduce sensitivity and increase coping capacity are keys to effective adaptation to climate change.

Adaptation is fundamentally a risk-management strategy. Risk is a combination of the magnitude of the potential consequences and the likelihood that they will occur. Managing risk in the context of adapting to climate change involves using the best available social and physical science to understand the likelihood of climate impacts and their associated consequences, then selecting and implementing the response options that seem most effective. Because knowledge about future impacts and the effectiveness of response options will evolve, policy decisions to manage risk can be improved if they incorporate the concept of "adaptive management," which implies monitoring of progress in real time and changing management practices based on learning and as a recognition of changing conditions is incorporated. It is important to be clear at the start about the problem to be managed. Without a shared understanding of the nature of the problem, the desired goals of the stakeholders, and the "decision context", collective risk management is not likely to be successful. In framing the problem, it is important to include the perspectives of individuals and interested parties whose voices and concerns might not otherwise be heard, those who will assume the responsibility of administering and implementing the adaptation and sustaining it over time, and those involved in monitoring success or failure against stated goals and objectives. In coping with uncertainty, it will be particularly important to separate relevant signals from random noise in the observations, carefully analyze new scientific information, and design "midcourse" adjustments based on lessons learned.

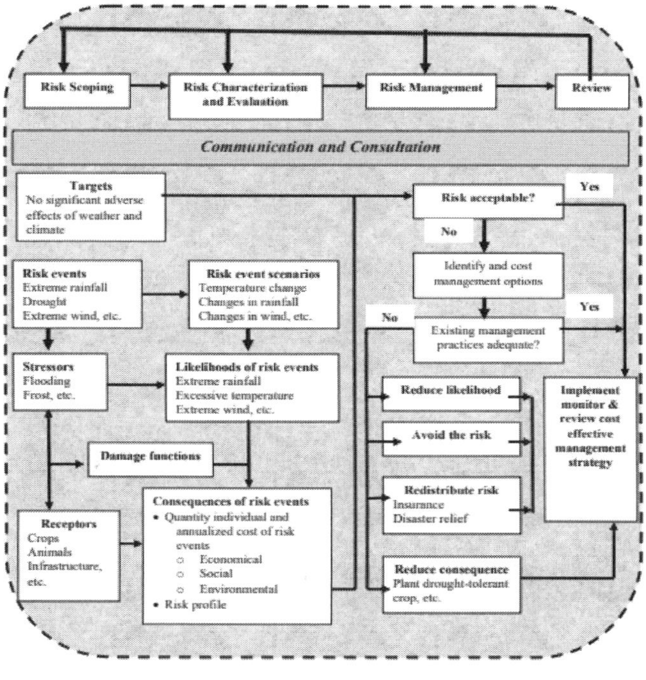

Fig. 3: Procedures for characterizing and managing risks and their application to the agricultural sector

17.2 AGRICULTURE AND GREENHOUSE GASES

One of the most important resources societies depends upon is agriculture. With so much irrefutable evidence demonstrating that global warming is indeed occurring and that ecosystems worldwide are already feeling the effects and having to adapt, the protection of agricultural resources is critical. Without the successful adaptation of agriculture and livestock to the long-term changes of climate due to global warming, future adaptation for anything would be difficult. Agricultural systems are influenced by several environmental factors—especially weather and climate. Agriculture and livestock depend on the health and wellbeing of soil conditions such as the presence and quality of organic matter and availability of adequate moisture. If the progression of global warming upsets the balance of any of these biophysical properties (precipitation, temperature, soil moisture, and organic matter), agriculture and livestock will be negatively affected. Global warming will have an effect on agriculture worldwide, but the outcomes will not be consistent. There may be positive effects in some areas and negative ones in others. As global warming influences temperatures and precipitation, it will play an important role in what types and amounts of crops can be produced. This will have a crucial effect on the world's food supply—what can be grown, how much, and where. Currently, much of the work done to understand global warming and determining how specific areas may change is through computer modeling. Modeling climate change is very complicated, however. Precipitation patterns and cloud formation and movement are currently very difficult to model because clouds are small-scale features that move and change rapidly; modeling is most successful on the mechanisms of climate that occupy larger scales and move slower, such as massive air masses like the global circulation patterns of jet streams, westerlies, and trade winds. Biophysical factors such as soil conditions and type and amount of organic matter help determine soil moisture and crop conditions. Climate change also reflects these ranges of conditions, which in turn affects crop production. It is expected that, agricultural production will experience a mixture of positive or negative effects that will depend on the geography and specific conditions of the area. The effect of increased temperature can have either a positive or negative effect, based on geography. For instance, for the colder high latitude locations such as Alaska and Canada, warmer temperatures would enable the existence of longer growing seasons. This could encourage the possibility of extended growing seasons as long as soil conditions are adequate. If soil conditions are not adequate and are not fertile enough (lacking proper nutrients and soil structure) to support the growth of crops, the length of the growing season will not matter because growing crops would not be successful. Adding large doses of fertilizer to increase fertility is also not a good option because they can have negative effects on the environment, such as being washed off into the drainage and entering and polluting the water cycle and biogeochemical systems. In more southern latitude locations, warmer temperatures would be beneficial if they eliminated the killing frosts that can currently destroy crops. This could potentially enable citrus crops to be grown farther north. The U.S. Department of Agriculture warns, however, that high temperatures could shorten the critical growing period interval, limiting the successful growth of large crop yields. Changes in precipitation may also have positive and negative effects. Increased precipitation in the traditionally arid areas such as the American Southwest would benefit them by enabling growing periods to exist. On the other hand, precipitation increase in areas where it results in flooding would be a disaster, causing erosion and loss of land and resources. In other geographical locations that experience a shortfall of precipitation, drought and water shortage would cause havoc with agricultural production. Accounting for greenhouse gas (GHG) emissions has emerged as an issue of considerable interest. While the scientific community is working for understanding of geochemical cycles, public policy is aiming to limit and decrease emissions and thereby to mitigate global climate change. The issues of monitoring and verification of international or sub-national commitments to reducing emissions are receiving increasing attention. Increasing levels of CO_2 in the atmosphere usually increases the net photosynthetic rates in plants, as well as reduces transpiration loss. Because the photosynthesis rate is enhanced in plants as atmospheric CO_2 increases, vegetation productivity rises. The specific reaction to increased CO_2 to each plant varies by plant species. It is well known that, the main commercial crops grown in the United States are wheat, rice, potatoes, barley, oats, and a wide variety of vegetables, which typically respond well to an increase in CO_2. With a modeled increase of a doubled level of CO_2, crops in the United States would be expected to generally increase quantities by 15 to 20 percent. On the other hand, the tropical vegetation species—or warm weather crops—such as sugarcane, sorghum, corn, and tropical grasses are much less responsive and estimates of only 5 percent are expected. The difficult aspect of this modeling is that some scientists believe crop productivity is also significantly difficult to predict because CO_2 is also influenced by other things such as nutrients and water availability, which makes it more difficult to predict the future. In computer modeling done by the IPCC, the estimated effects of global warming on agricultural resources will vary from crop to crop and geographic region to geographic region (Table 1).

Table 1: Estimated climate change effects on crop yields in Latin and North

Location of Sudy Site North America	
Canada	**Wheat:- 40% to +234%(results very due to geographic location)**
United Stated	**Corn:- 30% to -15%**
	Wheat:- 20% to -2%
	Soybean:- 40% to +15%
Latin America	
Argentina	**Corn:- 36% to 17%**
	Wheat: +3% to +48%
	Soybean:- 3% to 8%
Brazil	**Corn:- 25% to 2%**
	Wheat:- 50% to -15%
	Soybean:- 61% to -6%
Mexico	**Corn:- 61% to -6%**

General predictions have been made those areas that experience large temperature changes will experience more negative effects. Countries in the mid and high-latitude areas may experience an enhanced CO_2 effect accompanied by an increase in agricultural production; but yields in low-latitude countries such as Brazil may actually see a decline in agricultural production, especially with commodities such as corn and wheat, as drought becomes more likely. The commodity that seems to be the most resilient to climate change is rice. According to the IPCC, specific temperature ranges may also play an important role. For instance, areas that increase up to 3.6°F (2°C) are the areas that may experience positive crop yields. Areas that have an increase of 7.2°F (4°C), on the other hand, may actually see less production. Livestock are also greatly affected by global warming. They can be affected in two major ways:

- By the quantity of grazable land and whether there is enough to support the number of livestock on it;
- The quality of grazable land—the types and abundance of forage available for the health of the livestock living on the land. Global warming can seriously affect both of these variables.

The Pew Center on Global Climate Change has also identified some indirect effects of climate change on agriculture. One of the most significant is soil erosion and degradation. The effects of flooding, drought, and wind erosion can have devastating effects on crop yields. The positive and negative effects of global climate change as shown in Table 2.

17.3 STRATEGIES TO COMBAT THE IMPACT OF CLIMATIC CHANGES

The most important global debate of this century is on climate change. It is predicted that by 2050 there will be significant impacts including rising temperature, globally increased rainfall but increasing drought due to higher evaporation and changing rainfall distribution, and increased levels of CO_2 due to greenhouse and agriculture gas emissions. Climate change predictions over this century are for warmer (at least 1–2 °C) and drier conditions, with increased extreme weather events and increased CO_2 levels, in the regions where the principal temperate grain legumes of chickpea, lentil, faba bean and pea are mainly grown. However, rainfall may increase at higher and some tropical latitudes, and frost frequency may increase with drier conditions. Genetic adjustment of crops can help to mitigate the adverse effects on production, resulting from shorter crop cycles with warmer temperatures and reduced moisture, partly offset by growth response to increased CO_2. However, the negative impact of climate change will likely be far greater closer to the equator, in some of the world's poorest and most densely populated countries. Forecasts indicate that elevated CO_2 levels will have a fertilizing effect in some regions, although this will be negated by greater drought and heat stress in lower latitude areas.

Table 2: Positive and negative effects of global climate change

Positive Effects of Climate Change	Negative Effects of Climate Change
According to a report in National Geographic News (2002), as greehouse gas levels climb hight in the atmosphere, crop yields may increase i certain areas, such as fruits, vegetables, seeds, and grains. Although this may initially sound like a good thing, the positive effect is short-lived because it also results in a decrease in the crop's nutritional value	In fact, researcher Frank Mitloehner of the University of California has eight cows confined inside what he refers to as a "bio-bubble" (a big white tent) to answer an extermely important question: How much gas does a cow actually emit?

Positive Effects of Climate Change	Negative Effects of Climate Change
Peetr S. Curtis, an ecologist at Ohio State University, says, "But there's a trade-off between quantity and quality. While crops may be more productive, the resulting produce will be of lower nutritional value." Robert Mendelshon, an environmental economist at Yale University, says, "There is no doubt but that quality matters. If scientists can demonstrate a distinct loss of quality, this would be important and could change our impression of the global impact of climate change on agriculture from benign to harmful."	According to Mitloehner, one of the most surprising results of the experiment was that the most significant greenhouse has emissions were not a result of the manure. According to him, "We thought it was the waste that would lead to the majority of the [greenhouse gas] emissions, but it seemed to have been the animals. The major contributor to greenhouse gases was the ruminating process. During a cow's digestion process, as food eners the stomach, it mixes with bacteria, which breaks the food down and produces methane. Roughly 20 minutes later, the food returns to the cow's mouth as cud, which it chews, thereby releasing methane into the air. in addition to methane, it also release methanol and ethanol, as well as VOCs.
One of the key concepts Peter Curtis wanted to focus on was the overall reproductive traits of plants such as the number of seed and fruit, their size, and their nutritional quality because researchers had already confirmed that increasing CO_2 Levels did increase the growth of the plant's leaves, steams , and roots. The results of Cutis study showed that with a doubling of CO_2 total seed weight increased 25% the number of flowers increased 19% individual seed weight increased 4% and the number of seeds increased 16%. The study did have some unexpected results, however. According to Curtis, "The surprise is that nitrogen levels actually go down with elevated CO_2 levels, which reduces the nutritional value of these cases, nitrogen levels were 15-20% lower." Iraki Loladze, an ecologist at Princeton University, says, "The increase in crop productivity does not make up fo rthe fall in nutritional value of the crops-plants today provide 84% of the calories peopl eat worldwide and are also the source of major essential nutrients." He also noticed that levels of micrountrients (such as Fe, Zn, and 12) also dropped as CO_2 levels rose.	Another negative impact concerns manure management. The decomposition of animal waste in an anaerobic environment produces methane. According to the Environmental Protection Agency (EPA), manure storage and treatment systems are responsible for about 9% of the total methane emissions in the United Staes, and 31 percent of the methane emissions coming from the agricultural aector alone. Methane has an enormous greenhouse gas potential. It is about 21 times more effective at trapping heat than CO_2 In fact, 10% of the warming in teh United States caused by global warming is due to methane alone. More than 80 percent of the methane emission that originate from animal wases come from liquid-bassed manure management systems that include: anaerobic lagoons, holding tanks, and manure ponds. According to EPA, one solution that has been suggested to combat this problem is the use of anaerobic digester systes. An anaerobic digester is a container similar to a covered lagoon taht is designed to hold decomposing manure under warm, oxygen-free conditions that promote the growth of naturally occurring bactera. The pupose of the bacteria is to digest the manure, Producing methane and an effuent that farmers can use in place of untreated manure.

Positive Effects of Climate Change	Negative Effects of Climate Change
A study conducted by Ramakrishna Nemani (2003) at the University of Montana marked the first vegetation inventory and the effects of global warming as it relates to temperature and reciptation from a global perpective. They showed that global vegetaion had increased 6% from 1982 to 1999. In the Amazon region, Nemani says a major cause of vegetation growth is due to less cloud cover, resulting in increased solar radiation. The Amazon region accounts for 42% of the global increase.	The by-product (methane) produced by these digester is called "biogas," and it can now be used effectively as an energy source. Depending on how much they generate, they may also have the option to sell "extra electricity directly to local utility companies. Secondary benefits include a reduction of odors, methane emissions, and surface/groundwater contamination. Farmers can also sell the resulting high-quality ferilizer. These related business opportunities help the overall position of U.S. agribusiness.
The atmospheric scientist Dave Schimel, who works at the National Center for Atmoshpheric Research in Boulder, Colorado, keyed in on the impacts due to absence of cloud cover and its significance for increased vegetation According to Schimel, "Most studies of the effets of climate change have addressed temperature effets; some have also addressed water effects. But some of the most robust observed changes in climate have been in cloudiness and almost no studies have examined trends in solar radiation. So this is a really intersting new perpectibe."	Fertilizer management is another area of concern. Bassed on information collected by the EPA, the use of synthetic nitrogen and organic fertilizers has contributed to about 36% of total U.S. nitrous oxide (N_2O) emissions. Nitrous oxide is about 310 times more effective at trapping heat in the atmosphere than CO_2. While most N_2O is produced naturally through microbial proceses in the soil. it is the addition of syntheitc N_2O Introduced through fertlizers that has led to an increase in emissions from agricultural lands. According to the IPCC, if fertilizer applications are doubled, emissiosns of N_2O will double. Therefore, the IPCC recommended that N-fertilizer be used more efficently in order to reduce N_2O emissions.

To combat the impact of climate change in the years to come, so that world can survive, various approaches have been suggested differently worldwide. However, in the context of agriculture, solutions to climate change involve two aspects: adaptation which suggest how to maintain production under changed conditions, and mitigation which explain how to soften the impacts on the most vulnerable communities. Thus, various approaches have been mentioned to involve both the approaches. It could be summarized the strategies to combat the impact of climatic changes as follows:

i. Governmental policies Climate changes are occurring in all the continents with many countries having either documented changes or predicted them by climate modeling. Various ecological imbalances are being recorded and documented by national and international organizations around the world. In response to these changes and predictions governments have already initiated many policy decisions to combat the influence of climate change. If these initiatives are to be successful it is important that they receive high priority. For example, Japan, China, Australia, UK, Denmark, Bangladesh, Tunisia etc. have implemented policies aimed at mitigating the adverse impact of climate change. These government policies are essential otherwise it will be difficult to mitigate this impact globally.

ii. Reduction of CO_2 Emission Rising CO_2 levels are considered a precursor of climate change. CO_2 itself is the starting point for carbon fixation by the process of photosynthesis. It is well known from controlled-environment experiments that photosynthetic performance is enhanced by higher levels of CO_2 in the atmosphere because of the way the stomata operate. Field experiments have been set-up in several countries to examine the effects of elevated CO_2 during the cropping season and the approach is known as Free Air CO_2 enrichment (FACE). The general trend is that higher levels of CO_2 produce slightly higher yields, but different varieties may respond very differently. Also, the beneficial effects of higher CO_2 appear to be muted in drier environments (ICARDA, Caravan, 2008). The policies aimed at the development of emission reductions programmes and the provision of carbon offsets will create the biggest impact in tackling climate change, globally. Environmental and human needs mean real reductions need to happen now. However, it must be remembered that none of the IPCC (2007a, b) emission reduction scenarios allow us to avoid increases in global CO_2 levels. The successful implementation of CO_2 reduction emission needs following considerations around the world:

- Reducing large scale emissions in a business takes time, and offsetting alongside a reduction strategy has a greater immediate impact. Offsetting is the only way to deal with unavoidable large scale emissions and take responsibility for total emissions impact from day one. Many small scale point source emissions can be reduced without loss of productivity with existing technology (e.g. video link rather than driving) if there is a desire to do so.

- Developing a successful carbon market which helps allocate investment where it can make the greatest longer term impact. Businesses should support this. While there are still disagreements over the exact forms (cap and trade; baseline and credit) there is a need for an efficient global system.

- Funding the development of the new low carbon technologies which the world, especially the developing world, needs. This could create longer-term capacity, skills and knowledge, bringing wider benefits to local communities.

- Developing and promoting new management practices which reduce the carbon and nitrogen footprints by enhancing the long-term storage of carbon in managed land while simultaneously reducing the emission of the other greenhouse gases: nitrous oxide and methane.

Thus, it is important to halt the worsening of climate change effects by introducing limitations on greenhouse gas emissions at the international level by both the developing and developed world.

- Educating People and Strengthening National Organizationsn order to ensure successful mitigation of the impact of climate change on field crops in general and legume crops in particular, effective and well supported, institutional mechanisms should be established at national and international levels. Thus, effective means can be developed to take essential policy decisions around infrastructure development, resources mobilization, capacity building and implementation of technical programmes.
- Breeding Strategies and Development of Cultivars It is imperative to develop breeding approaches so new cultivars can withstand or take advantage of climate changes in respect of high and low temperature, moisture stress, increase CO2 levels, changes in seasonal temperature, light and rainfall distribution etc. To screen and identify resistant/tolerant donors it is essential to collect diverse germplasm lines including wild species and evaluate under the future probable environments. This could be achieved through the following ways:
- Focusing on dry lands and incorporation of drought Tolerance in crops
- Low and high temperature tolerance in crops
- Water use efficiency of crops
- Plant Modelling

Because climate change will result in altered growing conditions there is a need to gain as much information as possible in a short period, often for future environments. One year at a time experiments will not be the most efficient method for doing this, so there is a great need to use crop modeling to understand the future needs of the crop. Research of this nature is being carried out, however, there is a need to develop and apply increased use of plant modelling approaches that allow assessment of impacts of changes in climate, climatic variability and extreme weather events on pest, beneficial and endangered insects, arable plant communities, plant diseases and the sustainability of crop production.

- Integrated Production Management Technologies In the last 50 years, attempts to overcome biotic constraints of crop production have mainly focused on use of chemical pesticides and/or host- plant resistance. These single – factor management strategies to combat biotic and/or abiotic constraints have often been studied in isolation from each other. As a result, the yield losses caused by pests or diseases epidemics, along with poor agronomy, have remained alarming and significant. There is a greater opportunity to combine best technologies that combat insect-pests and disease with improved agronomical practices and emerge with integrated management packages. Integrated management packages for crops provide greater scope and need validation, upscaling and out-scaling with the involvement of farmers. Thus, combining of different technologies like desired agronomic management including weed and irrigation management, disease management, insect and pest integrated management, development and incorporation of high yielding resistant, widely adapted and quality cultivars are essential. Under changing climates, the implementation of integrated production management technologies is very important to sustain the productivity of crops globally. The understanding of regional impacts of climate change on crops is important in relation to knowing the predicted impact of climate change.

18

IMPACTS CLIMATE CHANGES ON PEST AND DISEASE PREVALENCE

18.1 INTRODUCTION

In a wonderful book about "Climate change and crop production" for Reynolds (2010) and in a chapter entitled "Scenarios of climate change within the context of agriculture" for Jarvis et al., they analyzed the effects of climate changes on agriculture and its scenarios. They stated that, global climatic change is also likely to impact agriculture through shifts in patterns of pests and diseases (organisms that range from weeds, certain herbivorous insects, arthropods and nematodes to fungi, bacteria and viruses). Rising temperatures and variations in precipitation, humidity and other abiotic factors are affecting the diversity and responsiveness of agricultural pests and diseases across diverse geographic ranges. Of all the factors that influence the productivity of agricultural pests and diseases, temperature is cited as the most important to insect ecology, epidemiology and distribution, while humidity and rainfall patterns and temperature are what define the responsiveness of plant pathogens.

There has been a remarkable scientific output on the topic of how climate change is likely to affect plant diseases. Climate change influences the occurrence, prevalence, and severity of plant diseases. Projected atmospheric and climate change will thus affect the interaction between crops and pathogens in multiple ways. This will also affect disease management with regard to timing, preference, and efficacy of chemical, physical, and biological measures of control and their utilization within integrated pest management (IPM) strategies. Prediction of future requirements in disease management is of great interest for agro-industries, extension services, and practical farmers.

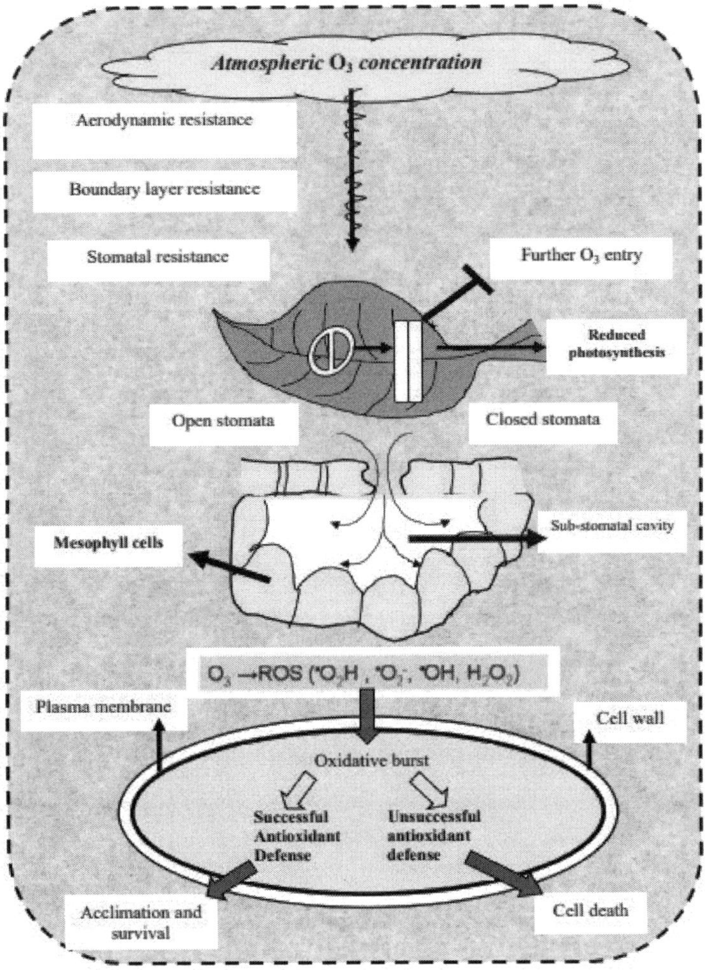

Fig. 1: Simplified scheme of O3 transfer, plant uptake and cell response

There are, however, few studies which quantify likely impacts on pest and disease prevalence due to the scarcity of system level studies that examine multitrophic complexities between causal and ancillary agents. For example, where crops are enriched by augmented atmospheric CO2 concentrations, pest/disease attacks are expected to negate optimization effects. Moreover, where positive impacts are highlighted (e.g. expanded range of cultivars, climatic fertilization enhancement), overriding negative consequences (in terms of biodiversity, yield, mitigation costs, etc.) are predicted to offset gains. While predictions are more certain on which specific pathogens and pests will thrive under greater variability in climate, what is difficult to say with certainty is what effects they will have on similarly climate stressed crops. Thus, a shroud of doubt still lingers as to the forecasting of climate warming on agriculture–plague interactions. This is because, aside from climate change effects, developing new pest species and the spread of existing ones are caused by:

- Natural expansion into unfilled ranges;
- Active dissemination on vehicles;
- Passive transport on traded plants and plant products; and
- Active flight (migrant species).

Numerous citations note the abundance and frequency of pests and diseases as likely to increase as local climates are adjusted outside their previously bounded norms, especially in situations where crops are moved to previously unsuitable areas. FAO (2008) cites Cannon (2008) who in a cursory report lists no less than 17 agricultural pests whose geographic coverage, species incidence and/or intensity will threaten to bring about impacts on agricultural production under climate change (Jarvis et al. 2010).

It is well established that, under climate change, characterized by increased temperatures and CO_2 levels, the fitness of plant herbivore pests is adjusting as their distributions and niches vary along with ambient conditions. In turn, their relationships with their natural enemies, phenologies (i.e. arrival and emergence times) and pressures from different pests and pathogens are noted in the scientific literature (Garrett et al. 2006). The physiological changes in plants growing under new extremes and farmers' adjusted management strategies will largely determine how these dynamics play out (i.e. it is difficult to say with certainty which groups of contaminants will increase or decrease and on which crops). Numerous studies are finding that herbivorous insect outbreaks are expected to increase in both frequency and intensity as global climate varies.

While the impacts of pests on yields and productivity are undoubtedly affected by a multitude of factors that are both biotic and abiotic, effective and proven biological controls in the form of pests' natural predators are not to beoverlooked as a factor that impacts agricultural potential. Climate change effects on herbivores and parasitoids of crops and cultivars and how disruptions in climate factors are adjusting fitness and competition for their natural enemies. Direct and indirect aspects of phenological modifications in plants are affecting the fecundity and abundance of herbivores, disadvantaging their natural predators. Increases in ambient CO_2 and temperature, and adjustments of humidity and precipitation rates are adjusting the availability of food resources for many pests, to their advantage and disadvantage, depending on the species involved.

In reference to biotic changes, climate alterations may affect microclimates around plants resulting in increased risk of infection from wetness and root diameter (Garrett et al. 2006). Moreover, elevated levels of CO_2 can bring about positive effects both indirectly (reduced expression of induced resistance in plants). Climate modifications affecting diseases are also linked to food safety concerns. For example the propensity for the spread of food borne pathogens from the greater temporal range of diseases during planting seasons has been noted (Gregory et al. 2009). It is well known that, preventing plant diseases has always been a major concern in agriculture and a cornerstone of breeding efforts to obtain higher yields. Although recent decades have seen major changes in ecosystems as a result of agriculture intensification, producing enough food for the growing population remains a major global challenge. A range of forces influence food systems and food security, but the global food supply needs to double by 2050, with the current world population of about 6.7 billion being projected to reach 9.5 billion by the mid-21st century (Borlaug, 2009). Human-induced climate change and increasing climate variability, resulting from the increase in the atmospheric concentration of greenhouse gases (GHGs), are recognized unequivocally. The fourth report of the Intergovernmental Panel on Climate Change (IPCC) established by the World Meteorological Organization (WMO) and the United Nations Environment Programme (UNEP) summarizes this evidence (IPPC, 2007). The most obvious effect is on the global mean temperature, which is expected to rise between 0.9 and 3.5°C by the year 2100. Cold days and nights and frost have become less frequent over most land areas, whereas hot days and nights are becoming more frequent. The melting of the ice caps and snow cover, resulting in rising sea levels, the variation in the frequency, timing and intensity of precipitation, leading to unusual floods mainly in coastal areas, and mid-term or severe drought in other regions, are well documented. Since plant diseases reduce crop performance and are considerably affected by environmental parameters, it is likely that major changes in eco-climatic conditions will lead to changes in plant disease frequency and severity, threatening the quantity and quality of agricultural products. Several reviews highlight the growing concern for the potential impact of climate change on plant diseases. New cropping practices, globalization and international trade have a rapid effect on the plant disease spectrum. Recent climatic changes undeniably observed worldwide give a new dimension to the evolution and distribution of plant pathogen populations resulting from crop intensification and long-term climate evolution. This makes it even more likely that plant disease evolution and its control will require increased emphasis in the future under changing climate scenarios. Although the epidemiology of many plant patho-systems is now better understood, it is difficult to separate climate change effects on the parameters affecting plant disease from normal seasonal variations. The effects of global climate change on plant diseases are subtle, progressive and difficult to document because of the scarcity of long-term data sets, resulting in uncertainty about possible future scenarios. Conclusions about a specific crop disease are often deduced from limited studies on one or a few specific physical variables (e.g. temperature, CO2 concentration and drought) conducted under controlled conditions, whereas multiple interactions occur in the context of climate change. Studying and understanding the drivers of change are essential if actions are to be implemented that prevent or reduce their impact.

18.2 MORE CO_2, INCREASE IN PESTS AND DISEASES

Elevated CO_2 can increase levels of simple sugars in leaves and lower their nitrogen content. These can increase the damage caused by many insects, who will consume more leaves to meet their metabolic requirements of nitrogen. Thus, any attack will be more severe. Higher temperatures from global warming, mainly due to elevated CO_2, will mean that more numbers of pests will survive the winter season. Elevated CO_2 will help in easier over-wintering of pathogens while higher temperatures will favour thermophilic fungi. Higher temperatures will lead to a poleward spread of many pests and diseases in both hemispheres. This will lead to more attacks over longer periods in the temperate climatic zone. Other possible effects of climate change need to be taken into account. On one hand, warmer temperature lowers the effectiveness of some pesticides but on the other hand, it favours insect carriers of many disease pathogens and natural enemies of pests and diseases. Thus, depending on the pest or pathogen, elevated CO_2 may act in a synergic or opposing manner with higher temperatures. Results of such interactions are difficult to be anticipated. Thus, one is obliged to wait for visual signs of appearance of a pest or disease for initiating action. Elevated CO_2 levels and higher temperatures will keep changing the composition and duration of infective stages of pests and diseases. The current agromet models for anticipation and control of crop pests and diseases will thus be ineffective.

When to initiate action?

As mentioned above, elevated carbon dioxide and higher temperature may act in a synergic or opposing manner depending on the pest or pathogen concerned. The result of these changes cannot be foreseen as yet and waiting for visual appearance of a pest or disease to initiate action, is the only remaining option. Organising manual surveys to cover all major crop pests and diseases will be a very costly and nearly impossible.

Under these circumstances, spore and insect trap data can assist in anticipating incidences of pests and diseases. A network spore and insect trapping centres to cover all major irrigated and rainfed crops and their pests and diseases needs to be set up. To ensure the network covers major irrigated crops with minimum stations, delineated homogenous rainfall zones and crop-climate zones need to be taken into account.

Summation of the mean air temperatures above a specified base value, known as accumulated degree-days, has been useful for anticipating the incidence of many crop pests and diseases. To overcome the unsuitability of calendar dates to begin calculations, the concept of Biofix is used. For a pest or a disease, Biofix date can mark the beginning of sustained trappimg of insects or disease spores.

It is crucial to measure the Economic Threshold Level (ETL)—the insect density at which control operations must begin to prevent crop loss from exceeding the cost of control operations. Economic Injury Level (EIL) is the insect density at which crop loss is more than the cost of control operations and varies with pests, crops and their growth stages. It is recommended to be taken as 75% of EIL.

Similarly in case of diseases, it is important to note the Critical Disease Level (CDL)—the point in time before which application of fungicides is not required and after which will be ineffective. Thus, for integrated management of pests and diseases using spore and insect trap data, it is necessary to laydown ETL and CDL criteria for important crop pests and diseases respectively.

Even within a season, there are variations in severity of pests and diseases. Thus, mid-seasonal advisories for resuming control action may be needed. In the case of some diseases, the above problem is sought to be overcome through the concept of Disease Severity Value, DSV. In this, after a specified initial period, DSVs are commenced to be accumulated as per a formula and each day a DSV is calculated. The ratings are accumulated to a threshold value that calls for initiation of action. After the action, DSV is set to zero and the accumulations commenced till they reach an assigned value calling for another spray and the process is repeated till end of the crop period. If adequate data is available, similar concept of Pest Severity Value, PSV can be developed. PSV values for important crop pests need to be developed. In perennial orchards the trees are attacked by more than one generation of a pest. Thus, after fixing the first Biofix date and criterion for initial spraying of pesticide, a watch must be kept for increase in number of trap catches to fix subsequent Biofix dates and spraying criteria

The increase in temperatures will be more at night than during daytime. Higher nocturnal temperature will reduce the duration of Leaf-Wetness and result in lesser disease incidence. Biological control of a pest or disease through introduction of their natural enemies from other regions will become more effective. Warm temperatures will favour their quick establishment and development.

Forewarnings based on current crop position and expected weather are most effective when there is a time lag between the onset of favourable conditions and the manifestation of pest or disease affliction. Secondly, an initial inoculum is detected in insect or spore traps and the organism's phenological development is amenable to calculation by the accumulated degree days approach above a base value. The effectiveness of forewarnings of pests and diseases is maximal over large mono-cropped areas and minimal in situations of variegated cropping.

18.3 EFFECT OF CLIMATE CHANGE ON PLANTS AND PLANT DISEASES

A major example for the devastating effects of climate change is floods caused by rising of sea level that can cause the disappearance of low-level lands and major crop losses. Another example is drought, where insufficiencies in water levels in the soil cause plants to lose their biological functions and even become more susceptible to diseases and pests. Climatic conditions contribute to the disease triangle, which involves the presence of a susceptible host, a pathogen, and suitable environmental conditions for infection to occur, and climate change affects environmental conditions whether it be in favor of the host or the pathogen. Examples of these conditions include dew, rain, relative humidity, temperature, aeration (wind), soil moisture, and sunlight intensity.

For any type of crop, the effect exerted by high temperature is highly dependent on the optimal growth and reproduction temperature of the crop. In certain regions, increased temperature may prove beneficial to the types of crops that usually grow there or permit farmers to switch to planting crops that grow in warmer areas. However, that is not always the case; if the higher temperature exceeds a crop's optimum temperature yields will decline, or worse, appearance and infestation of pathogens might occur. Crop yield can be affected by high levels of CO_2. A few laboratory experiments showed that high levels of CO_2 could positively affect growth. However, certain variables such as varying temperatures, water, ozone, and low nutrient levels may oppose these possible increases in yield. For instance, if the temperature is higher than a crop's optimal temperature requirement, if there are insufficiencies in water and nutrients, increase in yield may be low. Increased levels of CO_2 are linked to lower nitrogen and protein content in soybean and alfalfa plants, which results in a great quality reduction. Reduced forage and grain quality can reduce the ability of rangeland and pasture to support livestock which rely on grazing. Although rising CO_2 can stimulate plant growth, it also lowers the nutritional value of the majority of food crops. Rising levels of atmospheric carbon dioxide directly affect the concentrations of protein and essential minerals by reducing their content in a variety of plant species, which include rice, soybeans, and wheat. Therefore, the effect of rising CO_2 on the crops' nutritional value is considered a possible and indirect threat to human health as well. Moreover, due to the increased use and lowered efficiency of pesticides due to development of pest resistance, human health is additionally threatened by pesticide use as well as their residual toxicity in humans. More extreme temperatures and precipitation might decrease growth in certain crops. As previously mentioned, extreme events such as droughts and floods can decrease yield and damage crops. For instance, increased evening temperatures affected corn yield throughout the US Corn Belt. Additionally, premature budding due to a warmer winter instigated losses equivalent to 220$ million of cherries in Michigan in 2010 and 2012.

Moreover, drought has developed into a major problem in regions with increased summer temperatures as this causes dryness in soils. Even though higher irrigation may be possible in certain regions, water supplies may be also lowered in other locations, causing a lower availability of water for irrigation when more is needed. Many weeds, pests, and fungi thrive under warmer temperatures, wetter climates, and increased CO_2 levels. Currently, US farmers spend more than 11$ billion every year to control weeds, which compete with crops for nutritional resources. The ranges and distribution of weeds and pests are likely to increase with climate change. This could cause new problems for farmers' crops previously unexposed to these species.

Climate change parameters can have effects on both the host and the pathogen, for example, certain degrees of temperature promote pathogen growth, and certain temperatures can cause the host to have higher resistance to pathogenic infections. An example highlighting these events involves wheat and oats, which become more susceptible to rust diseases with increased temperature, while some forage species become more resistant]. Moreover, changes in temperature as limited as CO_2 changes could cause certain pests to undergo from 1 to 5 additional lifecycles per season, which increases the ability of the pests to overcome plant resistance.Certain mycotoxins such as fusarium mycotoxins (produced by *Fusarium* spp.) have increased concentrations at harvest due to high humidity and temperature. Humid conditions also increase proliferation of weeds, and weed biomass increases with increasing temperatures.

Certain parameters can have different effects depending on plant physiology, for example, increased CO_2 levels can cause a decrease in plant decomposition rates, which results in higher fungal inoculum levels, and these concentrations may induce the production of more fungal spores. On the other hand, high CO_2 concentrations may cause physiological changes to plants, causing them to acquire higher resistance to certain pathogens. Other extreme conditions may include low water levels and soil erosion which causes a decline in soil fertility and hence plant health.Fungicide activity is also a major determinant factor; climate change may highly affect fungicide efficiency. Highly frequent rainfalls greatly impact the efficiency of contact fungicides, as rain has the ability to sweep and eliminate contact fungicides from the hosts' surface, rendering them ineffective. However, plants with high metabolic rates have increased intake of fungicides and aren't highly affected by this parameter.

18.4 INTEGRATED PEST MANAGEMENT AND MITIGATING PEST MANAGEMENT

IPM stands for integrated pest management, according to the Food and Agricultural Organizations (FAO) in the United States, IPM is an ecosystem approach involving crop protection which combines different strategies and practices toward growing healthier crops and minimizing the use of pesticides to protect the environment. It is an analytical method used to analyze the agroecosystem and its different elements in order to optimally manage these elements to control and minimize pests while protecting the environment and the economic health. That is, the available methods of control (biological, cultural, chemical, and physical) should be considered and rationally applied by the farmers. However, IPM is more than just a tool and collection of control choices. It also comprises precaution techniques (which mainly include monitoring, prevention, early diagnosis, and forecasting) which assist in the control of pest populations as where data is collected, and preventive actions is recommended. A significant part in IPM techniques is proper decision-making for any interferences. Every decision made must be justifiable both ecologically and economically. Consequently, control programs which involve systematic application of chemicals which may harm the environment are unacceptable in IPM processes. As an alternative, precedence is given to alternative control techniques as well as preventive methods. IPM has been applied in various countries and areas that differ in their natural, social, and economic circumstances in addition to their levels of agricultural expansion. However, advancement in crop yield and safety may be realized in any existing conditions through the implementation of IPM. The practice of IPM is not a strict and simple form of submission to regulations and rules, but it rather involves taking actions with environmentally friendly approaches through principles and approaches that contribute to the reduction of the use of chemicals and increasing food security to achieve agricultural sustainability. In order to make IPM as effective as possible, it should be modified to local conditions.

18.5 DECISION SUPPORT SYSTEM (DSS)

In order to produce an efficient model, understanding the decision support system is essential. The decision support system is an informatics tool which uses mathematical models such as equations and statistics to help the decision-maker take action. The three main phases of a decision-making system include intelligence, design, and choice; two other subsequent phases include implementation of the decision and monitoring of its effect and outcomes. The decision-making process in IPM is highly complex and dynamic; it requires a high level of organization and constant update of operators; it requires the presence of databases and means to collect data and information as well as tools to handle data. The decision-making process generally provides the capability to identify when difficulties may occur and how to deal with these difficulties depending on data provided. The main properties of a successful DSS include ease of use, presentation format, system restrictiveness, decisional guidance, feedback, and interaction support:

18.5.1 Ease of use:

A DSS system is only beneficial if users perceive a DSS to be easy to use and that using it enhances their performance and productivity. The system should be easy to operate and interact with and requires minimal cognitive efforts; to sum up it should reduce mental effort and time consumed to analyze data and increase comfort of the user.

18.5.2 Presentation format:

The way information is presented through the program/system may influence the user's judgment/decision; therefore, the way information is presented through the decision support system should be focused on showing key data in an accurate and favorable format.

18.5.3 System restrictiveness and decisional guidance:

These refer to how much a DSS limits the options of the user and to which extent it guides the user toward a certain decision. A good DSS provides a reasonable amount of options within the scope of the topic/field at hand, as well as ample guidance to perform the optimal decision and achieve the most beneficial outcome.

18.5.4 Feedback:

The way the system provides messages and the wording of certain commands is important because they promote positive user experience and enhance the decision-making process.

18.5.5 Interaction support:

Interaction support means that users are permitted a particular level of interaction with a DSS. The DSS design is the determining factor on the presence as well as the level of interactivity between the user and the system. Users may have control over the system when a certain level of interaction support is present. The received control over the use of a system may have a motivational effect on its use. The system involves an integration of certain components and interacting factors with a common objective. In the case of pathogen monitoring, a pathogen monitoring system would basically receive input based on environmental factors and properties, and the output is expressed in the form of maps, information, and graphs. The system would then, based on information provided by threshold charts for pathogen favorable growth conditions, give out options that would determine the best possible course of actions in order to prevent disease outbreaks and crop loss. With this, an integrated model is illustrated. Although the systems are accurate, the main drawbacks include maintenance needs of meteorological stations, the requirement of different systems with different parameters depending on the type of plant, and disease as well as their growth conditions and application of these processes in farms which could be a difficult task due to old farming traditions.

18.5.6 Forecasting and Early Warning System:

Forecasting of pests and diseases appearance in plants is an additional application that demands a consistent and dependable stock of weather data. There are huge losses in yields due to pests and diseases prevalence which could have been controlled in several situations if the appropriate forecasting techniques were available. Consequently, the forecasting and early warning system, within the Plant Pathology Research Institute of the Agricultural Research Center, was established in the same timeframe of system and climate change applications in agriculture. Prediction of pest, disease prevalence, and progression based on weather data is extremely crucial when planning and implementing control measures. The idea of forecasting and early warning system has been introduced in 1926; with the appearance of computers and information technology, softwares were developed in order to produce convenient warning systems. Currently, early warning systems are being used to deal with pathogens of certain crops such as faba beans. The main concept involves the use of a mobile telemetry automated weather station system to monitor environmental conditions and softwares to interpret the input; based on the information provided, a decision can be made regarding the protection and maintenance of plant health.

19

EFFECTS OF CLIMATE CHANGE ON ANIMAL PRODUCTION AND LIVESTOCK

19.1 INTRODUCTION

- The IPCC report estimates a confidence level > 90% that there will be more frequent warm spells, heat waves and heavy rainfall and a confidence level > 66% that there will be an increase in drought, tropical cyclones and extreme high tides. The magnitude of the events will vary depending on the geographic zones of the World. The AR4 has been subjected to scientific criticism. It has been said that the report understates or overstates the dangers of climate change, and overstates the faults due to anthropogenic greenhouse gas concentrations. Nevertheless, the recognized scientists contributing to the IPCC report and the results of our analysis of the evolution of temperature–humidity index (THI) in the Mediterranean area in the last five decades of 20th century (data unpublished) convince us to follow the theory of global warming. The effects of global warming will not be adverse everywhere in the world.
- Despite the importance of livestock to poor people and the magnitude of the changes that are likely to befall livestock systems, the intersection of climate change and livestock in developing countries is a relatively neglected research area. Little is known about the interactions of climate and increasing climate variability with other drivers of change in livestock systems and in broader development trends. In many places in the tropics and subtropics, livestock systems are changing rapidly, and the spatial heterogeneity of household response to change may be very large. While opportunities may exist for some households to take advantage of more conducive rangeland and cropping conditions, for example, the changes projected will pose serious problems for many other households. Although, the

effects of global warming will not be adverse everywhere, a relevant increase of drought is expected across the world affecting forage and crop production. Hot environment impairs production (growth, meat and milk yield and quality, egg yield, weight, and quality) and reproductive performance, metabolic and health status, and immune response. The process of desertification will reduce the carrying capacity of rangelands and the buffering ability of agro-pastoral and pastoral systems. Other systems, such as mixed systems and industrial or landless livestock systems, could encounter several risk factors mainly due to the variability of grain availability and cost, and low adaptability of animal genotypes. Regarding livestock systems, it will be strategic to optimize productivity of crops and forage (mainly improving water and soil management), and to improve the ability of animals to cope with environmental stress by management and selection. To guide the evolution of livestock production systems under the increase of temperature and extreme events, better information is needed regarding biophysical and social vulnerability, and this must be integrated with agriculture and livestock components.

- A huge increase in the demand of animal production is expected in the next decades. Food and water security will be one of the other priorities for humankind in the 21st century. Over the same period the World will experience a change in the global climate that will cause shifts in local climate that will impact on local and global agriculture. As forecast a slight increase in crop productivity at mid to high latitude for an increased local mean temperature of 1–3 °C. Also in these areas, frosts, heat waves or heavy rainfall can cancel the advantages of the increase in temperature. A lower increase of temperature, around 1–2 °C, however, could worsen crop and cereal production at lower latitudes. The areas affected most will be in the boreal hemisphere, in particular North America, Northern Europe, Northern Asia and, at a lower latitude, the Mediterranean basin and West-Central Asia.
- Some countries in these areas have a very high farm animal density and animal production comes mostly from industrialized livestock systems, which rear highly selected pigs, poultry and dairy cows. The indirect effects of global warming such as soil infertility, water scarcity, grain yield and quality and diffusion of pathogens may impair animal production in these systems more than the direct effects. Indeed, in these systems the animals can cope better with the direct effects of high temperature, i.e., heat stress, with the help of diet, techniques of cooling or farm management. On the other hand, the employment of techniques to adapt air temperature of barns to the thermo-neutrality of the animals causes higher energy consumption and therefore, worsens global warming and increases general costs of animal production. Moreover, industrialized systems produce more manure than can be used as fertilizer on nearby cropland resulting in soil accumulation of phosphorous, nitrogen and other pollutants.

19.2 LIVESTOCK SYSTEMS IN DEVELOPING COUNTRIES

Livestock systems in developing countries are changing rapidly in response to a variety of drivers. Globally, human population is expected to increase from around 6.5 billion today to 9.2 billion by 2050. More than 1 billion of this increase will occur in Africa. Rapid urbanization is expected to continue in developing countries, and the global demand for livestock products will continue to increase significantly in the coming decades. The potential impact of these drivers of change on livestock systems and the resource-poor people who depend on them for their livelihoods is considerable. These impacts will be influenced by both supply-side shifts in natural resource use as well as marketed demand changes. Given the complexity of livestock (and in most cases crop–livestock) systems in developing countries, a mix of technological, policy and institutional innovations will be required. On the technology side, improvements will be linked to a combination of feed and nutrition, genetics and breeding, health and environmental management options, with different combinations appropriate to different systems. Significant changes in physical and biological systems have already occurred on all continents and in most oceans, and most of these changes are in the direction expected with warming temperature. For the future, there is considerable uncertainty, but recent "best estimates" of temperature increases from the IPCC in the Fourth Assessment Report (AR4) are in the range 1.8–4 °C in 2090–2099 relative to 1980–1999, depending on the scenario of future greenhouse-gas emissions that is used to drive the climate models. The impacts of temperature increase at even the lower end of this range will be far-reaching. At the lower end of the range of temperature rise (1–3 °C), global food production might actually increase but above this range would probably decrease. However, broad trends will be overshadowed by local differences, as the impacts of climate change are likely to be highly spatially variable. Climate change will alter the regional distribution of hungry people, with particularly large negative effects in sub-Saharan Africa. Smallholder and subsistence farmers, pastoralists and artisanal fisherfolk will suffer complex, localized impacts of climate change, due both to constrained adaptive capacity in many places and to the additional impacts of other climate related processes such as snow-pack decrease, particularly in the Indo-Gangetic Plain, and sea-level rise. Furthermore, changes in the frequency and severity of extreme climate events will have significant consequences for food production and food security; it is not only projected mean climate change that will have an impact. Increasing frequencies of heat stress, drought and flooding events are estimated to be likely, even though they cannot be modeled in any satisfactory way with current levels of understanding of climate systems, but these will have adverse effects on crop and livestock productivity over and above the impacts due to changes in mean variables alone.

The challenges for development are already considerable, and there is now general concern that climate change and increasing climate variability will compound these. However, there is only limited knowledge about the interactions of climate with other drivers of change in agricultural systems and on broader development trends. Such work is increasingly important for evaluating how farming systems may evolve in the future. It should be understood the likely impacts of climate change on vulnerable people through its effects in and on other sectors. These include impacts on water resources and other ecosystems goods and services, and human health and nutrition, for example. Enhanced understanding is needed of the likely impacts of climate change on the vulnerability of the resource-poor, so that resilience to current climate variability as well as to the risks associated with longer-term climate change can be gauged, and appropriate actions set in place to increase or restore resilience where this is threatened.

19.3 LIVESTOCK SYSTEMS BASED ON GRAZING AND THE MIXED FARMING SYSTEMS

It is expected that the livestock systems based on grazing and the mixed farming systems will be more affected by global warming than an industrialized system. This will be due to the negative effect of lower rainfall and more droughts on crops and on pasture growth and of the direct effects of high temperature and solar radiation on animals. These systems exist mainly in developing countries where the human demand for animal products is increasing due to the continuous growth in the population and per capita consumption. A loss of 25% of animal production by global warming is foreseen in these countries. A worse scenario is foreseen for Africa and some zones of Asia where extensive or pasture-based systems remain the norm. In any case at a World level, animal production has to increase in the next decades to satisfy the growing need. It is estimated that by 2050, global meat consumption will be twice that of today. How can we make animal production equal animal consumption in the next decades? The challenge will be how to better balance either the increase in the number of stock or the productivity per head, at the same time improving the sustainability of the livestock sector. This is an important task as today the billion land animals which are reared and slaughtered, either directly or indirectly contribute to total human induced greenhouse gas by 18% and total CO2 emission by 9%. The efficiency of water utilization will be another primary mission necessary to achieve sustainability of animal agriculture in expectation of increasing water scarcity and worsening quality. To save water means to grow plants and to rear animals in systems demanding less water. The question is where and under what conditions in the World will it be convenient to produce animal products? Animal production requires high volumes of water per unit of product. An example is beef that demands about 23 tons of water per kg of product. Animal agriculture will face hard challenges in many fields in the 21st century. Decision makers, research institutions and extension services have to support livestock activities to cope at best with the loss of production, worsening of animal products, enlargement of land desertification and the worsening of animal health under the effects of the climate change we expect in the next decades.

It could be analyzed some relevant effects of global warming on livestock production and on the forecast of the evolution of major livestock systems. It is not to foresee the global evolution of animal production and livestock systems under the effects of climate change owing to the enormous number of components under either direct or indirect climatic effects that shape livestock systems. It could be also in briefly reviewing some elements of the complex relationship between livestock and climate change in developing countries. Classification of livestock systems in developing countries and the impacts of climate change on livestock are be considered. Livestock have obvious impacts on water resources and biodiversity, as well as these things being affected by climate change and having impacts on livestock.

19.4 ANTHROPOGENIC CLIMATE CHANGE

Anthropogenic climate change has recently become a well-established fact and the resulting impact on the environment is already being observed. The greenhouse effect is a key mechanism of temperature regulation. Without it, the average temperature of the earth's surface would not be 15°C but -6°C. The earth returns energy received from the sun back to space by reflection of light and by emission of heat. A part of the heat flow is absorbed by so-called greenhouse gases, trapping it in the atmosphere. The principal greenhouse gases involved in this process include carbon dioxide (CO2), methane (CH4) nitrous oxide (N2O) and chlorofluorocarbons. Since the beginning of the industrial period anthropogenic emissions have led to an increase in concentrations of these gases in the atmosphere, resulting in global warming. The average temperature of the earth's surface has risen by 0.6 degrees Celsius since the late 1800s. Recent projections suggest that average temperature could increase by another 1.4 to 5.8 °C by 2100. Even under the most optimistic scenario, the increase in average temperatures will be larger than any century-long trend in the last 10 000 years of the present-day interglacial period. Ice-core based climate records allow comparison of the current situation with that of preceding interglacial periods. Global warming is expected to result in changes in weather patterns, including an increase in global precipitation and changes in the severity or frequency of extreme events such as severe storms, floods and droughts. Climate change is likely to have a significant impact on the environment. In general, the faster the changes, the greater will be the risk of damage exceeding our ability to cope with the consequences. Mean sea level is expected to rise by 9– 88 cm by 2100, causing flooding of low-lying areas and other damage. Climatic zones could shift poleward and uphill, disrupting forests, deserts, rangelands and other unmanaged ecosystems. As a result, many ecosystems will decline or become fragmented and individual species could become extinct. The levels and impacts of these changes will vary considerably by region. Societies will face new risks and pressures. Food security is unlikely to be threatened at the global level, but some regions are likely to suffer yield declines of major crops and some may experience food shortages and hunger. Water resources will be affected as precipitation and evaporation patterns change around the world. Physical infrastructure will be damaged, particularly by the rise in sea level and extreme weather events. Economic activities, human settlements, and human health will experience many direct and indirect effects. The poor and disadvantaged, and more generally the less advanced countries are the most vulnerable to the negative consequences of climate change because of their weak capacity to develop coping mechanisms.

19.5 LIVESTOCK GLOBALLY PLAY A CONSIDERABLE ROLE IN CLIMATE CHANGE,

Livestock globally play a considerable role in climate change, in terms of their contribution to greenhouse-gas emissions. Livestock activities have significant impact on virtually all aspects of the environment, including air and climate change, land and soil, water and biodiversity. The impact may be direct, through grazing for example, or indirect, such as the expansion of soybean production for feed replacing forests in South America. Livestock's impact on the environment is already huge, and it is growing and rapidly changing. Global demand for meat, milk and eggs is fast increasing, driven by rising incomes, growing populations and urbanization. As an economic activity, livestock production is technically extremely diverse. In countries or areas where there is no strong demand for food products of animal origin, subsistence and low in put production prevails, mainly for subsistence rather than for commercial purposes. This contrasts with commercial, high-input production in areas serving a growing or established high demand. Such diverse production systems make extremely diverse claims on resources. The diversity of production systems and interactions makes the analysis of the livestock–environment interface complex and sometimes controversial. The livestock sector affects a vast range of natural resources, and must be carefully managed given the increasing scarcity of these resources and the opportunities that they represent for other sectors and activities. While intensive livestock production is booming in large emerging countries, there are still vast areas where extensive livestock production and its associated livelihoods persist. Both intensive and extensive production requires attention and intervention so that the livestock sector can have fewer negative and more positive impacts on national and global public goods. Global agriculture will face many challenges over the coming decades and climate change will complicate these. A warming of more than 2.5°C could reduce global food supplies and contribute to higher food prices. The impact on crop yields and productivity will vary considerably. Some agricultural regions, especially in the tropics and subtropics, will be threatened by climate change, while others, mainly in temperate or higher latitudes, may benefit. The livestock sector will also be affected. Livestock products would become costlier if agricultural disruption leads to higher grain prices. In general, intensively managed livestock systems will be easier to adapt to climate change than will crop systems. Pastoral systems may not adapt so readily. Pastoral communities tend to adopt new methods and technologies more slowly, and livestock depend on the productivity and quality of rangelands, some of which may be adversely affected by climate change. In addition, extensive livestock systems are more susceptible to changes in the severity and distribution of livestock diseases and parasites, which may result from global warming.

19.6 THE UNITED NATIONS FRAMEWORK CONVENTION ON CLIMATE CHANGE (UNFCCC)

As the human origin of the greenhouse effect became clear, and the gas emitting factors were identified, international mechanisms were created to help understand and address the issue. The United Nations Framework Convention on Climate Change (UNFCCC) started a process of international negotiations in 1992 to specifically address the greenhouse effect. Its objective is to stabilize greenhouse gas concentrations in the atmosphere within an ecologically and economically acceptable timeframe. It also encourages research and monitoring of other possible environmental impacts, and of atmospheric chemistry. The direct warming impact is highest for carbon dioxide simply because its concentration and the emitted quantities are much higher than that of the other gases. Methane is the second most important greenhouse gas. Once emitted, methane remains in the atmosphere for approximately 9–15 years. Methane is about 21 times more effective in trapping heat in the atmosphere than carbon dioxide over a 100- year period. Atmospheric concentrations of CH_4 have increased by about 150 percent since pre-industrial times, although the rate of increase has been declining recently. It is emitted from a variety of natural and human-influenced sources. The latter include landfills, natural gas and petroleum systems, agricultural activities, coal mining, stationary and mobile combustion, wastewater treatment and certain industrial process (US-EPA, 2005). The IPCC has estimated that slightly more than half of the current CH_4 flux to the atmosphere is anthropogenic (IPCC, 2001b). Total global anthropogenic CH_4 is estimated to be 320 million tonnes CH_4/yr, i.e. 240 million tonnes of carbon per year. This total is comparable to the total from natural sources. Nitrous oxide, a third greenhouse gas with important direct warming potential, is present in the atmosphere in extremely small amounts. However, it is 296 times more effective than CO_2 in trapping heat and has a very long atmospheric lifetime (114 years) . Livestock activities emit considerable amounts of these three gases. Direct emissions from livestock come from the respiratory process of all animals in the form of carbon dioxide. Ruminants, and to a minor extent also monogastrics, emit methane as part of their digestive process, which involves microbial fermentation of fibrous feeds. Animal manure also emits gases such as CH_4, nitrous oxides, NH3 and CO_2, depending on the way they are produced (solid, liquid) and managed (collection, storage, spreading). Livestock also affect the carbon balance of land used for pasture or feed crops, and thus indirectly contribute to releasing large amounts of carbon into the atmosphere. The same happens when forest is cleared for pastures. In addition, greenhouse gases are emitted from fossil fuel used in the production process, from feed production to processing and marketing of livestock products. Some of the indirect effects are difficult to estimate, as land use related emissions vary widely, depending on biophysical factors as soil, vegetation and climate as well as on human practices.

19.7 CLASSIFICATION OF LIVESTOCK SYSTEMS IN THE DEVELOPING WORLD

It could be used the systems classification of Seré and Steinfeld (1996), whose methods were built on the concept of the agro-ecological zone. There are two parts to the classification. The agro-climatic part is based on the length of growing period (LGP), defined as the period in days during the year when the rainfed available soil moisture supply is greater than half the potential evapotranspiration (PET). It includes the period required to evapotranspire up to 100 mm of available moisture stored in the soil profile. Excluded are any time intervals with daily mean temperatures of less than 5 °C. Three categories are defined:

1. Arid/semi-arid, with a LGP of less than or equal to 180 days;

2. Humid/sub-humid, with a LGP greater than 180 days;

3. Tropical highlands/temperate. Temperate regions are defined as those with one month or more with monthly mean temperature, corrected to sea level, below 5 °C. Tropical highlands are defined as those areas with a daily mean temperature, during the growing period, of between 5 and 20 °C.

The second part of the classification distinguishes between solely livestock systems and mixed farming systems. Solely livestock systems are those in which more than 90% of dry matter fed to animals comes from rangelands, pastures, annual forages and purchased feeds and less than 10% of the total value of production comes from non-livestock farming activities. Mixed farming systems are livestock systems in which more than 10% of the dry matter fed to animals comes from crop by-products, stubble or more than 10 percent of the total value of production comes from non-livestock farming activities. The solely livestock systems are split into two:

- Grassland-based systems, in which more than 10% of the dry matter fed to animals is farm produced and in which annual average stocking rates are less than 10 temperate livestock units per hectare of agricultural land.
- Landless livestock production systems, in which less than 10% of the dry matter fed to animals is produced on the farm and in which annual average stocking rates are above 10 temperate livestock units per hectare of agricultural land. The landless systems are further split into two categories: landless monogastric systems, in which the value of production of the pig/poultry enterprises is higher than that of the ruminant enterprises; and landless ruminant systems, in which the value of production of the ruminant enterprises is higher than that of the pig/poultry enterprises.

The classification system of Seré and Steinfeld (1996) thus has eleven system types: livestock only, rangeland-based (LG), which may be arid/semiarid (LGA), humid/ sub-humid (LGH), or tropical highland/temperate (LGT); landless monogastric-based (LLM), and landless ruminant-based (LLR); mixed, rainfed systems (MR) by the three agro-ecological zones, and mixed, irrigated systems (MI), also by the three agro-ecological zones. The systems are tabulated in Figure 2.6. Clearly, these systems may be affected by climate change in different ways. Impacts on the arid/semiarid rangelands (LGA) of Latin America or sub-Saharan Africa, for example, could be expected to be associated more with effects on pasture (as the major feed in these systems) and water availability, compared with impacts on the tropical-highland mixed, rainfed systems (MRT) of sub-Saharan Africa or South Asia, where the primary effects may be more on crop residues (as a key feeding resource), for example. The following section presents an overview of the impacts of climate change on livestock in relation to these different systems.

19.8 CLIMATE CHANGE'S IMPACTS ON LIVESTOCK

The effects of climate change on the health of farm animals have not been studied in depth. However, it can be assumed that as in the case of humans, climate change, in particular global warming, is likely to greatly affect the health of farm animals, both directly and indirectly. Direct effects include temperature related illness and death, and the morbidity of animals during extreme weather events. Indirect impacts follow more intricate pathways and include those deriving from the attempt of animals to adapt to thermal environment or from the influence of climate on microbial populations, distribution of vector-borne diseases, host resistance to infectious agents, feed and water shortages, or food-borne diseases. It could be organised impacts of climate change on livestock under seven headings: feeds, quantity and quality; heat stress; water; livestock diseases and disease vectors; biodiversity; systems and livelihoods; and indirect impacts. It could be concluded that increasing of temperatures cause severe damage to the physiology, the metabolism and to the healthiness of animals. Modification of existing regimes of precipitation and the increase of aridity will have repercussions on the availability of feedstuff for animals. The increased difficulty in livestock production in the world will correspond to the increasing needs in animal products. There is still a great deal that is not well understood concerning the interactions of climate and increasing climate variability with other drivers of change in livestock systems and in broader development trends. Multiple and competing pressures are likely on tropical and subtropical livestock systems in the future, to produce food, to feed livestock, and to produce energy crops, for example (Table 1 and 2).

Table1 : Livestock systems classification

Generic	Specific	System
LG (livestock only)	LGA	Livestock only system, arid-semiarid (LGPP180 days)
	LGH	Livestock only system, humid-subhumid (LGPP180 days)
	LGT	Livestock only system, highland/temperate
MR (mixed rainfed)	MRA	Mixed rainfed crop/livestock systems, arid-semiarid (LGP < 180 Ddays)
	MRH	Mixed rainfed crop/livestock systems, humid-subhumid (LGPP180 days)
	MRT	Mixed rainfed crop/livestock systems, highland/temperate
MI (mixed irrigated)	MIA	Mixed rainfed crop/livestock systems, arid-semiarid (LPG < 180 days)
	MIH	Mixed rainfed crop/livestock systems, humid-subhumid (LGPP180 days)
LL (landless)	MIT	Mixed rainfed crop/livestock systems, highland/temperate
	LLM	Landless monogastric systems
	LLR	Landless ruminat systems

Table 2 : Some of the knowledge gaps of climate change impacts on livestock-based systems and livelihoods in the tropics and subtropics

Area	Gap
Feeds: quantity and quality	Rangelands (LG) primary productivity impacts, species distribution and change due to CO_2 and other competitive factors, estimation of carrying capacities.
	Mixed systems (MR, MI): localized impacts on primary productivity, harvest inexes and stover production
Heat stress water	What is the extent of the problem, in a development contextt?
	Suface and groundwater supply, and impacts on livestock (particularly LG systems)
	Effective ways to increase livestock water productivity
Diseases and disease vectors	How may the prevalence and intensity of key epizootic livestock diseases change in the future?
	How may climate change affect disease as systems intensify (particularly MR, MI, LL systems)?
Biodiversity	Ecological biodiversity: what will happen to numbers of species as systems change?
	Animal breed biodiversity: can the animal genetic resources that might be useful in the future be specified?
Livestock systems	Localized impacts on ivelibоods
	How will systems evolve in future?
	Magnitude and effects of systems changes on ecosystems goods and services
Indirect impacts	How do human helth impacts of climate change intertwine with livelihood systems and vulnerability?

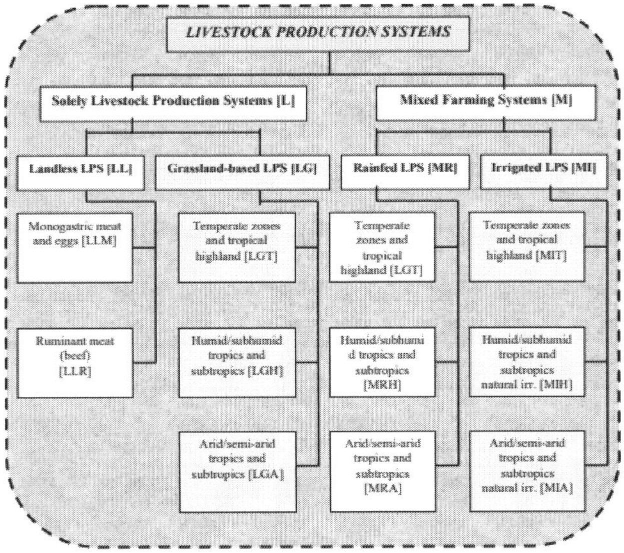

Fig. 6: Classification of livestock production systems (adapted sere and steinfeld, 1996)

The answers of the livestock systems to these requisites will be diverse. The grazing and mixed rain-fed systems, which count on the availability of pastures and farm crops, will be the most damaged by climate change. The positive trend both in the number of heads and in productivity that we observed in recent decades could slow down or even become negative, if an effort is not made to adapt. Livestock and livestock systems are substantial users of natural resources and globally they contribute significantly to global warming, while at the same time they make contributions of critical importance to the livelihoods of at least a billion poor people in rural households, almost all of whom are in developing countries. Damage could be considerable, since grazing and mixed rain-fed systems raise almost 70% of all ruminants in the world. Worldwide they produce almost 2/3 of the milk and meat from ruminants. Moreover, more that 50% of this production is raised in developing countries where the need of animal products will increase more. At the moment it seems difficult to evaluate whether or not the areas lost to desertification caused by climate change, will be compensated by the zones favored by the change in climate. The need for twice the amount of animal products in the next 4 decades should be satisfied essentially by the irrigated and industrial systems. How could this happen? An increase in heads and also better indices of productivity in these systems are predictable. Therefore, we will have more pigs and poultry, point that is already being confirmed by the growing trend in pork and poultry production. Furthermore, because of the difficulties of grazing and rain-fed systems, milk and beef production will shift to industrialized systems, even though this increase in industrialized production will be more moderate than poultry and pork. Scientific research can help the livestock sector in the battle against climate change. All animal scientists must collaborate closely with colleagues of other disciplines, first with agronomists then, physicists, meteorologists, engineers, economists, etc. The effort in selecting animals that up to now has been primarily oriented toward productive traits, from now on, must be oriented toward robustness, and above all adaptability to heat stress. In this way molecular biology could allow to directly achieve genotypes with the necessary phenotypic characteristics. Research must continue developing new techniques of cooling systems such as thermo-isolation, concentrating more than in the past on techniques requiring low energy expenditure. New indices that are more complete than THI to evaluate the climatic effects on each animal species must be developed and weather forecast reports must also be developed with these indices, to inform the farmers in advance. Above all to beat the climate change or in any case not to let the climate beat livestock systems, researchers must be very aware of technologies of water conservation. In the future we can profit, more than in the past, from the years of experience of the people living in arid zones by applying our scientific knowledge to useful traditional practices.

20

CLIMATE CHANGES AND LAND USE

20.1 INTRODUCTION: AN OVERVIEW

Land use describes use of the land surface by humans. Normally, use of land is defined in an economic context, so we think of land as it is used for agricultural, residential, commercial, and other uses. However, strictly speaking, we can seldom really see the use of the land, except on the very closest inspection, so we consider also land cover— the visible features of the Earth's surface— included in the vegetative cover, natural and as modified by humans, its structures, transportation and communications, and so on (Figure 1). As a practical matter, it must be considered land use and land cover together, while also recognizing the distinction between the two. Modern society depends on accurate land use data for both scientific and administrative purposes. They form an essential component of local and regional economic planning to ensure that various activities are positioned on the landscape in a rational manner. For example, accurate knowledge of land use patterns permits planning to avoid placing residential housing adjacent to heavy industry or in floodplains. Accurate land use information can ensure that residential neighborhoods are logically placed with respect to commercial centers and access to transportation services. In another context, land use is an important component of climatic and hydrologic modeling to estimate the runoff of rainfall from varied surfaces into stream systems. And accurate land use data are important for transportation planning, so traffic engineers can estimate the flow of vehicles from one region to another and design highways with appropriate capacities. Land use patterns also reflect the character of a society's intimate interaction with its physical environment. Although land use patterns within our own society seem self-evident and natural, other societies often organize themselves in contrasting patterns. This fact becomes obvious when it is possible to observe different economic and social systems occupying similar environments.

Fig. 1: Land use and land cover. Land use refers to the principal economic enterprises that characterize an area of land: agriculture, manufacturing, commerce, and residential. Land cover indicates the physical features that occupy the surface of the Earth, such as water, forest, and urban structures.

20.2 URBAN LAND USE AND LAND COVER

Urban land use and land cover (LULC) datasets are very important sources for many applications, such as socioeconomic studies, urban management and planning, and urban environmental evaluation. The increasing population and economic growth have resulted in rapid urban expansion in the past decades. Therefore, timely and accurate mapping of urban LULC is often required. Although many approaches for remote sensing image classification have been developed, urban LULC classification is still a challenge because of the complex urban landscape and limitations in remote sensing data. Conventional survey and mapping methods do not provide the necessary information in a timely and cost-effective manner. Remotely sensed data, with their advantages in spectral, spatial, and temporal resolution, have demonstrated their power in providing information about the physical characteristics of urban areas, including size, shape, and rates of change, and have been used widely for mapping and monitoring of urban biophysical features. Geographic information system (GIS) technology provides a flexible environment for entering, analyzing, and displaying digital data from various sources that are necessary for urban feature identification, change detection, and database development. The integration of remote sensing and GIS technologies has been applied widely and has been recognized as an effective tool in urban-related research. Because of the confusion of spectral signatures in some land cover types, such as between impervious surface and soil and between low-density residential area and forest, ancillary data have become an important source for improving urban LULC classification accuracy. However, in urban areas, the separation of different densities of residential areas, the distinction between residential areas and forest or grass, and the separation of commercial/industrial areas from residential areas are important. Census data, such as those on housing density distribution, are closely related to urban LULC patterns. This chapter explores, based on a case study of Indianapolis, Indiana, the use of housing information in different stages in the image classification procedure in order to identify a suitable method for improvement of urban LULC classification performance.

Climate change and land-use change are both key drivers of biodiversity change. Interactions between these drivers are complex and currently not well understood, and may have a greater overall impact on biodiversity change than either of these drivers operating in isolation. In spite of this, most biodiversity studies assess the impacts of climate change or land-use change and associated habitat fragmentation in isolation. Furthermore, only a small number of biodiversity studies include the effects of land-use change in contrast to the large number of studies of climate change. Calls have been made for studies that integrate both drivers however only a few such studies have been undertaken to date (de Chazal et al. 2009). An implication of the lack of integrated analysis is that studies of biodiversity change that examine the effect of either climate change or land-use change in isolation are likely to either over- or under-estimate the potential effects. Interactions between climate and land-use change may also lead to surprising outcomes. The individual and combined effects of climate change and land-use change on biodiversity are also determined by how these drivers as well as biodiversity are defined with different definitions resulting in a range of effects and interactions.

A much broader range of impacts of land-use/cover change on ecosystem goods and services were further identified. Of primary concern are impacts on biotic diversity worldwide, soil degradation, and the ability of biological systems to support human needs. Land-use/cover changes also determine, in part, the vulnerability of places and people to climatic, economic, or sociopolitical perturbations. When aggregated globally, land-use/cover changes significantly affect central aspects of Earth System functioning. All impacts are not negative though as many forms of land-use/cover changes are associated with continuing increases in food and fiber production, in resource-use efficiency, and in wealth and well-being.

20.3 UNDERSTANDING AND PREDICTING THE IMPACT OF SURFACE PROCESSES ON CLIMATE

Understanding and predicting the impact of surface processes on climate required long-term historical reconstructions – up to the last 300 years – and projections into the future of land-cover changes at regional to global scales. Quantifying the contribution of terrestrial ecosystems to global carbon pools and flux required accurate mapping of land cover and measurements of landcover conversions worldwide. Fine resolution, spatially explicit data on landscape fragmentation were required to understand the impact of land-use/cover changes on biodiversity. Predicting how land-use changes affect land degradation, the feedback on livelihood strategies from land degradation, and the vulnerability of places and people in the face of land-use/cover changes requires a good understanding of the dynamic human-environment interactions associated with land-use change). Sustainable land use refers to the use of land resources to produce goods and services in such a way that, over the long term, the natural resource base is not damaged, and that future human needs can be met. The time horizon of the concept covers several generations. Over the last few decades, numerous researchers have improved measurement of land-cover change, the understanding of the causes of land-use change, and predictive models of land-use/cover change, in part under the auspices of the Land-Use and Land-Cover Change (LUCC) project of the International Geosphere-Biosphere Programme (IGBP) and International Human Dimensions Programme on Global Environmental Change (IHDP). Their work, part of an international effort that has helped to propagate the emergence of "land-change science," has taken on the task of demonstrating the role of land change in its own right within the Earth System. An "integrated land science" has emerged, uniting environmental, human, and remote sensing/GIS sciences to solve various questions about land-use and landcover changes and the impacts of these changes on humankind and the environment. This science has demonstrated both the pivotal role of land change in the Earth System and its complexities that transcend such simplifications as unidirectional and permanent land-cover change caused by immediate population or consumption changes, replacing them by a representation of a much more complex process of land-use/cover change. The new Global Land Project is developing further land-change science based on the foundations generated by LUCC and by other projects on terrestrial ecosystems. Since time immemorial, humankind has changed landscapes in attempts to improve the amount, quality, and security of natural resources critical to its well-being, such as food, freshwater, fiber, and medicinal products. Through the increased use of innovation, human populations have, slowly at first, and at increasingly rapid pace later on, increased its ability to derive resources from the environment, and expand its territory. Several authors have identified three different phases – the control of fire, domestication of biota, and fossil-fuel use – as being pivotal in enabling increased appropriation of natural resources. Agriculture has been the greatest force of land transformation on this planet. Nearly a third of the Earth's land surface is currently being used for growing crops or grazing cattle. Much of this agricultural land has been created at the expense of natural forests, grasslands, and wetlands that provide valuable habitats for species and valuable services for humankind. It is estimated that roughly half of the original forests (ca. 8 000 years ago) have been lost. The pace of agricultural land transformation has been particularly rapid in the last 300 years. Much of the large-scale cultivation in 1700 was concentrated in the Old World, specifically in Europe, the Indo – Gangetic Plains, eastern China, and Africa. Roughly 2–3% of the global land surface was cultivated at that time. Since then, the rate of cropland expansion increased with scale, regional-scale trends are alarming. It is estimated that the United States of America paved over roughly 2.9 million ha of agricultural land between 1982 and 1997, and that ~30% of the increase in developed land during 1982–1997 occurred on prime farmland. China lost nearly 1 million ha of its cultivated land to expansion of infrastructure (both urban and rural) between 1988 and 1995. Some rough estimates indicate that 1 to 3 million ha of cropland may be taken out of production every year in developing countries to meet the land demand for housing, industry, infrastructure, and recreation. Before the 19th century, land in Sub-Saharan Africa was used largely for hunting, gathering, herding, and shifting cultivation. Some settled agriculture existed in Africa long before the imposition of colonial rule in the late nineteenth century, but in the pre-colonial period, demographic and economic needs allowed for land cleared for cultivation to be left fallow for long periods or abandoned as cultivators moved on and cleared new land. Estimates of cropland areas before ~1900 are variable, in part because of the lack of data and in part because "croplands" were part of a shifting cultivation rotation, where the distinctions between cropped areas and fallows are unclear. Shifting cultivation included annual clearing of 0.5–3.0 ha of forest per family creating a mosaic of cropped fields intermingled with fields 2–3 years old, fallows, and stands of secondary and mature forest. Clearing of previously cultivated areas (old fallows of 10–50 years) was generally preferred over clearing old-growth forests. Before ~1900, land use had probably been in a "quasi-equilibrium" for thousands of years. Changes included both increases and decreases as a result of wars, epidemics, famines, and slave trade (both intra-African and trans-Atlantic). In fact, populations are thought to have declined somewhat during the 19th century. Between 1850 and 1900 European colonization introduced changes; but with a few exceptions, the most rapid and dramatic changes occurred after 1930. The area under export crops expanded significantly because colonial governments needed the revenues, they provided to recover from worldwide depression of the 1930s. In addition, by the 1930s, the railroads and most of the other major transport routes were in place in colonial Africa, and it became feasible to begin development of areas that had hitherto been inaccessible. This combination of demographic pressure and economic incentive has continued to the present. Cropland area in Sub-Saharan Africa is estimated to have been 119 million ha in 1961 and 163 million ha in 2000, an increase of 37% in 40 years. The rate of forest clearing for long-term shifting cultivation has been even greater than the rate of clearing for permanent croplands in recent decades.

20.3 LAND USE PATTERNS AND CLIMATE CHANGE

Land use patterns change over time in response to economic, social, and environmental forces. The practical significance of such changes is obvious. For planners and administrators, they reveal the areas that require the greatest attention if communities are to develop in a harmonious and orderly manner. From a conceptual perspective, study of land-use changes permits identification of long-term trends in time and space and the formation of policy in anticipation of the problems that accompany changes in land use. Land use and land cover change is very simple in concept: Two maps representing the same region, prepared to depict land use patterns at different dates, are compared, point by point, to summarize differences between the two dates (Figure 2). In practice, compiling land use and land cover change requires mastery of several practical procedures that often reveal previously hidden difficulties. First, the two maps must use the same classification system, or at least two classification systems that are compatible. Compatible classifications mean that the classes can be clearly matched to one another without omission or ambiguity. For example, if one map represents forested land as a single category, "forest," and a second uses three classes— "coniferous," "deciduous"—and "mixed coniferous/deciduous," then the two maps are compatible in the sense that the three categories of the second maps can be combined into a single forested class that is logically compatible with the "forested" class on the first map. As long as the classes are matched at that level of generalization, one can be confident that the comparisons would reveal changes in the extent of forested land. However, if the second map employs classes such as "densely forested," "partially forested," and "sparsely forested," we could not reasonably match these classes to the "forested class" on this first map—the two sets of classes are not logically compatible.

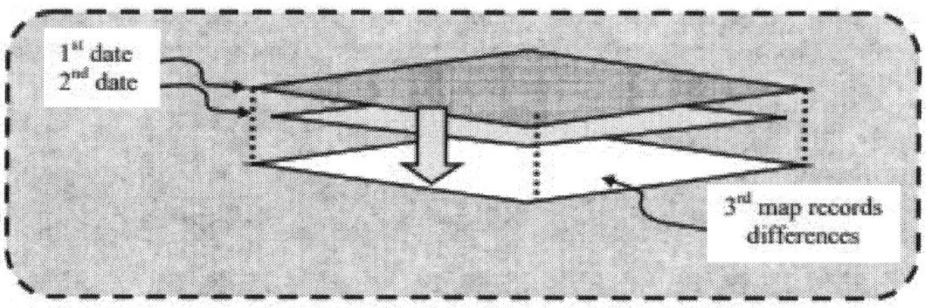

Fig. 2: Schematic representation of compilation of land use change using sequential aerial imagery

The meanings conveyed by these classes, as represented on the two maps, are not equivalent, so they cannot be compared to determine whether the differences represent true change or simply differences in the way the maps were prepared. Second, the two maps must be compatible with respect to scale, geometry, and level of detail. Can we match one point on one map with the corresponding point on the second and be sure that both points refer to the same place on the Earth's surface? If not, we cannot conduct a change analysis using the two maps because we cannot be sure that differences in the two maps reflect genuine differences in land use and not simply differences in the projection, scale, or geometric properties of the two maps. These two basic conditions must be satisfied before change data can be complied using any two maps or images: compatibility of the information portrayed and geometric compatibility. Often, land use maps are prepared to meet the very specific objectives of the sponsoring organization. Therefore, it may be very difficult to reuse that map in a land use change study many years later. Objectives may have changed, or the level of detail, the classification system, and so forth, reflect the needs of a previous era. It is for this reason that it is common in land use change studies for maps to be prepared at the time of the study, using consistent materials, methods, and procedures, even if earlier studies already exist and are available.

20.4 LAND USE AND ITS RELATIONSHIP WITH GLOBAL CLIMATE CHANGES

20.4.1 Recent Changes in Agricultural Areas

Historically, humans have increased agricultural output mainly by bringing more land into production. The greatest concentration of farmland is found in Eastern Europe, with more than half of its land area under crops. In the United Kingdom, about 70% of its area is classified as agricultural land (cropland, grassland/rough grazing), with agriculture and areas set aside for conservation or recreation intimately intertwined. Despite claims to the contrary, the amount of suitable land remaining for crops is very limited in most developing countries where most of the growing food demand originates. Where there is a large surplus of cultivable land, land is often under rain forest, permanent pastures, or in ecologically marginal areas.

20.4.2 Recent Changes in Pastoral Areas

Natural vegetation covers have given way not only to cropland but also to pasture, defined as land used permanently for herbaceous forage crops, either cultivated or growing wild (FAO, 2004). The distinction between pasture and natural savannas or steppes is not always clear. In many parts of the world, such landscapes are "multifunctional", making it difficult to classify them for inventories. Therefore, the LUCC-MA assessment did not deal with grazing land changes. Nevertheless, broad patterns can probably be derived from the FAO statistics, which show that most pastures are located in Africa (26% of the global total of ~35 million ha) and Asia (25%), and only a small portion is located in North America (8%) and Europe (2%) (FAO, 2004). Latin America and the Caribbean have 18% of the world's pastures, while the FSU nations have 10%, and Oceania has 12%. During the last decade, pastures increased considerably in Asia and the FSU (6.8% and 10%, respectively), whereas the largest decreases were seen in Europe and Oceania. Data suggest that pasture land has apparently decreased in eastern Africa; however, as eastern Africa recorded a large increase in head of cattle over this period (872 000 additional head of cattle per year between 1992 and 1999, according to FAO, 2004), it is likely that many pastoral areas in this part of Africa are classified as natural vegetation (Figure 3).

20.4.3 Recent Changes in Urbanization

In 2000, towns and cities housed more than 2.9 billion people, nearly half of the world population. Urban populations have been growing more rapidly than rural populations worldwide over the last two decades, particularly in developing countries. According to the UN Population Division (United Nations Population Division, 2002), the number of mega-cities, defined here as cities with more than 10 million inhabitants, has increased from one in 1950 (New York) to 17 in 2000, the majority of which are in developing countries. Urban form and function have also changed rapidly. Built-up or impervious areas are roughly estimated to occupy between 2% to 3% of the Earth's land surface. This relatively small area reflects high urban population densities: for example, in 1997, the 7 million inhabitants of Hong Kong lived on as little as 120 km2 of built-up land. However, urbanization affects land in rural areas through the ecological footprint of cities. This footprint includes, but is not restricted to, the consumption of prime agricultural land in peri-urban areas for residential, infrastructure, and amenity uses, which blurs the distinction between cities and countryside, especially in western developed countries. Urban inhabitants within the Baltic Sea drainage, for example, depend on forest, agriculture, wetland, lake, and marine systems that constitute an area about 1 000 times larger than that of the urban area proper.

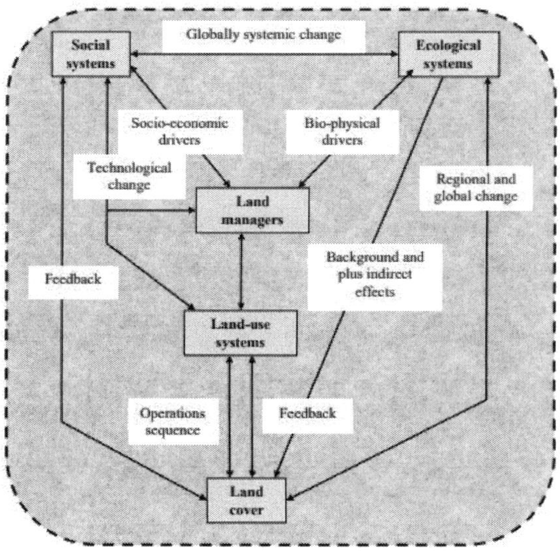

Fig. 3: Framework for understanding land-use/cover situations

20.3.4 Recent Forest-Cover Changes

Deforestation, one of the most commonly recognized forms of land-cover change, is nevertheless plagued by inconsistencies in definitions. The Food and Agriculture Organization (FAO) of the United Nations defines deforestation as occurring when tree canopy cover falls below 10% in natural forests (or when a forest is transformed to other land uses even if tree canopy cover remains higher than 10% – e.g., shifting cultivation). On the basis of this definition, and using country forest inventories, expert estimates, forest-plantation data, and an independent remote sensing survey, the Global Forest Resources Assessment 2000 estimated a net decrease in forest area of 9.4 million ha yr^{-1} from 1990 to 2000. This change was a result of a 12.5 million ha yr^{-1} net decrease in natural forests (comprising deforestation of 14.6 million ha yr^{-1}, conversion to forest plantation of 1.5 million ha yr^{-1}, and regeneration of 3.6 million ha yr–1), and 3.1 million ha yr–1 net increase in forest plantations (1.5 million ha yr–1 converted from natural forests, and 1.6 million ha yr^{-1} of afforestation). Most of the deforestation occurred in the tropics, while most of the natural forest re-growth occurred in Western Europe and eastern North America; the total net forest change was positive for the temperate regions and negative for the tropics.

20.3.5 Recent Changes in Drylands

Desertification is a difficult process to evaluate because of its varying definitions and perceptions. The United Nation's Convention to Combat Desertification (UNCCD) defines desertification as "land degradation in arid, semi-arid and dry sub-humid areas resulting from various factors, including climatic variations and human activities". Land degradation is defined as the decrease or destruction of the biological productivity of the land, including vegetation degradation, water and wind erosion, and chemical and physical deterioration, or a combination of these processes. The LUCC-MA synthesis of the main areas of degraded drylands was constrained by lack of reliable data. Most available data were heterogeneous in terms of the monitoring methods or the indicators used. The study found that the main areas of degraded dryland lie in Asia.

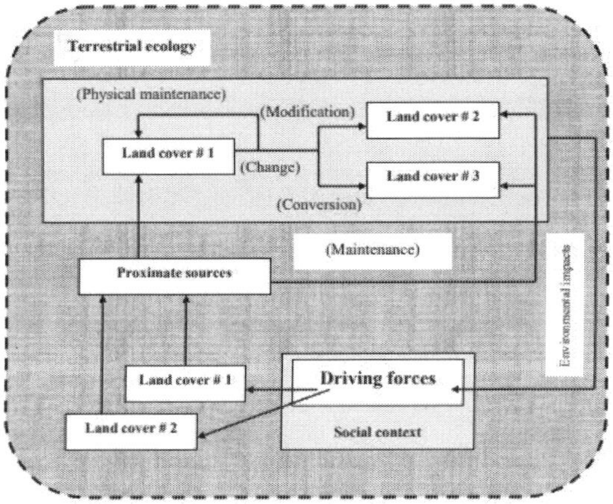

Fig. 4: Links between human activities and land use and land cover

The synthesis did not support the claim that the African Sahel is a desertification "hot spot" at the present time. However, it found major gaps in desertification studies, including around the Mediterranean Basin, in eastern Africa, in parts of South America (in northern Argentina, Paraguay, Bolivia, Peru and Ecuador) and in the United States of America. If dryland degradation data were available in compatible format for all the continents, the global distribution of the most degraded drylands could be different, but the patterns observed in Asia would most likely remain the same. As shown in Table 1, in the USA the most commonly used land cover classification system is the USGS land use and land cover classification system for use with remote sensor data. Most projects use the top classes of the Anderson scheme and define lower classes based on the needs of the specific project. There are many other classification schemes in use by programmes such as the USFWS National Wetlands Inventory, the USGS GAP Analysis Program and the NOAA Coast watch Change Analysis Program. Most of these classification systems have subclasses that can still be aggregated into the upper classes of the USGS Anderson Classification System (Figures 3 and 4).

Table 1: Land Use and Land cover Classification System for use with Remote Sensor Data

Level 1	Level 2
1 Urban or Built-Up Land	11 Residential 12 Commercial and services 13 Industrial
2 Agricultural Land	14 Transportation, communications and utilities 15 Industrial and commercial complexes 16 Mixed urban or buil-up land
3 Rangeleand 4 Forest Land 4 Forest Land 5 Water Wetland 7 Barren Land 8 Tundra	17 Other urban or built-up Land 21 Cropland and pasture 22 orehards, groves, vineyards, nurseris and omamental horticultural areas 23 Confined feeding operations 24 Other agricultural land 31 Serbaceous rangeland 32 Shrub-brush land rangeland 33 Mixed rangeland 41 Deciduous forset land 42 Evergreen forset land 43 mixed forset land 51 Streams and canals 52 lakes 53 Rservous 54 Bays and esruaries 61Forested wetland 62 Non-forsted wetland 71 Dry salt flats
9 Perennial snow or ice	72 Behes 73 Sandy areas other than beaches 74 Bare exposed rock 75 Strip mines, quarries, and gravel pits

Level 1	Level 2
	76 Trasnitional areas
	77 Mixed
	81 Shrub and brush tundra
	82 Herbaceous
	83 Bare ground tundra
	84 Wet tundra
	86 Mixed tundra
	91 Perennial snofields
	92 Perennial Snow or

20.4 CLIMATE CHANGES AND BIOENERGY

Bioenergy is a composite word, combining "biomass" and "energy". Bioenergy is the production of heat, power and liquid fuels (termed biofuels) from biomass. Biomass is plant matter or derived from plant matter, and can be sourced from processing and post-consumer residues, in-field residues from forestry and agriculture, and purpose-grown energy crops. It could be considered that the impacts on soil health of the latter two sources of biomass. The potential for bioenergy to mitigate climate change is a major driver of the recent policy development and industry expansion. Following the Intergovernmental Panel on Climate Change (IPCC) definition, it could be defined climate change as a statistically significant variation in either the mean state of the climate or in its variability, persisting for an extended period (typically decades or longer). Climate change may be due to natural internal processes or external forcings, or to persistent anthropogenic changes in the composition of the atmosphere or in land use. Traditional biofuels (e.g., wood, crop residues, animal dung) have been used as sources of household energy (such as for cooking and heating) since the dawn of human civilization. Modern biofuels go back to 1896 when Henry Ford's first car was designed to run on ethanol. They achieved prominence during the oil embargo of the 1970s. However, the current interest in biofuels is driven by high energy demands and the impacts of fossil fuel combustion on emissions of CO2, CH4, and N2O with the attendant changes in climate. Global energy consumption is expected to grow by 50% between 2005 and 2030. From its humble beginning in the late 19th and early 20th centuries, fuel ethanol production in the United States grew to 310 million liters (ML) in 1980, 2.3 billion liters (BL) in 1985, 2.8 BL in 1990, 4.9 BL in 1995, 6.3 BL in 2000, 15.4 BL in 2005, and 25.9 BL in 2007.

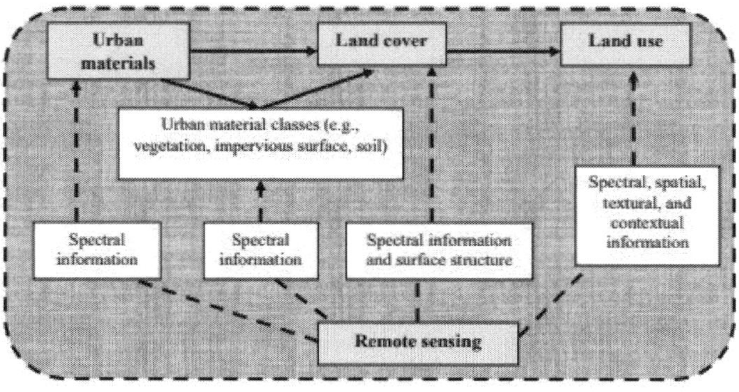

Fig. 5: Relationship among remote sensing of urban materials, land cover, and land use, a conceptual framework.

As the world searches for low-carbon renewable energy, there will be significant opportunities for biomass to provide heat, electricity and transport fuels. Bioenergy can contribute to climate change mitigation through increasing sequestration of carbon (e.g. in forests) and through reducing use of fossil fuels. With various feedstocks and process routes and multiple products, there are many bioenergy options that can produce not only energy but also other products, such as biochar, that may enhance the economic, social and environmental role of bioenergy. Bioenergy could supply greater than 30% of world primary energy demands by 2050. However, questions remain concerning the capacity and sustainability of large-scale intensive biomass energy systems, particularly over implications for soil health (Muller, 2009). The major policy driver for the current promotion of bioenergy is its potential to contribute to climate change mitigation. Bioenergy systems can displace the use of fossil fuel, leaving the fossil fuel "in the ground", and may sequester carbon (through storage of carbon in the soil or vegetation). The potential benefit from increasing biomass energy (stationary and transport) and avoiding fossil fuel use is significant. The IPCC estimates that by 2030, increased energy generation efficiency and switching to alternative low-carbon sources including bioenergy could reduce emissions by approximately 1.6–2.5 Gt CO_2-e per annum for light transport, and between 2.0 and 4.2 Gt CO_2-e per annum for stationary energy production. Also important are the production-chain emissions, due to: fossil fuel use, for example, in cultivation, harvest, transport, fertiliser manufacture; non-CO_2 emissions, such as nitrous oxide (N_2O) from applied fertiliser; and change in biomass and soil carbon. Co-products can also have a significant impact on the GHG balance of some bioenergy systems: biochar, a co-product of pyrolysis, can be used as a soil amendment improving soil health and yielding mitigation benefits; residues from ethanol and biodiesel production can be used as animal feeds, saving emissions associated with feed production. Land use change (LUC), both direct and indirect, can significantly add to the production-chain emissions of bioenergy systems. While climate change is a major driver for the industry, mitigation value must be balanced with other objectives. This becomes a challenge for policy development and implementation, as there are often trade-offs: maximizing carbon sequestration and biomass production may reduce potential biodiversity benefits, or adversely impact hydrology. However, there are potential "win–win" scenarios where mitigation, conservation and production objectives can be addressed simultaneously, for example, the development of a mallee industry for bioenergy production in Western Australia. To make a significant contribution to climate change mitigation, a huge quantity of biomass will be required. For example, it is estimated that for the USA to produce 30% of their domestic transport fuel demand would require the growing and processing of approximately one billion tonnes of biomass per annum. The potential impact on GHG emissions of large-scale bioenergy systems will require carefully developed policy to meet multiple objectives and optimize production.

Bioenergy generally costs significantly more than existing fossil fuel-based energy sources, but it is promoted for various reasons including:

- Energy security;
- Environmental benefits, particularly mitigation of climate change;
- Rural and regional development;
- Human health.

The development of bioenergy is encouraged by policies including renewable energy targets, industry development grants and emissions trading. However, there are concerns about potential impacts of the expansion and intensification of large-scale bioenergy systems including:

- The "food versus fuel" debate – it has been asserted that utilization of food for fuel [particularly corn (*Zea mays*) for ethanol] has led to rising food prices, but others claim that this impact has been exaggerated or can be avoided by integrating food, feed and biomass production. Nevertheless, the stronger competition for resources as the human population grows could be a significant challenge for future bioenergy development.
- The actual net GHG balance of bioenergy systems – some systems may deliver minimal GHG savings, when production-chain emissions and impacts on terrestrial carbon stocks are accounted.
- Indirect land use change (iLUC) – use of agricultural land for biomass production may lead indirectly to deforestation, with implications for soil health and climate change.
- Local environmental impacts – for example, impacting air quality due to increased emissions in production of energy from biomass crops and biofuel refining; reduced water supply where production of biofuels uses significant water; and soil health.

20.5 MORE UNCERTAINTY FOR AGRICULTURE UNDER CLIMATE CHANGE

The issues of agricultural production are complicated by increasing climate uncertainty. The relationship between agriculture and climate change is complex. Agriculture contributes to global warming through emissions of methane and nitrous oxide. Changes in land use practices (management of cropland and grazing land) are considered to be the best mitigation options. Agriculture is also extremely sensitive to climate change, and it is anticipated that large areas of croplands, in particular in semi-arid zones, will need to adapt to new conditions with lower precipitation. Climate change is expected to alter hydrologic regimes and patterns of freshwater resource availability, with impacts on both rainfed and irrigated agriculture. Severe reductions in river runoff and aquifer recharge are expected in the Mediterranean basin and in the semi-arid areas of Southern Africa, Australia and the Americas, affecting water availability for all uses. The projected increase in the frequency of droughts and floods will hurt crop yields and livestock, with greater impacts coming earlier than previously predicted. While climate change does not seem to threaten global food production; it will alter the distribution of agricultural potential. Most of the increase in cereal production will be concentrated in the Northern Hemisphere, while more frequent and severe droughts and floods will hurt local production, especially in subsistence sectors at low latitudes. Several densely populated farming systems in developing countries are at risk from the impacts of climate change. A combination of reduced river base flows increased flooding and rising sea levels is expected to impact highly productive irrigated systems that help maintain the stability of cereal production. The production risks will be amplified in alluvial plains dependent on glacier melt (Colorado, Punjab) and, in particular, in lowland deltas (the Ganges and Nile).

21

THE TOXICOLOGY OF CLIMATE CHANGES

21.1 INTRODUCTION

The U. N. Intergovernmental Panel on Climate Change (IPCC) has completed four assessments covering the evidence, impacts, and mitigation of climate change. They report unequivocal global warming with evidence of increases in global mean air and ocean temperatures, widespread snow and ice melt, and rising global sea level. Temperature is projected to increase 1.8– 4.0 °C by the end of the century under a range of probable greenhouse gas emission scenarios with the greatest warming expected at high latitudes. In addition to global warming, some regions, such as North and South America, northern Europe, and northern and central Asia are projected to experience increased precipitation, while others, including southern Africa and Asia and the Mediterranean, are expected to experience substantial droughts. Heat waves, precipitation and storm events are predicted to be more frequent and intense. Oceanic acidification linked to increasing atmospheric carbon dioxide levels is a growing threat to marine organisms and ecosystems. Climate change will have a powerful effect on the environmental fate and behavior of chemical toxicants by altering physical, chemical, and biological drivers of partitioning between the atmosphere, water, soil/sediment, and biota, including: air-surface exchange, wet/dry deposition, and reaction rates (e.g., photolysis, biodegradation, oxidation in air). Temperature and precipitation, as altered by climate change, are expected to have the largest influence on the partitioning of chemical toxicants. In addition, an array of important processes, such as snow and ice melt, biota lipid dynamics, and organic carbon cycling, will be altered by climate change potentially producing significant increases in fugacity (thermodynamic measure of substance tendency to prefer one phase over another) and contaminant concentrations. The IPCC projects that climate change is likely to affect the health of millions of people, and that the effects will be mostly. The elderly, infants, children, and urban poor are expected to be most vulnerable to the rapidly changing climate).

21.2 ADVERSE CONSEQUENCES OF CLIMATE CHANGE ON HUMAN HEALTH

Notable adverse consequences of climate change on human health include increased death and injury associated with more severe and frequent heat waves, extreme weather events, and enhanced vector-borne and allergic disease transmission. While adverse health outcomes are projected to be greatest in low-income countries, more severe, frequent, and widespread heat waves and storm events will also impact developed countries unprepared to cope with these events.

Table 1: Climate change-induced effects of contaminants on human health

Climate change induced effect	Relationships/Interactions
Increased cardio-respiratory disease	Temperature exacerbates the adverse effects or ozone and PM The elderly and individuals with preexisting cardio-respiratory disease may be more vulnerable to these effects
Altered exposure and risk	Some populations may experience increases or decreases m POP exposures and health risks depending on the region and diet of exposed individuals Pesticides may impair mechanisms of temperature regulation especially during times of thermal stress
Increased susceptibility to pathogens	Toxicants can suppress immune function and climate-induced shifts in disease vector range will result in novel pathogen exposure human vulnerability to climate shifts in pathogens Immune ssystem impairment linked to toxicants may increase Low-income populations. infants, children. and the chronically ill may be more susceptible
Increased alk-rvemcm, potential	Air pollution and allergen exposures linked to climate change can exacerbate allergic disease and asthma incidences Climate chance enhanced allergen production coupled with POP exposures may sensitize individuals to alkrvx disease Low-income populations. III rants. children. and the chronically ill may he more susceptible

There continues to be a lack of data describing the effects of contaminant exposures on human health and vulnerable subpopulations under projected climate change scenarios. However, a number of studies suggest that the toxicity of ozone and particulate matter (PM) will be exacerbated with global warming, and some of these data support that older adults will be especially vulnerable. Other potential interactions between climate change and toxicant exposure include increased susceptibility to pathogens and aeroallergens.

21.3 CLIMATE APPLICATIONS AND AGRICULTURE

Climate variability creates risk in rainfed farming. Risk in turn discourages investment by farmers, governments and development agencies. For instance, in dry regions recurrent droughts debilitate and destabilize poor, agricultural-based societies, and contribute to land degradation by reducing vegetative cover and water supplies. Drought triggers the exploitation of diminishing resources in order to survive. Climate change caused by global warming is likely to increase the frequency of climatic extremes in the future and result in changes in cropping practices and patterns over time and space. If climate variability could be predicted in advance, it would help societies prepare for and cope with the resultant shocks. As well, since drought is a trigger for desertification, better drought prediction and monitoring could help prevent land degradation (Shapiro et al. 2007). In light of population growth and climate change, investment in agriculture is the only way to avert wide-scale food shortages or, in the worst-case scenario, catastrophic human suffering. Assuming investment is forthcoming, maintaining food security will require crop scientists to integrate and apply a broad range of strategies. These include tried and tested technologies such as conventional breeding and agronomy as well as new approaches such as molecular genetics and conservation agriculture. Examples of successful applications as well as future prospects of how each discipline can be expected to evolve over the next 30 years are presented. The Intergovernmental Panel on Climate Change indicates that rising temperatures, drought, floods, desertification and weather extremes will severely affect agriculture, especially in the developing world. While the convergence of population growth and climate change threatens food security on a worldwide scale, the opportunity also exists to address the pernicious threat of famine. Indeed, the prerequisites to develop a globally coordinated effort to ensure long-term food security are available for the first time in human history. Namely: (i) the realization that agricultural problems worldwide have a common scientific basis; (ii) a vast and expanding database encompassing all disciplines that impinge on agricultural productivity; (iii) a de facto network of agricultural scientists working in almost every country in the world; and (iv) unprecedented opportunities for communication, data analysis and investment. These elements, the indisputable fruits of an industrialized global economy, were not available to our predecessors, which is probably why climate change in history spelt death. Agricultural researchers worldwide are, therefore, working to mitigate these and other effects of climate change to increase productivity within a finite natural resource basis. Assuming investment is forthcoming, maintaining food security in the face of population growth and climate change will require a holistic approach that includes stress-tolerant germplasm, coupled with sustainable crop and natural resource management as well as sound policy interventions. There will be duplication of effort as regions struggle with parallel challenges; however, judicious public investment can reduce redundancy of effort permitting local organizations to focus on adaptive research.

21.4 ADAPTATION: LIVING WITH CLIMATE CHANGES

Adaptation is an automatic or planned response to change that minimizes the adverse effects and maximizes any benefits. It is one of the two possible means of coping with human-induced climate change and sea-level rise. The other option is to reduce the magnitude of human-induced climate change by reducing greenhouse gas emissions. This is called mitigation. Adaptation is essential to cope with the climate change and sea-level rise that we cannot avoid now and in the near future, while mitigation would limit the extent of future climate change. Adaptation is essentially a local challenge, while mitigation is essentially a global process that will only be achieved by international cooperation and a common commitment. Adaptation is necessary because climate change is already happening, and the long lag times in the climate system make further climate change, and especially sea-level rise, inevitable. Further climate change is already built into the system by past greenhouse gas emissions, that is, we are already committed to it. The effect happens decades to centuries after the cause and cannot readily be stopped. Additional to the existing commitment to climate change, even more climate change is made inevitable by the long time, at least several decades, between deciding to reduce emissions, and the socioeconomic system changing enough to actually reduce greenhouse gas emissions sufficiently to stop making the situation worse. Very substantial reductions in greenhouse gas emissions will be necessary before greenhouse gas concentrations stop going up. In fact, for centuries to come, atmospheric CO2 concentrations will not fall much below whatever maximum levels are reached, even after major reductions in emissions unless we actively withdraw greenhouse gases out of the atmosphere. This is due to the large reservoirs of CO2 in the ocean, soil and biosphere, which are in equilibrium with the atmosphere on a time-scale of decades. The permanent sinks are much slower to act. In short, we cannot simply turn off climate change, so we must learn to live with it. That is why adaptation is essential.

Studies indicate that Africa's agriculture is negatively affected by climate change. Adaptation is one of the policy options for reducing the negative impact of climate change. Common adaptation methods in agriculture include use of new crop varieties and livestock species that are better suited to drier conditions, irrigation, crop diversification, adoption of mixed crop and livestock farming systems, and changing planting dates.

A handbook on methods for climate change impacts assessment and adaptation strategies has been developed by the United Nations Environment Program. The handbook discusses the principles and strategies for adaptation. These can be summarized in eight alternatives but not exclusive strategies:

- **Bear losses:** This is the baseline response of 'doing nothing'. Bearing loss occurs when those affected have failed to act until it is too late, or have no capacity to respond in any other way (for example, in extremely poor communities) or where the costs of adaptation measures are considered to be high in relation to the risk or the expected damages. The big problem with this solution is that losses may become unbearable.

- **Share losses:** This involves a wider community in sharing the losses. Sharing takes place in traditional societies and in the most complex, high-tech societies. In traditional societies, mechanisms include sharing losses with extended families, villages or similar small-scale communities. In societies organized on a larger-scale, losses are shared through emergency relief, rehabilitation, and reconstruction paid for by government funds or public appeals, or through private insurance. However, insurance usually applies only when the risk is considered random and uncertain for the individual insured, not when it is predictable. Even with shared losses, the accumulated loss to society may eventually become unacceptable, at which point other actions must be taken.
- **Modify the threat:** For some risks, it is possible to exercise a degree of control over the specific environmental threat. For 'natural' events such as a flood or drought, possible measures include flood control works (dams, dikes, levees) or water storages. For climate change, attempts to modify the threat through such measures may quickly become too expensive, and the more sensible modification to reduce the threat is to slow the rate of climate change by reducing global greenhouse gas emissions and eventually stabilizing greenhouse concentrations in the atmosphere. (Note, however, that in Intergovernmental Panel on Climate Change (IPCC) terminology, measures that reduce climate change are referred to as 'mitigation' of climate change, in distinction to 'adaptation', which is reserved for an optimal response to a given climate change.)
- **Prevent effects:** A frequently used set of adaptation measures involves steps to prevent the effects of climate change and variability. Examples for agriculture would be changes in crop management practices such as increased irrigation, additional fertilizer, and pest and disease control.
- **Change use:** Where the threat or reality of climate change makes the continuation of an economic activity impossible or extremely risky, consideration can be given to changing the use. For example, a farmer may choose to switch to crop varieties more adapted to lower soil moisture. Similarly, agricultural land may be returned to pasture or forest or other uses may be found such as recreation, tourism, wildlife refuges, or national parks.
- **Change location:** A more extreme response is to change the location of economic activities. For example, major crops and farming regions could be relocated away from areas of increased aridity and heat to areas that are currently cooler and which may become more attractive for some crops in the future. This may be

possible in some countries, but not in others where migration to cities or other countries may be the only alternatives.

- Research: Possibilities for adaptation can also be opened up by research into new technologies and methods of adaptation, such as greater water use efficiency, cheap water desalinization, or new crop cultivars.
- Educate, inform, and encourage behavioral change: Dissemination of knowledge through education and public information campaigns can lead to adaptive behavioral change. Such activities have been little recognized and given little priority in the past, but are likely to assume increased importance as the need to involve more communities, sectors, and regions in adaptation becomes apparent. Water conservation and fire prevention campaigns and regulations are already major adaptive trends in countries such as Australia. Discouragement of maladaptive trends such as development in low-lying coastal areas is another useful strategy. This may involve planning rules or other ‘carrots and sticks.

21.5 REMOTE-SENSING AND GIS APPLICATIONS IN AGROMETEOROLOGY

A picture is worth a thousand words. Is this true, and if so, why? Pictures concisely convey information about positions, sizes, and interrelationships between objects. By their nature, they portray information about things that we can recognize as objects. These objects in turn can convey deep levels of meaning. Because humans possess a high level of proficiency in deriving information from such images, we experience little difficulty in interpreting even those scenes that are visually complex. We are so competent in such tasks that it is only when we attempt to replicate these capabilities using computer programs; for instance, that we realize how powerful our abilities are to derive this kind of intricate information. Each picture, therefore, can truthfully be said to distill the meaning of at least a thousand words. It is well known that, geographic information system (GIS) is a computer technology that uses a geographic information system as an analytic framework for managing and integrating data; solving a problem; or understanding a past, present, or future situation. GIS is, therefore, about modelling and mapping the world for better decision making. Remote sensing is the acquisition of information about an object or phenomenon, without making physical contact with the object. In modern usage, the term generally refers to the use of aerial sensor technologies to detect and classify objects on Earth (both on the surface, and in the atmosphere and oceans) by means of propagated signals (e.g., electromagnetic radiation emitted from aircraft or satellites). The Earth's climate is now clearly changing and getting warmer. Many components of the climate system are changing at rates and in patterns that are not natural and are best explained by the increased atmospheric abundances of greenhouse gases and aerosols generated by human activity during the 20th century. These changes include temperatures of the atmosphere, land and ocean; the extent of sea ice and mountain glaciers; the sea level; the distribution of precipitation; and the length of the seasons (AGU, 2008). The Earth has warmed consistently and unusually over the past few decades in a manner that can be explained only when a greenhouse process is overlaid on orbital variation, solar variation, volcanic eruptions and other natural disturbances. Observational evidence, complex modelling and simple physics all confirm this. Whatever the proportion of human-induced rise in global temperature versus natural rise, there is no doubt that the temperature and the sea level are rising, the Greenland and Antarctic ice sheets are disintegrating, and major weather patterns and ocean currents are shifting. The Earth's warming is already causing severe droughts and flooding, major vegetation transformations in deserts and forests, massive tundra methane releases and the degradation of the Amazon rainforest and Saharan vegetation. It is also starting to impact the Indian Ocean Monsoon; the Atlantic Conveyor Belt and El Niño weather patterns. The economic impacts of droughts in the USA alone cause $68 billion in losses per year.

Remote sensing is now a mature enough technology to answer some of the fundamental questions in global environmental change science, namely:

- How and at what pace is the Earth system changing and what are the forces causing these changes?
- How does the Earth system respond to natural and human-induced changes?
- How well can we predict future perturbations to the Earth system and what are the consequences of change to human civilization?

Ice sheets, ocean currents and temperatures, deserts and tropical forests each have somewhat different remote sensing requirements. For instance, ocean temperatures are measured by thermal infrared sensors, while ocean currents, winds, waves, and sea level require various types of radar instruments on satellites. Most ocean features are large and require spatial resolutions of kilometers, while observations of desert or forest changes may require resolutions of tens of meters and many bands within the visible and near-infrared region of the electromagnetic spectrum. Monitoring of coral bands pooled in the short wavelength visible spectrum. Fortunately, by the turn of this century, most of these requirements had been met by NASA (the National Aeronautics & Space Administration) and NOAA (the National Oceanic & Atmospheric Administration) satellites and aircraft, the European Space Agency (ESA) and the private sector. Furthermore, new satellites are being launched, carrying imagers with fine spatial (0.64 m) and spectral (200 narrow bands) resolutions, as well as other environmental sensors. These provide a capability to detect changes in both the local and the global environment even more accurately. For the first time, constellations of satellites are being launched with the sole aim of quantifying aspects of the Earth's climate synergistically. With such technology available, governments are no longer alone in being able to monitor the extent of tropical forests and coral reefs, the spread of disease and the destruction caused by war. The technology of remote sensing evolved gradually into a scientific subject after World War II. Its early development was driven mainly by military uses. Later, remotely sensed data became widely applied for civil applications. The range of remote sensing applications includes archaeology, agriculture, cartography, civil engineering, meteorology and climatology, coastal studies, emergency response, forestry, geology, geographic information systems, hazards, land use and land cover, natural disasters, oceanography, water resources, and so on. Most recently, with the advent of high spatial resolution imagery and more capable techniques, urban and related applications of remote sensing have been rapidly gaining interest in the remote sensing community and beyond. Advances in the application of Geographical Information Systems (GIS) and the Global Positioning System (GPS) help to incorporate geo-coded ancillary data layers in order to improve the accuracy of satellite image analysis. When these techniques for generating, organizing, sorting and analyzing spatial information are combined with mathematical climate and ecological models, scientists and managers can improve their ability to assess and predict the impact of global environmental changes and trends. To handle the vast quantities of information being generated by today's Earth observation programmes, there have been significant advances made in the use of the Internet to store and disseminate geospatial data to scientists and the public. The Internet is set to play an even greater role in the handling of products delivered by future missions. It well established that, the Earth's climate is getting warmer and the patterns of the weather and ocean currents are changing. Severe droughts and flooding are becoming more prevalent and the ice sheets of Greenland and at the poles are disintegrating. The global sea level is rising by about 2 to 3 mm per year, threatening to inundate many coastal areas by the end of this century. Remote sensors on satellites offer an effective way for monitoring environmental trends on a global scale. They can detect physical and biological changes in the atmosphere, in the oceans and on land. Satellite systems have become the defining technology in our ability to quantify global change. The accuracy and applicability of satellite imagery is constantly improving due to technological advances, such as finer spectral/spatial resolution, more powerful computers, the Global Positioning System (GPS) and Geographical Information Systems (GIS). When these techniques for generating, organizing and analyzing spatial information are combined with mathematical and environmental models, scientists and managers have a means for assessing and predicting the impact of global environmental changes. It could be explained the following items:

21.6 REMOTE SENSING CONCEPTS

The field of remote sensing has been defined many times (Table 2). Examination of common elements in these varied definitions permits identification of the topic's most important themes. From a cursory look at these definitions, it is easy to identify a central concept: the gathering of information at a distance. This excessively broad definition, however, must be refined if it is to guide us in studying a body of knowledge that can be approached in a single course of study. This definition of Campbell and Wynne (2011) serves as a concise expression of the scope of this volume. It is not, however, universally applicable, and is not intended to be so, because practical constraints limit the scope of this volume. So, although this text must omit many interesting topics (e.g., meteorological or extraterrestrial remote sensing), it can review knowledge and perspectives necessary for pursuit of topics that cannot be covered in full here. The scope of the field of remote sensing can be elaborated by examining its history to trace the development of some of its central concepts. A few key events can be offered to trace the evolution of the field (Table 2.16 and Figure 2.12). Remote sensing refers to the activities of recording, observing, and perceiving (sensing) objects or events in far-away (remote) places. In remote sensing, the sensors are not in direct contact with the objects or events being observed. Electromagnetic radiation normally is used as the information carrier in remote sensing. The output of a remote sensing system is usually an image representing the scene being observed. A further step of image analysis and interpretation is required to extract useful information from the image. In a more restricted sense, remote sensing refers to the science and technology of acquiring information about the earth's surface (i.e., land and ocean) and atmosphere using sensors onboard airborne (e.g., aircraft or balloons) or space borne (e.g., satellites and space shuttles) platforms. Depending on the scope, remote sensing may be broken down into

(1) satellite remote sensing (when satellite platforms are used),

(2) photography and photogrammetry (when photographs are used to capture visible light),

(3) thermal remote sensing (when the thermal infrared portion of the spectrum is used),

(4) radar remote sensing (when microwave wavelengths are used), and

(5) LiDAR remote sensing (when laser pulses are transmitted toward the ground and the distance between the sensor and the ground is measured based on the return time of each pulse). Remote sensing is primarily concerned with collecting and interpreting information about an object or landscape from a remote vantage point. The platform can be anywhere, ranging from a balloon just above the surface of the Earth to a satellite hundreds of kilometres away in space. Examples of remote sensing include aerial photography, satellite imagery, radar altimetry and laser bathymetry. Coupled with ground measurements, remote sensing can provide valuable information about the surface of the land, the oceans and the atmosphere. Techniques for acquiring aerial photographs were already developed in the 1860s; however, color and color-infrared (CIR) aerial photographs were not widely used until the 1940s and 1950s. The 1950s and 1960s marked the appearance of remote sensing applications for airborne radar and video technologies. A significant event in terms of land remote sensing was the 1972 launch of the first Landsat satellite, originally called ERTS-1. The satellite was designed to provide frequent broadscale observations of the Earth's land surface. Since 1972, additional Landsat satellites have been put into orbit. Other countries, including France, Japan, Israel, India, Iran, Russia, Brazil, China,

and perhaps North Korea, have also launched satellites whose onboard sensors provide digital imagery on a continuous basis. The appearance of geographic information systems (GIS) in the mid- 1960s reflects the progress in computer technology and the influence of quantitative revolution in geography. GIS has evolved dramatically from a tool of automated mapping and data management in the early days into a capable spatial data Shandling and analysis technology and, more recently, into geographic information science (GISc). The commercial success since the early 1980s has gained GIS an increasingly wider application. Therefore, to give GIS a generally accepted definition is difficult nowadays. GIS today is far broader and harder to define. Many people prefer to define its domain as geographic information science and technology (GIS & T), and it has become imbedded in many academic and practical fields. The GIS & T field is a loose coalescence of groups of users, managers, academics, and professionals all working with geospatial information. Each group has a distinct educational and "cultural" background. Each identifies itself with particular ways of approaching particular sets of problems. GIS has been called or defined as an enabling technology because of the breadth of uses in the following disciplines as a tool. Disciplines that traditionally have studied the earth, particularly its surface and near surface in either physical or human aspect, include geology, geophysics, oceanography, agriculture, ecology, biogeography, environmental science, geography, global science, sociology, political science, epidemiology, anthropology, demography, and many more.

Table 2.: Remote sensing: some definitions)

Definition
Remote sensor has been variously defined but basically it is the art or science of telling something about an object without touching it.
Remote sensing is the acquisition of physical data of an object without touch or contact.
Imagery is acquired with a sensor other than (or in addition to) to a conventional camerahrough which a scene is recorded, such as by electronic scanning, microwave, radar, thermal, the normal visual range of the film and camera special techniques are applied to process and interpret remote sensing imagery for the purpose of producing conventional maps, thematic maps, resources surveys, etc., in the fields of agriculture, archaeology, forestry, geography, geology, and others.
Remote sensing is the observation of a target by a device separated from it by some distance.
The term "remote sensing" in its broadest sense merely means "reconnaissance at a distance." Remote sensing, though not precisely defined, includes all methods of obtaining pictures or other forms of electromagnetic records of the Earth's suface from a distance, and treatment and processing of the picture data. Remote sensing then in the widest sense is concerned with detecting and recording electromagnetic radiation from the target areas in the field of view of the sensor instrument. This radiation may have originated directly from separate components of the target area; it may be solar energy reflected from them; or it may be reflections of energy transmitted to the target area from the sensor itself.
"Remote sensing" is the term currently used by a number of scientists for the study of remote objects (earth lunar, and planetary surfaces and atmospheres, stellar and galactic phenomena, etc.) from great distances. Broadly defined, remote sensing denotes the joint effects of employing modern sensors, data-processing equipment, information theory and processing methodology, communications theory and devices, space and airborne vehicles, and large-system theory and practice for the purposes of carrying out aerial or space surveys of the earth's surface.
Remote sensing is the science of deriving information about an object from measurements made at a distance from the object, i.e., without actually coming in contact with it. The quantity most frequently measured in present-day remote sensing systems is the electromagnetic energy emanating from objects of Internet, and although there are other possibilities (e.g., seismic waves, sonic waves, and gravitational force), energy.
Remote sensing is the practice of deriving information about the Earth's land and water surfaces using images acquired from an overhead perspective, using eletromagnetic radiation in one or more regions of the eletromagnetic spectrum, relflected or emitted from the Earth's surface.

Table 3: Milestones in the history of remote sensing (adapted from Campbell and Wynne, 2011)

Year	Events
1800	Discovery of infrared by Sir William Hershel
1839	Beginning of practice of photography
1847	Infrared spectrum shown by A. H. L. Fizeau and J. B. L. Foucault to share properties with visible light
1850-18601	Photography from ballons
1873	Theory of electromagnetic energy developed by James Clerk Maxwell
1909	Photography from airplanes
1914-1918	Worl War I: aerial reconnaissance
1920-1930	Development and initial applications of aerial photography and photogrammetry
1929-1939	Economic depression generates environmental crises that lead to governmental applications of aerial photography
1930-1940	Development of radars in Germany, United States, and United Kingdom
1939-1945	World War II: applications of nonvisible portions of electromagnetic spectrum; training of persons in acquistion and interpretation of airphotos
1950-1960	Military research and development
1956	Colwell's research on plant disease detection with infrared photography
1960-1970	First use of term remote sensing
	TIROS weather satellite
	Skylab Remote sensing observations from space
1972	Launch of Landsar I
1970-1980	Rapid advances in digital image processing
1980-1990	Landsat 4: new generation of Landsat sensors
1986	SPOT French Earth observation satellite
1980s	Development of hyperspectral sensors
1990s	Global remote sensing systems, LiDAR

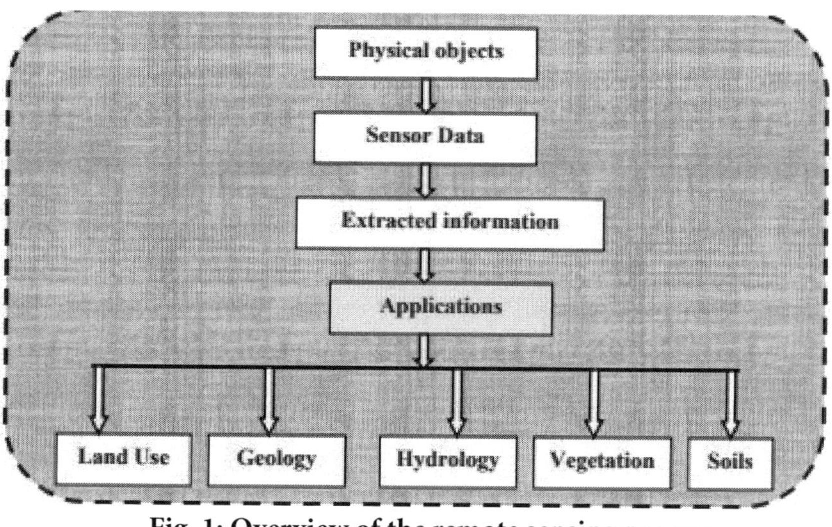

Fig. 1: Overview of the remote sensing process

Table 4: Units of Length and Frequencies Used in Remote Sensing

Unit	Distance	Unit	Frequency (cycles per second)
Kilometer (km)	1,00 m	Hertz (Hz)	1
Meter (m)	1.0 m	Kilohertz (kHz)	10^3 (=1,000)
Centimeter (cm)	0.01 m= 10^{-2} m	Megahertz (MHz)	10^6 (=1,000,000)
Millmeter) (mm)	0.001 m = 10^{-3} m	Gigahertz (GHz)\	10^9 (=1,000,000,000)
Micrometer (μm)	0.000001 m = 10^{-6} m		
Nanometer (nm)	10^{-9} m		
Angstrom unit (A)	10^{-10} m		

21.7 REMOTE SENSORS AND SYSTEMS

It is well known that aerial photography started approximately in 1858 when the famous photographer, Gaspard Tournachon, obtained the first aerial photographs from a balloon near Paris. Since then, aerial photography has advanced, primarily during wartime, first to include color infrared films (for camouflage detection) and later to use sophisticated digital cameras. Aerial photography and other remote sensing techniques are now used successfully in agriculture, forestry, land use planning, fire detection, mapping wetlands and beach erosion, and many other applications. For example, in agriculture it has been used for land use inventories, soil surveys, crop condition estimates, yield forecasts, acreage estimates, crop insect/pest/disease detection, irrigation management and, more recently, precision agriculture. Remote sensing has been used to describe a new field of information collection that includes aircraft and satellite platforms carrying electro-optical and antenna sensor systems since the 1960s. Up to that time, camera systems dominated image collection and conventional photographic media dominated the storage of the spatially varying visible (VIS) and near-infrared (NIR) radiation intensities reflected from the Earth. Beginning in the 1960s, electronic sensor systems were increasingly used for collection and storage of the Earth's reflected radiation and satellites were developed as an alternative to aircraft platforms. Advances in electronic sensors and satellite platforms were accompanied by an increased interest and use of electromagnetic radiant energy, not only from the VIS and NIR wavelength regions, but also from the thermal infrared (TIR) and microwave regions. For instance, the thermal infrared region is used for mapping sea surface temperatures and microwaves (radar) are used for measuring sea surface height, currents, waves and winds on a global scale. Remote sensors can be classified by application, wavelength or active/passive mode. Under applications we have imagers, which produce two-dimensional images and can be used for map-making; radiometers measure the radiant energy in a few specific bands very accurately; while spectrometers provide the energy distribution across a spectral continuum or many spectral bands. Profilers, such as radar and LiDAR, measure the distance to features, allowing us to determine the topography or bathymetry of an area. Radar and LiDAR are primarily 'active' devices, since they provide their own pulse of energy. Most other sensors are 'passive', because they use electromagnetic energy provided by the sun or the Earth.

21.8 MONITORING CHANGES IN GLOBAL VEGETATION COVER

Some of the earliest applications of civilian remote sensing have been in agriculture and forestry. The reason for this rich lineage is that chlorophyll, and the plants that contain it, have a distinct, readily detectable spectral signature in the visible and near-infrared portions of the electromagnetic spectrum. Space-and airborne remote sensing systems have therefore been used for many decades to estimate crop health, predict agricultural yield and vegetation vitality, map deforestation and quantify the damage inflicted by disease and insect infestation on crops and forests. Since vegetation is also sensitive to changes in climate, including temperature and rainfall, it can be used as an indicator of global climate change.

Table 5: Classification of remote sensors

Classification by application	Classfieation by wavelength	Classification by mode
Imagers (mappers)	Visible	Active
Photographic (film)	(array or film)	LiDAR
Multispectral (array)		Radar
Radar Side-scan sonar	Near infrared (reflected)	† Sonar
Radiometers	Thermal infrared (emitted)	Passive Visible infrared
Spectrometers	Microwave	Microwave
Profilers (rangers)	† Sound waves	
LiDAR, Radar	† Seismic waves	

† Not electromagnetic waves

Plants, by virtue of their chlorophyll content, have a very distinct spectral signature which differs markedly from other land cover types such as soil, water and bare earth. The vegetation spectrum has two primary chlorophyll absorption bands in the blue and red regions which are due to pigments such as chlorophyll a, chlorophyll b and β-carotene, which reside in the upper layers of the leaf (palisade parenchyma). By contrast, there is strong reflectance at near-infrared wavelengths, caused by scattering in the spongy mesophyll, the deeper layers of the leaf. In the middle infrared region, three water absorption bands express themselves as a function of how hydrated the leaf tissue is. Thus, if the plant is stressed or drying up, there will be less absorption by the chlorophyll and water bands, as well as less reflection in the near-infrared. The ability of the near-infrared to penetrate deep into a leaf canopy makes it also useful for estimating plant biomass. Figure 2suggests that it is easy to distinguish vegetated areas from bare ones;

however, since the spectra of various plants are somewhat similar, it is not a trivial task to identify plants to the species level on the basis of their spectral signature alone.

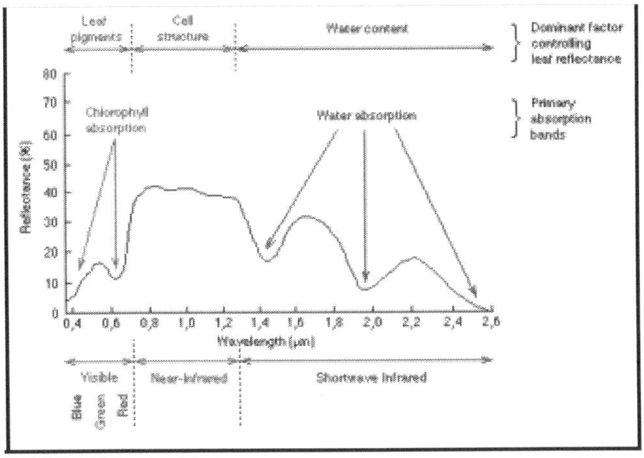

Fig. 2: Spectral response characteristics of green vegetation. Chlorophyll contained in a leaf has strong absorption at 450 nm and 670 nm and high reflectance in the near-infrared (700–1,200 nm). In the shortwave-IR, vegetation displays three absorption features that can be related directly to the absorption spectrum of water (blue line) contained within the leaf.

The remote sensing of vegetation also benefits from the fact that the wavelengths of visible light to which chlorophyll is sensitive occupy an atmospheric 'window', i.e. the light passes relatively unhindered through the atmosphere to the Earth's surface. Even if positioned outside the Earth's atmosphere, any remote sensing system that is tuned to image these wavelengths will enjoy a relatively high signal, and this is the reason that so many sensing systems utilize the visible spectrum. More than 45 per cent of the radiant energy down welling on the Earth's surface during the day is in the visible wavelength region, and it is also no coincidence that this is the portion of the spectrum that plants have evolved to initiate photosynthetic reactions.

All photosynthetic organisms have one or more pigments capable of absorbing visible radiation. Two portions of the electromagnetic spectrum are strongly absorbed by the chlorophyll pigment. These correspond to blue (450 nm) and red light (670 nm), and the magnitude of the absorption is directly related to the amount of radiant energy utilized by the plant. The upshot is a small peak at 500–600 nm (visible green), which explains why our eyes perceive most vegetation to be green. As shown by Figure 3, the near-infrared region (700– 1,200 nm) acts inversely, displaying a pronounced peak in reflectance. It is a fairly unique characteristic of green vegetation that the reflectance in the near-IR is several times as large as that in the visible band. Since the transition from low reflectance in the red to high in the near-infrared is very abrupt, it is often termed the 'red edge'. The red edge position (REP) shifts according to changes of chlorophyll content, leaf area index (LAI), biomass and hydric status, age, plant health levels and seasonal patterns.

For healthy plants with high chlorophyll content and high LAI, the red edge position shifts toward the longer wave lengths; when a plant suffers from disease or chlorosis and low LAI, it shifts toward the shorter wavelengths (Figure 3). Many techniques have been developed to study quantitatively and qualitatively the status of vegetation from satellite images. The vegetation index concept was developed as a method of reducing the number of variables present in multi-spectral measurements down to one unique parameter. Vegetation indices are combinations of spectral channels, blended in such a way that they strengthen the spectral contribution of green vegetation. This is achieved by minimizing the disturbing influences of soil background, irradiance, solar position, yellow vegetation and atmospheric attenuation variations.

Climate change will cause shifts in areas suitable for cultivation of a wide range of crops. It could be used current and projected future climate data for ~2055, and the different models to predict the impact of climate change on different areas. Most detrimentally affected in terms of reduction of suitable areas for a range of crops will be sub-Saharan Africa and the Caribbean, areas with the least capacity to cope. Conversely, Europe and North America will see an increase in area suitable for cultivation. These regions have the greatest capacity to manage climate change impacts. Altered and unpredictable weather patterns can increase crop vulnerability to pests, diseases and the effects of extreme climate events such as high temperatures, droughts and torrential rains. Sequential extremes, such as droughts followed by intense flooding rains, are catastrophic in themselves and are compounded by ecological effects such as the expansion of the ranges of pathogens, diseases, and pests that affect human populations and agricultural production.

The impact of climate change on production of various crops varies markedly depending mainly on the region, growing season, the crops and their temperature thresholds. Cereals, oilseed and protein crops depend on temperature and, in many cases, day length, to reach maturity. Temperature increase may shorten the length of the growing period for these crops and, in the absence of compensatory management responses, reduce yields and change the area of cultivation by rendering unsuitable some currently cultivated areas and suitable, others not currently cultivated.

The challenges that climate change presents to humanity require an unprecedented ability to predict the responses of crops to environment and management. GIS and crop simulation models are two powerful and highly complementary tools that are increasingly used for such predictive analyses. The role of both technologies in predicting future situations centres around extrapolation. For GIS, extrapolation from the past based on correlation in a very loose sense plays an important role. The inherently spatial aspects of climate and climate change make them readily amenable for incorporation into a GIS-based analysis system. It is becoming ever more apparent that climatic changes are occurring non-uniformly across regions or agroecosystems. GIS provides a useful tool to capture this spatial heterogeneity and provides powerful ways in which to visualize and communicate the actual or potential changes that are occurring. A GIS-based framework has been the fundamental element of several major assessments of the potential impact of climate change on agriculture.

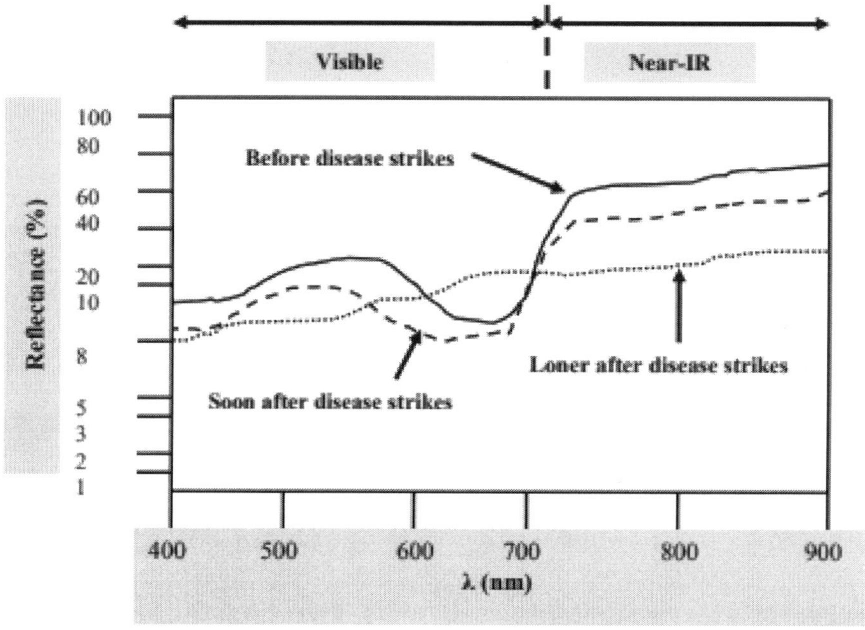

Fig. 3: Spectral reflectance changes in some plants impacted by diseases, drought, pollution or insect infestations.

The flexibility of GIS-based analysis systems to handle differing scenarios in a rapid and efficient manner is another important factor. The suite of advanced global general circulation models (GCMs) that inform major assessments such as those of the IPCC (e.g., Solomon et al. 2007), and the accompanying emission scenarios developed for the IPCC assessments (IPCC, 2000) form the basis of many climate change assessments. Increasingly, the outputs of the GCMs under differing emissions scenarios are available in data formats suitable for direct use in GIS-based systems. The availability of multiple GCM outputs, coupled to GIS based systems, has permitted increasing opportunities for analysis of spatial convergence or divergence of GCM outputs at global or regional scales. Common tasks within a GIS include the input, storage, manipulation, analysis and display (often in the form of maps or graphs) of georeferenced data. Mapping is a key output of any GIS, but it is certainly not the only functionality. Common data inputs include data in either vector or raster (gridded) formats. The latter are particularly useful for the representation of continuous data (e.g. climatic variables) and cell-by-cell modelling.

Globally, GIS is applied to disciplines ranging from managing utility networks to health, archaeology and ecology. Increasingly, it is a common component of climate change assessments. The geographic aspect of GIS makes it an interesting option for application to agricultural problems and priority setting because so many of the environmental and socio-economic factors that impact agriculture or agricultural research vary greatly over regions. Typical examples would include rainfall patterns, soil variability, disease and pest distribution, market locations, crop distributions, land-use patterns and human demographics (Table 2.19).

Hodson and White (2010) reported about the different uses for GIS as follows:

- GIS has seen widespread use for delineation of suitability zones and agroecological zonation. The ability to combine multi-thematic data based on common geography has been an extremely powerful tool. Common approaches to general crop suitability mapping have included the geographical overlay and intersection of key factors such as optimal temperature and moisture ranges, soil types and topographic features. GIS is perfectly suited for undertaking such analysis.
- Similar approaches have been undertaken to determine environmental niches in which wild relatives of crops are most likely to occur.
- Location-based climatic factors have also formed the basis of zonation's for targeting germplasm of major food crops.
- Climate-based mapping of potential pest and disease probability occurrence zones is also relevant to the scope of this review. This has been undertaken using very similar approaches to those described for wild relatives and crops.
- Abiotic stresses, such as drought, have also been assessed using GIS-based analytical approaches. Again, climate is a key driver and GIS captures the spatial variation that is essential to interpretation in an agricultural context.

Table 6: Examples of typical application themes for GIS/spatial technologies

Thematic area	Comment
Rainfall/climate patterns	Interpolated raster (gridded) surfaces derived from meteorological station data
Soil variability Disease and pest distribution Market Locations/accessibility Crop distributions y Land-use patterns Abiotic stresses Identification of wild species collection Site/suitability zones Crop suitability zones and agroecological zonation	Spatial variation of major soil types and derived soil properties Actual distributions and climatic suitability zones Accessibility surfaces based on least cost distance travel times Spatial allocation of reported agricultural census data into most likely areas using land use and suitability Satellite-derived land-conver estimates on varying spatial and temporal scales Gridded surfaces of human population density Modelled spatial distributions of key strreses, e.g. drought, heat Actual distributions and modeled ecological niches for important wild relatives crop species Climate, soil and landform-based agroecological zones for major crops

These examples illustrate how GIS has been the key technology applied to a range of differing agroecological themes. Spatial integration of multi-thematic data sets was a common element, but so too was use of climate data. This pivotal role of GIS in agroclimatic analysis is relevant whether the analysis is based on current or historical climate data or predicted future climate data. The increasing availability of outputs from a range of GCMs under varying emission scenarios is permitting a range of GIS-based assessments of the potential characteristics of future crop production zones, and associated abiotic and biotic stresses. Several illustrative case studies will be described in succeeding sections that build upon the examples and themes already outlined. The application of GIS-based systems to agroclimatic analysis under current climate conditions has already been outlined. The availability of a range of GCM outputs run under a suite of emission scenarios is now permitting similar approaches for potential future climates. With any such approaches it should always be borne in mind that outputs from the GCMs are not precise and variation occurs between different models and scenarios. In addition, for agricultural assessments downscaled GCM results are usually required and this introduces another set of uncertainty. Despite these caveats, the results of such studies can provide useful indications of the potential magnitude of change and the spatial variation that may occur. Selected examples are given below in order to illustrate the range of approaches being undertaken.

Most vegetation indices utilize the red and near-infrared bands. For example, based on the reflectance difference that green vegetation displays between the visible region (10 percent) and the near-infrared region (50 per cent) of the electromagnetic spectrum (e.g. in channels 1 and 2 of the AVHRR images of the NOAA satellites), the Normalized Difference Vegetation Index (NDVI) can be expressed as:

$$NDVI = (NIR - R)/(NIR + R)$$

Where: R is the red band and NIR is the near-infrared band reflectance. The range of values obtained by the NDVI is between −1 and +1. Only the positive values correspond with the vegetated zones, and the higher the index, the greater the chlorophyll content of the target. Negative values, generated by a higher reflectance in the visible region than in the infrared region, are due to clouds, snow, bare soil and rock. The value of the NDVI can change depending on the land use, the season, the hydric situation and the climate of the area (Figure4).

To quantify pigments in vegetation, hyper-spectral data with high radiometric and spatial resolution are required. To this end, airborne instruments such as the AVIRIS (Airborne Visible/Infrared Imaging Spectrometer) and HyMap have been used, as well as more recent experimental space borne imaging spectrometers such as the Compact High Resolution Imaging Spectrometer (CHRIS) and Hyperion. Before performing image analysis for thematic land cover or vegetation mapping, one must choose or develop a classification system which meets the needs of the problem to be addressed. A workable classification system must meet at least the following criteria, i.e. it must be:

- Exhaustive, i.e. all possible land covers or conditions must be assigned to a class;
- Mutually exclusive, i.e. any land cover must be assignable to one and only one class;
- Hierarchical, i.e. begin at a general level and divide each class into subclasses at the next lower level; ¾ Flexible, i.e. allow the production of many different specialty maps (e.g. data layers of a GIS) and be able to respond to changing needs.

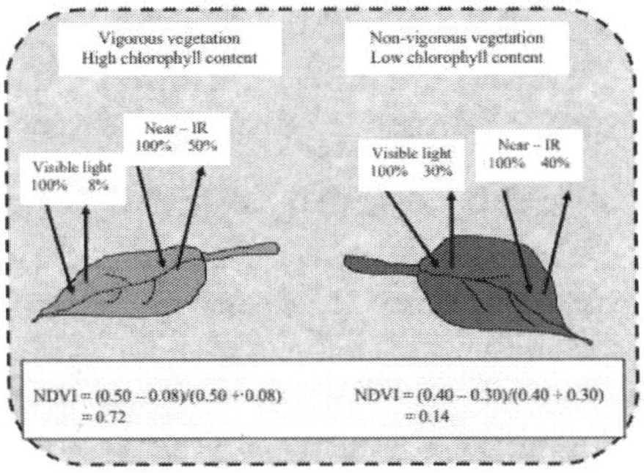

Fig. 4: The NDVI algorithm capitalizes on the fact that green vegetation reflects less visible light and more near-IR, whereas sparse or less green vegetation reflects a greater portion of the visible and less near-IR. By combining these reflectance characteristics in a ratio, NDVI provides a useful index of photosynthetic activity.

RELATED TERMINOLOGY

Abatement: Abatement is the word which is used to denote the result of decreased Greenhouse Gases Emission. This can also be taken as an activity to lessen the effects of Greenhouse Effect. 2. Aboveground Biomass: All living biomass above the soil including the stem, stump, branches, bark, seeds and foliage is known as aboveground biomass.

Abrupt Climate Change: Sudden (on the order of decades), large changes in some major component of the climate system, with rapid, widespread effects.

Absolute Humidity: The quantity of water vapour in a given volume of air expressed by mass is known as absolute humidity.

Absolute Risk: A quantitative or qualitative prediction of the likelihood and significance of a given impact is known as absolute risk. In the Voluntary Carbon Standard (VCS), the level of absolute risk can be calculated using the 'likelihood × significance' methodology. The calculated risk can then be converted into a risk classification.

Accelerated Erosion: Erosion is a natural phenomenon which results in soil losses and water quality degradation. Accelerated erosion is generally caused by activities that disturb or expose the soil to the erosive forces of gravity and rainwater. Climatic or weather conditions combined with human activity can accentuate soil erosion. For example, severe and intense storm events may increase the rate of accelerated erosion.

Adaptation to Climate Change: An initiatives and measures to reduce the vulnerability of natural and human systems against actual or expected Climate Change effects is known as adaptation to climate change. Crucial to reducing vulnerability to Climate Change, understanding how individuals, groups and natural systems can prepare for and respond to changes in climate is known as adaptation.

Adaptation: Adjustment or preparation of natural or human systems to a new or changing environment which moderates harm or exploits beneficial opportunities.

Adaptive Capacity: The ability of a system to adjust to climate change (including climate variability and extremes) to moderate potential damages, to take advantage of opportunities, or to cope with the consequences.

Aerosols: A collection of airborne solid or liquid particles, with a typical size between 0.01 and 10 mm that reside in the atmosphere for at least several hours is known as aerosols. Aerosols may be of either natural or anthropogenic origin. Aerosols may influence climate in two ways: directly through scattering and absorbing radiation and indirectly through acting as condensation nuclei for cloud formation or modifying the optical properties and lifetime of clouds.

Aerosols: Small particles or liquid droplets in the atmosphere that can absorb or reflect sunlight depending on their composition.

Afforestation: Planting of new forests on lands that historically have not contained forests.

Agroecology: The application of ecological principles to the production of food, fuel, fiber and pharmaceuticals is known as agroecology. The term encompasses a broad range of approaches and is considered a science, a movement and a practice.

Agroforestry: An ecologically based natural resource management system in which trees are integrated in farmland and rangeland is known as agroforestry.

Albedo: The amount of solar radiation reflected from an object or surface, often expressed as a percentage.

Alternative Energy: Energy derived from non-fossil-fuel sources or energy sources other than the traditional forest product and commercial energy items is known as alternative energy. They are tidal power, wind power, solar power etc.

Alternative Energy: Energy derived from nontraditional sources (e.g., compressed natural gas, solar, hydroelectric, wind). [1]

Anthropogenic: Made by people or resulting from human activities. Usually used in the context of emissions that are produced as a result of human activities. [3]

Atmosphere: The gaseous envelope surrounding the Earth. The dry atmosphere consists almost entirely of nitrogen (78.1% volume mixing ratio) and oxygen (20.9% volume mixing ratio), together with a number of trace gases, such as argon (0.93% volume mixing ratio), helium, radiatively active greenhouse gases such as carbon dioxide (0.035% volume mixing ratio), and ozone.

Atmospheric Lifetime: Atmospheric lifetime is the average time that a molecule resides in the atmosphere before it is removed by chemical reaction or deposition. In general, if a quantity of a compound is emitted into the atmosphere at a particular time, about 35 percent of that quantity will remain in the atmosphere at the end of the compound's atmospheric lifetime. This fraction will continue to decrease in an exponential way, so that about 15 percent of the quantity will remain at the end of two times the atmospheric lifetime, etc.

Bio-Carbon Fund: The Bio-Carbon Fund provides carbon finance for projects that sequester or conserve greenhouse gases in forests, agro- and other ecosystems. Through its focus on bio-carbon, or 'sinks', it delivers carbon finance to many developing countries that otherwise have few opportunities to participate in the Clean Development Mechanism (CDM), or to countries with economies in transition through Joint Implementation (JI).

Biofuels: Gas or liquid fuel made from plant material. Includes wood, wood waste, wood liquors, peat, railroad ties, wood sludge, spent sulfite liquors, agricultural waste, straw, tires, fish oils, tall oil, sludge waste, waste alcohol, municipal solid waste, landfill gases, other waste, and ethanol blended into motor gasoline.

Biogeochemical Cycle: Movements through the Earth system of key chemical constituents essential to life, such as carbon, nitrogen, oxygen, and phosphorus.

Biological Hazard: Process or phenomenon of organic origin or conveyed by biological vectors including exposure to pathogenic micro-organisms, toxins and bioactive substances that may cause loss of life, injury, illness or other health impacts, property damage, loss of livelihoods and services, social and economic disruption or environmental damage is known as biological hazard.

Biomass: Materials that are biological in origin, including organic material (both living and dead) from above and below ground, for example, trees, crops, grasses, tree litter, roots, and animals and animal waste.

Biosphere: The part of the Earth system comprising all ecosystems and living organisms, in the atmosphere, on land (terrestrial biosphere) or in the oceans (marine biosphere), including derived dead organic matter, such as litter, soil organic matter and oceanic detritus.

Black Carbon (BC): A climate forcing agent formed through the incomplete combustion of fossil fuels, bio-fuel and biomass and is emitted in both anthropogenic and naturally occurring soot is known as black carbon. It has recently emerged as a major contributor to global climate change, now attributed as the second largest contributor to global warming.

Black Carbon Aerosol: Black carbon (BC) is the most strongly light-absorbing component of particulate matter (PM), and is formed by the incomplete combustion of fossil fuels, biofuels, and biomass. It is emitted directly into the atmosphere in the form of fine particles ($PM_{2.5}$).

Borehole: Any exploratory hole drilled into the Earth or ice to gather geophysical data. Climate researchers often take ice core samples, a type of borehole, to predict atmospheric composition in earlier years.

Carbon Capture and Sequestration: Carbon capture and sequestration (CCS) is a set of technologies that can greatly reduce carbon dioxide emissions from new and existing coal- and gas-fired power plants, industrial processes, and other stationary sources of carbon dioxide. It is a three-step process that includes capture of carbon dioxide from power plants or industrial sources; transport of the captured and compressed carbon dioxide (usually in pipelines); and underground injection and geologic sequestration, or permanent storage, of that carbon dioxide in rock formations that contain tiny openings or pores that trap and hold the carbon dioxide.

Carbon Cycle: All parts (reservoirs) and fluxes of carbon. The cycle is usually thought of as four main reservoirs of carbon interconnected by pathways of exchange. The reservoirs are the atmosphere, terrestrial biosphere (usually includes freshwater systems), oceans, and sediments (includes fossil fuels). The annual movements of carbon, the carbon exchanges between reservoirs, occur because of various chemical, physical, geological, and biological processes. The ocean contains the largest pool of carbon near the surface of the Earth, but most of that pool is not involved with rapid exchange with the atmosphere.

Carbon Dioxide Equivalent: A metric measure used to compare the emissions from various greenhouse gases based upon their global warming potential (GWP). Carbon dioxide equivalents are commonly expressed as "million metric tons of carbon dioxide equivalents ($MMTCO_2Eq$)." The carbon dioxide equivalent for a gas is derived by multiplying the tons of the gas by the associated GWP.

Carbon Dioxide Fertilization: The enhancement of the growth of plants as a result of increased atmospheric CO_2 concentration. Depending on their mechanism of photosynthesis, certain types of plants are more sensitive to changes in atmospheric CO_2 concentration.

Carbon Dioxide: A naturally occurring gas, and also a by-product of burning fossil fuels and biomass, as well as land-use changes and other industrial processes. It is the principal human caused greenhouse gas that affects the Earth's radiative balance

Carbon Footprint: The total amount of greenhouse gases that are emitted into the atmosphere each year by a person, family, building, organization, or company. A persons carbon footprint includes greenhouse gas emissions from fuel that an individual burns directly, such as by heating a home or riding in a car. It also includes greenhouse gases that come from producing the goods or services that the individual uses, including emissions from power plants that make electricity, factories that make products, and landfills where trash gets sent.

Carbon Sequestration: Terrestrial, or biologic, carbon sequestration is the process by which trees and plants absorb carbon dioxide, release the oxygen, and store the carbon. Geologic sequestration is one step in the process of carbon capture and sequestration (CCS), and involves injecting carbon dioxide deep underground where it stays permanently.

Chlorofluorocarbons: Gases covered under the 1987 Montreal Protocol and used for refrigeration, air conditioning, packaging, insulation, solvents, or aerosol propellants. Since they are not destroyed in the lower atmosphere, CFCs drift into the upper atmosphere where, given suitable conditions, they break down ozone. These gases are being replaced by other compounds: hydrochlorofluorocarbons, an interim replacement for CFCs that are also covered under the Montreal Protocol, and hydrofluorocarbons, which are covered under the Kyoto Protocol. All these substances are also greenhouse gases.

Climate Change: Climate change refers to any significant change in the measures of climate lasting for an extended period of time. In other words, climate change includes major changes in temperature, precipitation, or wind patterns, among others, that occur over several decades or longer.

Climate Feedback: A process that acts to amplify or reduce direct warming or cooling effects.

Climate Lag: The delay that occurs in climate change as a result of some factor that changes only very slowly.

Climate Model: A quantitative way of representing the interactions of the atmosphere, oceans, land surface, and ice. Models can range from relatively simple to quite comprehensive.

Climate Sensitivity: In Intergovernmental Panel on Climate Change (IPCC) reports, equilibrium climate sensitivity refers to the equilibrium change in global mean surface temperature following a doubling of the atmospheric (equivalent) CO2 concentration.

Climate System (or Earth System): The five physical components (atmosphere, hydrosphere, cryosphere, lithosphere, and biosphere) that are responsible for the climate and its variations.

Climate : Climate in a narrow sense is usually defined as the "average weather," or more rigorously, as the statistical description in terms of the mean and variability of relevant quantities over a period of time ranging from months to thousands of years. The classical period is 3 decades, as defined by the World Meteorological Organization (WMO). These quantities are most often surface variables such as temperature, precipitation, and wind. Climate in a wider sense is the state, including a statistical description, of the climate system.

Coal Mine Methane: Coal mine methane is the subset of coalbed methane that is released from the coal seams during the process of coal mining.

Coalbed Methane: Coalbed methane is methane contained in coal seams, and is often referred to as virgin coalbed methane, or coal seam gas.

Co-Benefit: The benefits of policies that are implemented for various reasons at the same time including climate change mitigation acknowledging that most policies designed to address greenhouse gas mitigation also have other, often at least equally important, rationales (e.g., related to objectives of development, sustainability, and equity).

Cold Waves: Weather phenomenon that is distinguished by a cooling of the air is known as cold waves.

Concentration: Amount of a chemical in a particular volume or weight of air, water, soil, or other medium.

Conference of the Parties: The supreme body of the United Nations Framework Convention on Climate Change (UNFCCC). It comprises more than 180 nations that have ratified the Convention. Its first session was held in Berlin, Germany, in 1995 and it is expected to continue meeting on a yearly basis. The COP's role is to promote and review the implementation of the Convention. It will periodically review existing commitments in light of the Convention's objective, new scientific findings, and the effectiveness of national climate change programs.

Convention of Biological Diversity (or biodiversity): Known informally as the Biodiversity Convention, is an international legally binding treaty. It is one of the three environmental conventions to arise from the UN Earth Summit in Rio de Janeiro in 1992. Its objective is to develop national strategies for the conservation and sustainable use of biological diversity.

Coral Bleaching: The process in which a coral colony, under environmental stress expels the microscopic algae (zooxanthellae) that live in symbiosis with their host organisms (polyps). The affected coral colony appears whitened.

Cryosphere: One of the interrelated components of the Earth's system, the cryosphere is frozen water in the form of snow, permanently frozen ground (permafrost), floating ice, and glaciers. Fluctuations in the volume of the cryosphere cause changes in ocean sea level, which directly impact the atmosphere and biosphere.

Deforestation: Those practices or processes that result in the conversion of forested lands for non-forest uses. Deforestation contributes to increasing carbon dioxide concentrations for two reasons: 1) the burning or decomposition of the wood releases carbon dioxide; and 2) trees that once removed carbon dioxide from the atmosphere in the process of photosynthesis are no longer present.

Desertification: Land degradation in arid, semi-arid, and dry sub-humid areas resulting from various factors, including climatic variations and human activities.

Disaster Risk Management: The systematic process of using administrative directives, organizations, and operational skills and capacities to implement strategies, policies and improved coping capacities in order to lessen the adverse impacts of hazards and the possibility of disaster is knows as disaster risk management. This term is an extension of the more general term "risk management" to address the specific issue of disaster risks. Disaster risk management aims to avoid, lessen or transfer the adverse effects of hazards through activities and measures for prevention, mitigation and preparedness. 166.

Disaster Risk Reduction: The concept and practice of reducing disaster risks through systematic efforts to analyze and manage the causal factors of disasters including through reduced exposure to hazards, lessened vulnerability of people and property, wise management of land and the environment and improved preparedness for adverse events is known as disaster risk reduction.

Dryland Farming: A technique that uses soil moisture conservation and seed selection to optimize production under dry conditions.

Early Warning System: The set of capacities needed to generate and disseminate timely and meaningful warning information to enable individuals, communities, and organizations threatened by a hazard to prepare and to act appropriately in sufficient time to reduce the possibility of harm or loss is known as early warning system.

Earth Summit: Conference held in Rio de Janeiro (June 3-14, 1992) to reconcile worldwide economic development with environmental protection is known as earth summit. It was the largest gathering of world leaders in history with 117 heads of state and representatives of 178 countries.

Eccentricity: The extent to which the Earth's orbit around the Sun departs from a perfect circle.

Eco Tax: An eco tax is designated for the purpose of influencing human behavior (specifically economic behavior) to follow an ecologically benign path.

Ecological Footprint: A measure of human demand on the ecosystem is known as ecological footprint. It compares human consumption of natural resources with the planet's ecological capacity to regenerate them.

Ecosystem: Any natural unit or entity including living and non-living parts that interact to produce a stable system through cyclic exchange of materials.

El Niño - Southern Oscillation (ENSO): El Niño in its original sense, is a warm water current that periodically flows along the coast of Ecuador and Peru, disrupting the local fishery. This oceanic event is associated with a fluctuation of the intertropical surface pressure pattern and circulation in the Indian and Pacific Oceans, called the Southern Oscillation. This coupled atmosphere-ocean phenomenon is collectively known as El Niño-Southern Oscillation.

Emissions Factor: A unique value for scaling emissions to activity data in terms of a standard rate of emissions per unit of activity (e.g., grams of carbon dioxide emitted per barrel of fossil fuel consumed, or per pound of product produced).

Emissions: The release of a substance (usually a gas when referring to the subject of climate change) into the atmosphere.

Endangered: Species whose population is so small that it is in danger of becoming extinct is known as endangered.

Endemic: It is restricted or peculiar to a locality or region. With regard to human health, endemic can refer to a disease or agent present or usually prevalent in a population or geographical area at all times.

Energy Efficiency: Using less energy to provide the same service.

Energy Star: A U.S. Environmental Protection Agency voluntary program that helps businesses and individuals save money and protect our climate through superior energy efficiency.

Enhanced Greenhouse Effect: The concept that the natural greenhouse effect has been enhanced by increased atmospheric concentrations of greenhouse gases (such as CO_2 and methane) emitted as a result of human activities. These added greenhouse gases cause the earth to warm.

Enteric Fermentation: Livestock, especially cattle, produce methane as part of their digestion. This process is called enteric fermentation, and it represents one third of the emissions from the agriculture sector.

Eustatic Sea Level Rise: Change in global average sea level brought about by an alteration to the volume of the world ocean is known as eustatic sea level rise.

Evaporation: The process by which water changes from a liquid to a gas or vapor.

Evapotranspiration: The combined process of evaporation from the Earth's surface and transpiration from vegetation.

Extinction: The complete disappearance of an entire species is known as extinction.

Feedback Mechanisms: Factors which increase or amplify (positive feedback) or decrease (negative feedback) the rate of a process. An example of positive climatic feedback is the ice-albedo feedback.

Fluorinated Gases: Powerful synthetic greenhouse gases such as hydrofluorocarbons, perfluorocarbons, and sulfur hexafluoride that are emitted from a variety of industrial processes. Fluorinated gases are sometimes used as substitutes for stratospheric ozone-depleting substances (e.g., chlorofluorocarbons, hydrochlorofluorocarbons, and halons) and are often used in coolants, foaming agents, fire extinguishers, solvents, pesticides, and aerosol propellants. These gases are emitted in small quantities compared to carbon dioxide (CO_2), methane (CH_4), or nitrous oxide (N_2O), but because they are potent greenhouse gases, they are sometimes referred to as High Global Warming Potential gases (High GWP gases).

Fluorocarbons: Carbon-fluorine compounds that often contain other elements such as hydrogen, chlorine, or bromine. Common fluorocarbons include chlorofluorocarbons (CFCs), hydrochlorofluorocarbons (HCFCs), hydrofluorocarbons (HFCs), and perfluorocarbons (PFCs).

Forcing Mechanism: A process that alters the energy balance of the climate system, i.e. changes the relative balance between incoming solar radiation and outgoing infrared radiation from Earth. Such mechanisms include changes in solar irradiance, volcanic eruptions, and enhancement of the natural greenhouse effect by emissions of greenhouse gases.

Forecast: Definite statement or statistical estimate of the likely occurrence of a future event or conditions for a specific area.

Forest Carbon Partnership Facility: The Forest Carbon Partnership Facility (FCPF) which became operational in June 2008, is a global partnership focused on reducing emissions from deforestation and forest degradation, forest carbon stock conservation, sustainable management of forests and enhancement of forest carbon stocks (REDD+).

Fossil Fuel: A general term for organic materials formed from decayed plants and animals that have been converted to crude oil, coal, natural gas, or heavy oils by exposure to heat and pressure in the earth's crust over hundreds of millions of years.

Fuel Switching: In general, this is substituting one type of fuel for another. In the climate-change discussion it is implicit that the substituted fuel produces lower carbon emissions per unit energy produced than the original fuel, e.g., natural gas for coal.

Geosphere: The soils, sediments, and rock layers of the Earth's crust, both continental and beneath the ocean floors.

Glacier: A multi-year surplus accumulation of snowfall in excess of snowmelt on land and resulting in a mass of ice at least 0.1 km2 in area that shows some evidence of movement in response to gravity. A glacier may terminate on land or in water. Glacier ice is the largest reservoir of fresh water on Earth, and second only to the oceans as the largest reservoir of total water. Glaciers are found on every continent except Australia.

Global Average Temperature: An estimate of Earths means surface air temperature averaged over the entire planet.

Global Warming Potential:A measure of the total energy that a gas absorbs over a particular period of time (usually 100 years), compared to carbon dioxide

Habitat Fragmentation: A process during which larger areas of habitat are broken into a number of smaller patches of smaller total area, isolated from each other by a matrix of habitats unlike the original habitat.

Halocarbons:Compounds containing either chlorine, bromine or fluorine and carbon. Such compounds can act as powerful greenhouse gases in the atmosphere. The chlorine and bromine containing halocarbons are also involved in the depletion of the ozone layer.

Heat Island: An urban area characterized by temperatures higher than those of the surrounding non-urban area. As urban areas develop, buildings, roads, and other infrastructure replace open land and vegetation. These surfaces absorb more solar energy, which can create higher temperatures in urban areas.

Heat Waves: A prolonged period of excessive heat, often combined with excessive humidity.

Ice Core: A cylindrical section of ice removed from a glacier or an ice sheet in order to study climate patterns of the past. By performing chemical analyses on the air trapped in the ice, scientists can estimate the percentage of carbon dioxide and other trace gases in the atmosphere at a given time. Analysis of the ice itself can give some indication of historic temperatures.

Indirect Emissions: Indirect emissions from a building, home or business are those emissions of greenhouse gases that occur as a result of the generation of electricity used in that building. These emissions are called "indirect" because the actual emissions occur at the power plant which generates the electricity, not at the building using the electricity.

Industrial Revolution: A period of rapid industrial growth with far-reaching social and economic consequences, beginning in England during the second half of the 18th century and spreading to Europe and later to other countries including the United States. The industrial revolution marks the beginning of a strong increase in combustion of fossil fuels and related emissions of carbon dioxide.

Infrared Radiation: Infrared radiation consists of light whose wavelength is longer than the red color in the visible part of the spectrum, but shorter than microwave radiation. Infrared radiation can be perceived as heat. The Earths surface, the atmosphere, and clouds all emit infrared radiation, which is also known as terrestrial or long-wave radiation. In contrast, solar radiation is mainly short-wave radiation because of the temperature of the Sun.

Intergovernmental Panel on climate Change (IPCC): The IPCC was established jointly by the United Nations Environment Programme and the World Meteorological Organization in 1988. The purpose of the IPCC is to assess information in the scientific and technical literature related to all significant components of the issue of climate change. The IPCC draws upon hundreds of the world's expert scientists as authors and thousands as expert reviewers.

Landfill: Land waste disposal site in which waste is generally spread in thin layers, compacted, and covered with a fresh layer of soil each day.

$MMTCO_2Eq$ = (million metric tons of a gas) * (GWP of the gas)

Municipal Solid Waste (MSW): Residential solid waste and some non-hazardous commercial, institutional, and industrial wastes. This material is generally sent to municipal landfills for disposal.

Non-Methane Volatile Organic Compounds (NMVOCs): Organic compounds, other than methane, that participate in atmospheric photochemical reactions.

Ocean Acidification: Increased concentrations of carbon dioxide in sea water causing a measurable increase in acidity (i.e., a reduction in ocean pH). This may lead to reduced calcification rates of calcifying organisms such as corals, mollusks, algae and crustaceans.

Ozone Depleting Substance (ODS): A family of man-made compounds that includes, but are not limited to, chlorofluorocarbons (CFCs), bromofluorocarbons (halons), methyl chloroform, carbon tetrachloride, methyl bromide, and hydrochlorofluorocarbons (HCFCs). These compounds have been shown to deplete stratospheric ozone, and therefore are typically referred to as ODSs.

Ozone Layer: The layer of ozone that begins approximately 15 km above Earth and thins to an almost negligible amount at about 50 km, shields the Earth from harmful ultraviolet radiation from the sun. The highest natural concentration of ozone (approximately 10 parts per million by volume) occurs in the stratosphere at approximately 25 km above Earth. The stratospheric ozone concentration changes throughout the year as stratospheric circulation changes with the seasons. Natural events such as volcanoes and solar flares can produce changes in ozone concentration, but man-made changes are of the greatest concern.

Ozone Precursors: Chemical compounds, such as carbon monoxide, methane, non-methane hydrocarbons, and nitrogen oxides, which in the presence of solar radiation react with other chemical compounds to form ozone, mainly in the troposphere.

Ultraviolet Radiation (UV): The energy range just beyond the violet end of the visible spectrum. Although ultraviolet radiation constitutes only about 5 percent of the total energy emitted from the sun, it is the major energy source for the stratosphere and mesosphere, playing a dominant role in both energy balance and chemical composition.

Vulnerability: The degree to which a system is susceptible to, or unable to cope with, adverse effects of climate change, including climate variability and extremes. Vulnerability is a function of the character, magnitude, and rate of climate variation to which a system is exposed; its sensitivity; and its adaptive capacity.

Wastewater: Water that has been used and contains dissolved or suspended waste materials.

Water Vapor: The most abundant greenhouse gas, it is the water present in the atmosphere in gaseous form. Water vapor is an important part of the natural greenhouse effect.

REFERENCES

Abadin, H. G.; C. H. S. J. Chou; F. T. Llados (2007). Health effects classification and its role in the derivation of minimal risk levels: immunological effects. Regul Toxicol Pharm; Vol. 47 (3): 249–56.

ABS (2008). Australian Water Account. Australian Bureau of Statistics, Canberra.

Abu-Zeid, M.; A. K. Biswas (1991). Some major implications of climatic fluctuations on water management. Water Resources Development, Vol. 7: 74-81.

Adams, J. (2011). Vegetation-Climate Interaction: How Plants Make the Global Environment, 2nd. Praxis Publishing Ltd, Chichester, UK, Sperger Berlin Heidelberg, NY.

Adams, J. B.; M. E. Mann; C. M. Ammann (2003). Proxy evidence for an El Nino like response to volcanic forcing. Nature, 426, 274-278.

Adams, R. M.; C. Rosenzweig; R. M. Peart; J. T. Ritchie; B. A. McCarl; J. D. Glyer; et al. (1990). Global climate change and US agriculture. Nature, Vol. 345: 219–224.

Adamson, D. A.; F. Gasse; F. A. Street; M. A. J. Williams (1980). Late quaternary history of the Nile. Nature, Vol. 228: 50–55.

Adhya, T. K.; P. D. Sharma; A. K. Gogoi (2009). Mitigating Greenhouse Gas Emission from Agriculture. In: S. N. Singh (ed.), Climate Change and Crops, Environmental Science and Engineering, Springer-Verlag Berlin Heidelberg.

Agrawala, S.; A. Moehner; M. El Raey; D. Conway; M. van Aalst; M. Hagenstad; J.

Smith (2004). Development and climate change in Egypt: Focus on coastal resources and the Nile. Environment Directorate, Environment Policy Committee,

Working Party on Global and Structural Policies, Working Party on Development Co-operation and Environment. COM/ENV/EPOC/DCD/DAC(2004)1/FINAL.

Ainsworth, E. A.; A. Rogers (2007). The response of photosynthesis and stomatal conductance to rising [CO2]: Mechanisms and environmental interactions. Plant, Cell and Environment, Vol. 30: 258–270.

Ainsworth, E. A.; S. P. Long (2005). What have we learned from 15 years of free-air CO2 enrichment (FACE)? A meta-analytic review of the responses of photosynthesis, canopy properties and plant production to rising CO2. New Phytol, Vol. 165: 351–372.

Alcamo, J.; K. Kok; G. Busch; J. Priess; B. Eickhout; M. D. A. Rounsevell; D.

Rothman; M. Heistermann (2006). Searching for the future of land: scenarios from the local to global scale. In: Land-use and Land Cover Change: Local Processes, Global Impacts. Global Change IGBP Series, Springer-Verlag, Berlin, pp. 137–156.

Alcamo, J.; J. M. Moreno; B. Nováky; M. Bindi; R. Corobov; R. J. N. Devoy; C. Giannakopoulos; E. Martin; J. E. Olesen; A. Shvidenko (2007). Europe. Climate change 2007: impacts, adaptation and vulnerability. In: Parry ML, Canziani OF, Palutikof JP, van der Linden PJ, Hanson CE (eds.) Contribution of working group II to the fourth assessment report of the intergovernmental panel on climate change. Cambridge University Press, Cambridge, pp 541–580.

Aldaya, M. M.; A. Y. Hoekstra; J. A. Allan (2008). Strategic Importance of Green Water in International Crop Trade. Value of Water Research Report Series No. 25. UNESCO-IHE Institute for Water Education, Delft.\

Allan, J. A. (1997). Virtual Water: A Long Term Solution for Water Short Middle Eastern Economies? SOAS Water Issues Study Group, School of Oriental andAfrican Studies/King's College-London (Occasional Paper 3).

Allan, J. A. (2009). Nile basin asymmetries: a closed fresh water resource, soil waters potential, the political economy, and Nile transboundary hydropolitics. In: H. J.

Dumont (ed.), The Nile. Monographiae Biologicae, Vol. 89: 749–770. Springer, Dordrecht.

Allen, D. E.; B. P. Singh; R. C. Dalal (2011). Soil Health Indicators under Climate Change: A Review of Current Knowledge. In: Soil Health and Climate Change (Eds. Singh, B. P.; A. L. Cowie; K. Y. Chan). Soil Biology Series, Vol. 29.

Altieri, M. A.; and C. I. Nicholls (2004). Biodiversity and pest management in agroecosystems. Second Edition. The Haworth Press, Inc., Food Products Press, Binghamton, NY.

Alongi, D. M. (2002). Present state and future of the world's mangrove forests. Environmental Conservation, Vol. 29: 331–349.

Anderson, J. R., Hardy, E. E., Roach, J. T. & Witmer, R. E. (1976). A land use and land cover classification system for use with remote sensor data. US Geological Survey Professional Paper, 964, 28. Washington, DC.

Anwar, M. R., O'Leary, G., McNeil, D., Hossain, H. and Nelson, R. (2007). Climate change impact on rainfed wheat in south-eastern Australia. Proceedings of the 13th Australian Society of Agronomy Conference 2006. Field Crops Research 104, 139–147.

Araujo, M. B.; M. Cabeza; W. Thuiller; L. Hannah; P. H. Williams (2004). Would climate change drive species out of reserves? An assessment of existing reserveselection methods. Global Change Biology, Vol. 10: 1618–1626.

Assad, E. D.; H. S. Pinto; J. Jr. Zullo; A. M. H. Ávila (2004). Impacto das mudanças climáticas no zoneamento agroclimático do café no Brasil. Pesquisa Agropecuária Brasileira, 39, 1057–1064.

Attri, S. D., L. S. Rathore, M. V. K. Sivakumar, S. K. Dash (2011). Challenges and opportunities in agrometeorology. Springer-Verlag Berlin Heidelberg.

AGU (2008). Human impacts on climate. American Geophysical Union position statement. EOS, 89, 41.

Baertsch-Ritter, N., Keller, J., Dommen, J., Prevot, A. S. H. (2004). Effects of variousBmeteorological conditions and spatial emission resolutions on the ozone concentration and ROG/NOx limitation in the Milan area (I). Atmos. Chem. Phys. 4, 423–438.

Bailey, R. G. (1986). The Zaire River System. In B. R. Davies & K. F. Walker (eds), The Ecology of River Systems. Monographie Biologicae, Vol. 60: 201–214. Junk, The Hague.

Bard, S. M. (1999). Global transport of anthropogenic contaminants and the consequences for the Arctic marine ecosystem. Mar Pollut Bull; Vol. 38 (5):356– 79.

Barnabás, B.; K. Järgen; A. Fehér (2008). The effect of drought and heat stress on reproductive processes in cereals. Plant, Cell and Environment, Vol. 31: 11–38.

Bartlein, P. J.; C. Whitlock; S. L. Shafter (1997). Future climate in the Yellowstone National Park region and its potential impact on vegetation. Conservation Biology, Vol. 11: 782–792.

Barker, T., Bashmakov, I., Bernstein, L., Bogner, J.E., Bosch, P.R., Dave, R., Davidson, O.R., Fisher, B.S., Gupta, S., Halsnaes, K., Heij, G.J., Kahn-Ribeiro, S., Kobayashi, S., Levine, M.D., Martino, D.L., Masera, O., Metz, B., Meyer, L.A., Nabuurs, G.-J., Najam, A., Nakicenovic, N., Rogner, H.-H., Roy, J., Sathaye, J., Schock, R., Shukla, P., Sims, R.E.H., Smith, P., Tirpak, D.A., UrgeVorsatz, D. and Zhou, D. (2007). Technical summary. In: Metz, B., Davidson, O.R., Bosch, P.R., Dave, R. and Meyer, L.A. (eds) Climate Change 2007: Mitigation. Contribution of Working Group III to the Fourth Assessment Report of the Intergovernmental Panel on Climate Change. Cambridge University Press, Cambridge, UK, pp. 620–690.

Barrett, C. B. (2010). Measuring food insecurity. Science, Vol. 327 (5967): 825–828.

Batjes, N. (2009). Harmonized soil profile data for applications at global and continental scales: updates to the WISE database. Soil Use and Management 25, 124–127.

Bates, B. C.; Z. W. Kundzewicz, S. Wu; J. P. Palutikof (2008). Climate Change and Water, IPCC Technical Paper VI, Geneva: Intergovernmental Panel on Climate Change.

Beadle, L. C. (1981). The Inland Waters of Tropical Africa. An Introduction to Tropical Limnology. Second edition. Longman, London/New York, 475 pp.

Beatley, T. (1991). Protecting biodiversity in coastal environments – introduction and overview. Coastal Management, Vol. 19: 1–19.

Beaudry, R. M. (1993). Effect of carbon dioxide partial pressure on blueberry fruit respiration and respiratory quotient. Postharvest Biology and Technology, Vol. 3: 249–258.

Bell, M. L.; R. Goldberg; C. Hogrefe; P. L. Kinney; K. Knowlton; B. Lynn; et al. (2007). Climate change, ambient ozone, and health in 50 US cities. Climatic Change; Vol. 82 (1–2):61–76.

Bell, M. J.; G. R. Stirling; C. E. Pankhurst (2007). Management impacts on health of soils supporting Australian grain and sugarcane industries. Soil Till Res 97:256– 271.

Bender, R. J.; J. K. Brecht; C. A. Campbell (1994). Responses of 'Kent' and 'Tommy Atkins' mangoes to reduced O2 and elevated CO2. Procedures of the Florida State Horticultural Society, 107: 274–277.

Bengtsson, L.; K.I. Hodges, and N. Keenlyside (2009). Will extratropical storms intensify in a warmer climate?. J Climate 22, 2276–2301.

Bennett, L. T.; M. A. Adams (2004). Assessment of ecological effects due to forest harvesting: approaches and statistical issues. J. Appl. Ecol., 41, 585-598.

Bindi, M.; L. Fibbi; F. Miglietta (2001). Free air CO2 Enrichment (FACE) of grapevine (Vitis vinifera L.): II. Growth and quality of grape and wine in response to elevated CO2 concentrations. European Journal of Agronomy, Vol. 14: 145–155.

Bicheron, P., Leroy, M., Brockmann, C., Krämer, U., Miras, B., Huc, M., Niño, F., Defourny, P., Vancutsem, C., Arino, O., Ranera, F., Petit, D., Amberg, V., Berthelot, B. and Gross, D. (2006). Globcover: a 300 m global land cover product for 2005 using ENVISAT MERIS time series. In: Sobrino, J.A. (ed.) Proceedings of the Second International Symposium on Recent Advances in Quantitative Remote Sensing. Servicio de Publicaciones, Universitat de Valencia, Valencia, Spain, pp. 538–542.

Bishai, H. M.; S. A. Abdel-Malek; M. T. Khalil (2000). Lake Nasser. Publication of the

National Biodiversity Unit No. 11, Cairo, 577 pp.

Bomhard, B.; D. M. Richardson; J. S. Donaldson; G. O. Hughes; G. F. Midgley; D. C. Raimondo; A. G. Rebelo; M. Rouget; W. Thuiller (2005). Potential impacts of future land-use and climate change on the Red List status of the Proteaceae in the Cape Floristic Region, South Africa. Global Change Biology, Vol. 11 (9): 1452– 1468.

Borengasser, M.; W. S. Hungate; R. Watkins (2008). Hyperspectral remote sensing: principles and applications. Taylor & Francis Group, LLC. Taylor & Francis Series in Remote Sensing Applications Series.

Borlaug, N. (2009). Foreword. In: Strange, R. (ed.) Food Security. The Science and Economics of Food Production and Access to Food. Springer, Dordrecht, The Netherlands, p. 1.

Bradshaw, B.; H. Dolan; B. Smit (2004). Farm-level adaptation to climatic variability and change: crop diversification in the Canadian prairies. Climatic Change, Vol. 67: 119–141.

Brassard, J. P.; P. Singh (2008). Impacts of climate change and CO2 increase on agricultural production and adaptation options for Southern Quebec, Canada. Mitig Adapt Strategies Glob Chang, Vol. 13: 241–265.

Braun, H.-J.; G. Atlin and T. Payne (2010). Multi-location Testing as a Tool to Identify Plant Response to Global Climate Change. In: Reynolds, M. P. (Ed.). Climate change and crop production. CABI series in climate change; Vol. 1. CAB International.

Brinkman, R.; W. G. Sombroek (1996). The effects of global change on soil conditions in relation to plant growth and food production. In: F. Bazzaz and W.G. Sombroek, Eds. Global Climate Change and Agricultural Production. John Wiley & Sons, Chichester, United Kingdom, pp. 49–63.

Brook, B. W. (2008). Synergies between climate change, extinctions and invasive vertebrates. Wildlife Research, Vol. 35 (3): 249–252.

Brooks, N; W. N. Adger; P. M. Kelly (2005). The determinants of vulnerability and adaptive capacity at the national level and the implications for adaptation' inNGlobal Environmental Change, 15, pp. 151–63

Buckland, S. M.; K. Thompson; J. G. Hodgson; Grime, J. P. (2001). Grassland Ninvasions: effects of manipulations of climate and management. Journal of Applied Ecology, Vol. 38: 301–309.

Bullock, P. (2005). Climate change impacts. In: Encyclopedia of Soils in the Environment. (Eds. Hillel, D.; J. L. Hatfield; K. M. Scow; D. S. Powlson; M. J. Singer; C. Rosenzweig; D. L. Sparks). Elsevier Ltd., The Boulevard, Langford Lane, Kidlington, Oxford, UK.

Burney, J. A., S. J. Davis; D. B. Lobell (2010). Greenhouse gas mitigation by agricultural intensification. P. Natl. Acad. Sci. USA 107:12052–12057.

Burton, P. J.; A. C. Balisky; L. P. Coward; S. G. Cumming; D. D. Kneeshaw (1992). The value of managing for biodiversity. Forestry Chronicle, Vol. 68: 225–237.

Bush, M. B. (1996). Amazonian conservation in a changing world. Biological Conservation, Vol. 76: 219–228.

Camberlin, P. (2009). Nile basin climates. In: The Nile origin, environments, limnology and human use. Henri J. Dumont (Ed.). Monographiae Biologicae Series, Vol. 89: 307–333. Springer, Dordrecht.

Campbell, J. B.; R. H. Wynne (2011). Introduction to remote sensing. — 5th ed. The Guilford Press, Division of Guilford Publications, Inc. New York, NY.

Canadell, J. G.; C. Le Que´re; M. R. Raupach; C. B. Field; E. T. Buitenhuis; P. Ciais; T. J. Conway; N. P. Gillett; R. A. Houghton; G. Marland (2007). Contributions to accelerating atmospheric CO2 growth from economic activity, carbon intensity, and efficiency of natural sinks. Proceedings of the National Academy of Sciences of the United States of America, 104: 18866–18870.

Canadell, J. G.; D. E. Pataki; L. F. Pitelka (2007). Terrestrial Ecosystems in a Changing World. Springer-Verlag Berlin Heidelberg. Global Change – The IGBP Series.

CAPMAS (Central Agency for Public Mobilization and Statistics) (2006). The Central Agency for Public Mobilization and Statistics (Cairo: Government Printing).

Cardelino, C.A., Chameides, W. L. (1990). Natural hydrocarbons, urbanization, and urban ozone. J. Geophys. Res. 95, 13,971–13,979.

Casper, J. K. (2010). Greenhouse gases: worldwide impacts. Global Warming Series. Facts On File, Inc. An imprint of Infobase Publishing, New York NY.

Center for International Earth Science Information Network (CIESIN), Columbia University and Centro Internacional de Agricultura Tropical (CIAT) (2005).

Gridded Population of the World Version 3 (GPWv3). Palisades, NY: Socioeconomic Data and Applications Center (SEDAC), Columbia University. Available at: ht tp: / / sedac . c ies in. columbia.edu/gpw (accessed 8 August 2009).

Challinor, A. J.; T. R. Wheeler (2008). Crop yield reduction in the tropics under climate change: Processes and uncertainties. Agricultural and Forest Meteorology, Vol. 148: 343–356.

Challinor, A. J.; T. R. Wheeler; J. M. Slingo (2005). Simulation of the impact of high temperature stress on the yield of an annual crop. Agric For Meteorol, Vol. 135: 180–189.

Chambers, L. E.; L. Hughes; M. A. Weston (2005). Climate change and its impact on Australia's avifauna. Emu, 105: 1–20.

Chen, S.; X. Ren; J. Mao; C. Z.; Brune, W. H.; Lefer, B.; Rappenglück, B.; Flynn, J.; Olson, J.; J. H. Crawford (2010). A comparison of chemical mechanisms based on TRAMP-2006 field data. Atmos. Environ., 44, 4116-4125.

Chenini, F. (2010). Implications of Water Quality on Irrigation Practices under Water Scarcity. In: Sabater, S.; D. Barceló (Eds.). Water scarcity in the Mediterranean: perspectives under global change. The Handbook of Environmental Chemistry Series (Vol. 8), Springer-Verlag Berlin, Heidelberg, Germany.

Chornesky, E. A.; A. M. Bartuska; G. H. Aplet; K. O. Britton J. Cummings-Carlson; F. W. Davis; J. Eskow; D. R. Gordon; K. W. Gottschalk; R. A. Haack; A. J. Hansen; R. N. Mack; F. J. Rahel; M. A. Shannon; L. A. Wainger; T. B. Wigley (2005). Science priorities for reducing the threat of invasive species to sustainable forestry. Bioscience, Vol. 55: 335–348.

Christensen, L.; M. B. Coughenour; J. E. Ellis; Z. Z. Chen (2004). Vulnerability of the Asian typical steppe to grazing and climate change. Climatic Change, Vol. 63: 351–368.

Collingham, Y.C.; B. Huntley (2000). Impacts of habitat fragmentation and patch size upon migration rates. Ecological Applications, Vol. 10: 131–144.

Confalonieri, U.; B. Menne; R. Akhtar; K. L. Ebi; M. Hauengue; R. S. Kovats; et al. (2007). Human health - Climate Change 2007: Impacts, adaptation and vulnerability contribution of Working Group II to the Fourth Assessment Report of the Intergovernmental Panel on Climate Change. Cambridge, UK: Cambridge University Press.

Conway, D. (2002). Extreme Rainfall Events and Lake Level Changes in East Africa: Recent Events and Historical Precedents. In: E.O. Odada and D. O. Olago (eds.) The East African Great Lakes: Limnology, Palaeolimnology and Biodiversity. Advances in Global Change Research Vol. 12, Kluwer, Dordrecht. Pp. 63-92.

Conway, D. (2005). From headwater tributaries to international river: observing and adapting to climate variability and change in the Nile Basin. Glob Environ Change, Vol. 15(2): 99–114.

Conway, D.; A. Persechino; S. Ardoin-Bardin; H. Hamandawana; C. Dieulin; G. Mahe (2007). Rainfall and water resources variability in Sub-Saharan Africa during the 20th century. Journal of Hydrometeorology (in press).

Conway, D.; M. Hulme (1993). Recent fluctuations in precipitation and runoff over the Nile sub-basins and their impact on Main Nile discharge. Climatic Change, Vol. 25: 127-151.

Conway, D.; M. Hulme (1996). The impacts of climate variability and future climate change in the Nile Basin on water resources in Egypt. Int. J. Water Resources Development, Vol. 12: 261-280.

Connor, R.; J.-M. Faurès; J. Kuylenstierna; J. Margat; P. Steduto; D. Vallée; W. van der Hoek (2009). Water in a changing world. The United Nations World Water Development Report 3, The United Nations Educational, Scientific and Cultural Organization UNESCO), Earthscan, UK.

Cooper, P.J.M., Dimes, J., Rao, K.P.C., Shapiro, B., Shiferaw, B. and Twomlow, S. (2008). Coping better with current climatic variability in the rainfed farming systems of sub-Saharan Africa: an essential first step in adapting to future climate change? Agriculture, Ecosystems and Environment 126, 24–35.

CropLife International (2009). Facts and Figures – the status of global agriculture (2008-2009). CropLife International, Brussels, Belgium, 12 pp.

Crozier, L.; R. W. Zabel (2006). Climate impacts at multiple scales: evidence for differential population responses in juvenile Chinook salmon. Journal of Animal Ecology, Vol. 75: 1100–1109.

D'Amato, G.; G. Liccardi; M. D'Amato; M. Cazzola (2002). Outdoor air pollution, climatic changes and allergic bronchial asthma. Eur Respir J; Vol. 20: 763–76.

Da Fonseca, G. A. B., W. Sechrest; J. Oglethorpe (2005). Managing the matrix. In: Lovejoy, T.E., Hannah, L. (Eds.), Climate Change and Biodiversity. Yale University Press, New Haven, pp. 346–358.

Da Matta F. M.; A. Grandis; B. C. Arenque; M. S. Buckeridge (2010). Impacts of climate changes on crop physiology and food quality. Food Research International, Vol. 43: 1814–1823.

de Chazal, J.; M. D. A. Rounsevell (2009). Land-use and climate change within assessments of biodiversity change: A review. Global Environmental Change, Vol. 19: 306–315.

de Dios, V. R.; C. Fischer; C. Colinas (2007). Climate change effects on Mediterranean forests and preventive measures. New Forests, Vol. 33: 29–40.

de Fraiture, C.; D. Wichelns; J. Rockström; E. Kemp-Benedict; N. Eriyagama; L. J. Gordon; M. A. Hanjra; J. Hoogeveen; A. Huber-Lee; L. Karlberg (2007). Looking ahead to 2050: scenarios of alternative investment approaches. In: Molden, D. (Ed.), Comprehensive Assessment of Water Management in Agriculture, Water for Food, Water for Life: A Comprehensive Assessment of Water Management in Agriculture. International Water Management Institute, London: Earthscan, Colombo, pp. 91–145 (Chapter 3).

Del Valle-Tascon, S.; J. L. Carrasco-Rodriguez (2004). Impact of ozone on crops. In: R. Dris and S. M. Jain (eds.), Production Practices and Quality Assessment of Food Crops, Vol. 1, "Pre-harvest Practice", pp. 189–208. Kluwer Academic Publishers. Printed in the Netherlands.

Dale, V. H.; H. M. Rauscher (1994). Assessing impacts of climate change on forests – the state of biological modeling. Climatic Change, Vol. 28: 65–90.

Dasgupta, S.; B. Laplante, C. Meisner, D. Wheeler, and J. Yan (2007). The Impact of Sea Level Rise on Developing Countries; A Comparative Analysis. World Bank Policy Research Working Paper 4136, Washington: Development Research Group, World Bank.

Denman, K.L., G. Brasseur, A. Chidthaisong, P. Ciais, P.M. Cox, R.E. Dickinson, D. Hauglustaine, C. Heinze, E. Holland, D. Jacob, U. Lohmann, S Ramachandran, P.L. da Silva Dias, S.C. Wofsy and X. Zhang, (2007): Couplings Between Changes in the Climate System and Biogeochemistry. In: Climate Change 2007: The Physical Science Basis. Contribution of Working Group I to the Fourth Assessment Report of the Intergovernmental Panel on Climate Change [Solomon, S., D. Qin, M. Manning, Z. Chen, M. Marquis, K.B. Averyt, M.Tignor and H.L. Miller (eds.)]. Cambridge University Press, Cambridge, United Kingdom and New York, NY, USA.

Diaz-Sanchez, D.; L. Proietti; R. Polosa (2003). Diesel fumes and the rising prevalence of canopy: an urban legend? Curr Allergy Asthm R; Vol. 3: 146–52.

Dixon, R. K.; J. A. Perry; E. L. Vanderklein; F. H. Hiol (1996). Vulnerability of forest resources to global climate change: case study of Cameroon and Ghana. Climate Research, Vol. 6: 127–133.

Doblas-Reyes, F J, Hagedorn R, Palmer T N (2006). Developments in dynamical seasonal forecasting relevant to agriculture. Climate Res 33(1):19–26.

Dominici, F.; R. D. Peng; M. L. Bell; L. Pham; A. McDermott; S. L. Zeger; et al. (2006). Fine particulate air pollution and hospital admission for cardiovascular and respiratory diseases. J Amer Med Assoc; Vol. 295 (10):1127–34.

Donald, P. F.; A. D. Evans (2006). Habitat connectivity and matrix restoration: the wider implications of agri-environment schemes. Journal of Applied Ecology, Vol. 43: 209–218.

Donnelly, A.; T. Lawson; J. Craigon; C. R. Black; J. J. Colls; G. Landon (2001). Effects of elevated CO2 and O3 on tuber quality in potato (Solanum tuberosum L.). Agriculture, Ecosystems and Environment, Vol. 87(3): 273–285.

Dumont, J. H. (2009). The Nile – origin, environments, limnology and human use. In: A Description of the Nile basin, and a synopsis of its history, ecology, biogeography, hydrology, and natural resources. (Ed. Henri J. Dumont). Monographiae Biologicae Series, Vol. 89: 1 – 21. Springer, Dordrecht.

Duraiappah, A.; S. Naeem; T. Agardi; N. Ash; D. Cooper; S. Díaz; et al. (2005). Ecosystems and Human Well-being: Biodiversity Synthesis. Island Press, Washington, DC, 100 pp.

Dyer, J. M. (1994). Implications of habitat fragmentation on climate change-induced forest migration. Professional Geographer, Vol. 46: 449–459.

Dynesius, M.; C. Nilsson (1994). Fragmentation and flow regulation of river systems in the Northern third of the World. Science, Vol. 266: 753–762.

Easterling, W. (2010). Modes of agricultural adaptation to climate change. In: Handbook of Climate Change in Agroecosystems (World Scientific Publishing, Singapore).

Easterling, W.; M. Apps (2005). Assessing the consequences of climate change for food and forest resources: a view from the IPCC. Climate Change, Vol. 70: 165–189.

Easterling, W. E.; P. K. Aggarwal; P. Batima; L. M. Brander; L. Erda; S. M. Howden; et al. (2007). Food, fibre and forest products. In: M. L. Parry, O. F. Canziani, J. P.

Palutikof, P. J. van der Linden, & C. E. Hanson (Eds.), Climate change 2007: Impacts, adaptation and vulnerability. Contribution of Working Group II to the fourth assessment report of the intergovernmental panel on climate change (pp. 273–313). Cambridge: Cambridge University Press.

Ebi, K. L.; D. M. Mills; J. B. Smith; A. Grambsch (2006). Climate change and human health impacts in the United States: an update on the results of the U.S. national assessment. Environ Health Persp; Vol. 114 (9): 1318–24.

EEA (Environmental European Agency, 2008). Impacts of Europe's changing climate–2008 indicator – based assessment. EEA Report 4/2008 & JRC Report JRC47756, Copenhagen.

Eeley, H. A. C.; M. J. Lawes; S. E. Piper (1999). The influence of climate change on the distribution of indigenous forest in KwaZulu-Natal, South Africa. Journal of Biogeography, Vol. 26: 595–617.

Eggermont, H.; K. Van Damme; J. M. Russell (2009). Rwenzori mountains (Mountains of the Moon): headwaters of the White Nile. In: H. J. Dumont (ed.), The Nile – origin, environments, limnology and human use. Monographiae Biologicae, Vol. 89: 243–261. Springer, Dordrecht.

El- Marsafawy, Samia M., M. A. M. Ibrahim and N. G. Ainer (2007). Vulnerability and adaptation studies on sugarcane crop in Egypt under climate change conditions and their effects on farm net return. Annals of Agric. Sci. Moshtohor, Vol. 45 (2).

El- Marsafawy, Samia M. (2007). Impact of climate change on sunflower crop production in Egypt. Egypt J. Agric. Res., 85 (5).

El-Marsafawy, Samia M.; M. K. M. El- Samanody (2009). Economic Impacts of Future Climatic Changes on Sunflower Crop in Egypt. The Fifth Inter. Conf. of Sustain. Agric. Develop., Fac. of Agric., Fayoum Univ., 21-23 Dec., 2009, Fayoum, Egypt.

El Raey, M. (1999). Impact of Climate Change on Egypt. Environmental Software and Services. http://www.ess.co.at/GAIA/CASES/EGY/impact.html (accessed July 14, 2009).

El Raey, M. (2009). The Cost of Coastal Vulnerability to Climate Change. Conference on climate change and coastal cities, beaches and the Delta, Cairo, Egypt.

El Raey, M. (2011). Mapping Areas Affected by Sea-Level Rise due to Climate Change in the Nile Delta Until 2100. In: H.G. Brauch et al. (eds.), Coping with Global Environmental Change, Disasters and Security, Hexagon Series on Human and Environmental Security and Peace Vol. 5, Springer-Verlag Berlin Heidelberg. Pages: 733 – 788.

El Raey, M.; K. Dewidar; El Hattab M. (1999). Adaptations to the impacts of sea level rise in Egypt. Climate Research, 117-128.

El-Shaer, M. H., C. Rosenzweig, A. Iglesias H. M. Eid and D. Hellil (1997). Impact of climate change on possible scenarios for Egyptian agriculture in the future. Mitigation and Adaptation Strategies for Global Change 1: 233- 250.

El Shamy, M. E.; I. A. Seierstad; A. Sorteberg (2009). Impacts of climate change on Blue Nile flows using bias corrected GCM scenarios. In: Hydrological Earth Systems Science, 13; 551-565.

El Sharkawy, H.; H. Rashed, I. Rached (2009). Climate Change: The Impacts of Sea Level Rise on Egypt. The impacts of SLR on Egypt, 45th ISOCARP Congress.

Epstein, P. R. (2005). Climate change and human health. New Engl J Med; Vol. 353(14): 1433–6.

Erasmus, B. F. N.; A. S. Van Jaarsveld; S. L. Chown; M. Kshatriya; K. J. Wessels (2002). Vulnerability of South African animal taxa to climate change. Global Change Biology, Vol. 8: 679–693.

Estrella, N., Spark, T.H. and Menzel, A. (2007). Trends and temperature response in the phenology of crops in Germany. Global Change Biology 13, 1737–1747.

Evenson, R. E.; D. Gollin (2003). Assessing the impact of the Green Revolution 1960 to 2000. Science, Vol. 300: 758–762.

Falkenmark, M. (2007). Shift in thinking to address the 21st century hunger gap: moving focus from blue to green water management. Water Resources Management, Vol. 21 (1): 3–18.

Fahrig, L. (2003). Effects of habitat fragmentation on biodiversity. Annual Review of Ecology Evolution and Systematics, Vol. 34: 487–515.

FAOSTAT (2009). FAO Database. Food and Agriculture Organization of the United Nations. Available at: http://faostat.fao.org/site/339/ default.aspx (accessed 22 September 2009).

FAO (2006). FAO Nile Basin Dataset (virtwatnile.xls). Database prepared by FAO Land and Water Unit for the report at hand, Rome, Italy.

FAO (2008a). Food safety and climate change. Technical background document from the expert consultation held in Rome, 26-28 February 2008. Contribution to the High Level Conference on World Food Security: The Challenges of Climate Change and Bioenergy, 3-5 June 2008, Rome.

FAO (2008b). Climate Change, Water and Food Security. Technical background document from the expert consultation held in Rome, 26-28 February 2008. Contribution to the High Level Conference on World Food Security: The Challenges of Climate Change and Bioenergy, 3-5 June 2008, Rome.

FAO (2010). Challenges and opportunities for carbon sequestration in grassland systems. Integrated Crop Management, Vol. 9-2010. Rome: FAO.

Fazey, I.; J. Fischer; D. B. Lindenmayer (2005). What do conservation biologists publish? Biological Conservation, Vol. 124 (1): 63–73.

Fedoroff, N. V.; D. S. Battisti; R. N. Beachy; P. J. M. Cooper; D. A. Fischhoff; C. N. Hodges; V. C. Knauf; D. Lobell; B. J. Mazur; D. Molden; M. P. Reynolds; P. C.

Ronald; M. W. Rosegrant; P. A. Sanchez; A. Vonshak; J.-K. Zhu (2010). Radically rethinking agriculture for the 21st century. Science, Vol. 327 (5967): 833–834.

Felzer, B. S.; T. Cronin; J. M. Reilly; J. M. Melillo; X. Wang (2007). Impacts of ozone on trees and crops. Compters Rendus Geoscience, Vol. 339: 784–798.

Fiala, J.; L. Cernikovsky; F. de Leeuw; P. Kurfuerst (2003). Air pollution by ozone in Europe in summer 2003. Overview of exceedances of EC ozone threshold values during the summer season April–August 2003 and comparisons with previous years. Copenhagen, Denmark: European Environment Agency and the European Topic Centre on Air and Climate Change.

Fischer, G.; F. N. Tubiello; H. van Velthuizen; D. A. Wiberg (2007). Climate change impacts on irrigation water requirements: effects of mitigation, 1990–2080.

Technological Forecasting & Social Change, Vol. 74 (7): 1083–1107.

Fischlin, A.; G. F. Midgley; Price J. T.; R. Leemans; B. Gopal; C. Turley; M. D. A. Rounsevell; O. P. Dube; J. Tarazona; A. A Velichko. (2007). Ecosystems, their properties, goods and services. In: Parry, M. L., Canziani, O. F., Palutikof, J. P., van der Linden, P.J. & Hanson, C.E. (eds.), Climate change 2007: Impacts, Adaptation and Vulnerability. Contribution of Working Group II to the Fourth Assessment Report of the Intergovernmental Panel of Climate Change (IPCC). Cambridge University Press, Cambridge, UK, pp. 211–272.

Folke, C.; Å. Jansson; Larsson J, Costanza R (1997). Ecosystem appropriation by cities. Ambio 26(3):167–172.

Foley, J.A., DeFries, R., Asner, G.P., Barford, C., Bonan, G., Carpenter, S.R., Chapin, F.S. and Coe, M.T. (2005). Global consequences of land use. Science 309, 570– 574.

Fonseca, S. C.; F. A. R. Oliveira; J. K. Brecht (2002). Modelling respiration of fresh fruits and vegetables for modified atmosphere packages: A review. Journal of Food Engineering, Vol. 52: 99–119.

Forster, P.; V. Ramaswamy; P. Artaxo; T. Berntsen; R. Betts; D. W. Fahey; J.

Haywood; J. Lean; D. C. Lowe; G. Myhre; J. Nganga; R. Prinn; G. Raga; M. Schulz; R. Van Dorland (2007). Changes in atmospheric constituents and in radiative forcing. In: Solomon S, Qin D, Manning M, Chen Z, Marquis M, Averyt KB, Tignor M, Miller HL (eds.) Climate change 2007: the physical science basis.

Contribution of working group I to the fourth assessment report of the intergovernmental panel on climate change. Cambridge University Press, Cambridge.

Franklin, J. F.; F. J. Swanson; M. E. Harmon; D. A. Perry; T. A. Spies; V. H. Dale; A. McKee; W. K. Ferrell; J. E. Means; S. V. Gregory; J. D. Lattin; T. D. Schowalter; D. Larsen (1992). Effects of global climatic change on forests in northwestern North America. In: Peters, R.L., Lovejoy, T.E. (Eds.), Global Warming and Biodiversity. Yale University Press, New Haven, pp. 244–254.

FTF, Facing the future (2002). From 'Facing the future': a report to the International Institute for Environment and Development, p. 52, available at http://www.iied.org/mmsd.

Fu, G. B.; S. L. Chen; C. M. Liu; et al. (2004). Hydro-climatic trends of the Yellow River basin for the last 50 years. Climate Change, Vol. 65: 149–178.

Fuhrer, J. (2003). Agroecosystem responses to combinations of elevated CO2, ozone, and global climate change. Agriculture, Ecosystems and Environment, Vol. 97: 1– 20.

Fukámi, T.; D. A. Wardle (2005). Long-term ecological dynamics: reciprocal insights from natural and anthropogenic gradients. Proceedings of the Royal Society B – Biological Sciences, Vol. 272: 2105–2115.

Gao, Y.; G. Y. Qiu; H. Shimizu; K. Tobe; B. P. Sun; J. Wang (2002). A 10-year study on techniques for vegetation restoration in a desertified salt lake area. J. Arid Environ., 52, 483-497.

Geist, H. J. (2005). The causes and progression of desertification. Ashgate Publishing, Aldershot, 258 pp.

Geist, H.; W. McConnell; E. F. Lambin; E. Moran; D. Alves; T. Rudel (2006). Causes and Trajectories of Land-Use/Cover Change. In: Local Processes and Global Impacts: Land-Use and Land-Cover Change. Eric F. Lambin and Helmut Geist (Eds.). Springer-Verlag Berlin Heidelberg. Global Change – The International Geosphere-Biosphere Programme (IGBP) Series.

Giannakopoulos, C.; P. Le Sager; M. Bindi; M. Moriondo; E. Kostopoulou; C. M. Goodess (2009). Climatic changes and associated impacts in the Mediterranean resulting from global warming. Global Planet Change, Vol. 68: 209–224.

Gil, S. V.; A. Becker; C. Oddino; M. Auaz; A. Marinelli G. March (2009a). Field assessment of biological, chemical and physical responses of soil to tillage intensity, fertilization and grazing. Environ Manag 44:378–386.

Gil, S. V.; J. Meriles; C. Confornta; G. Figoni; M. Basanta; E. Lovera; G. J. March (2009b). Field assessment of soil biological and chemical quality in response to crop management practices. World J Microbiol Biotechnol 25:439–448.

Gillespie, A. (2006). Climate Change, Ozone Depletion and Air Pollution: Legal Commentaries within the Context of Science and Policy. Koninklijke Brill NV incorporates the imprints Brill Academic Publishers, Martinus Nijhoff Publishers, and VSP.

Gillson, L.; K. J. Willis (2004). 'As earth's testimonies tell': wilderness conservation in a changing world. Ecology Letters, Vol. 7: 990–998.

Gleick, P. H. (1998). The World's Water: The Biennial Report on Freshwater Resources, 1998-1999. Washington, DC: Island Press.

Gleick, P. H. (2003). Global freshwater resources: soft-path solutions for the 21st century. Science, Vol. 302 (28): 1524–1528.

Goesch, T.; A. Hafi; M. Oliver; S. Page; D. Ashton; S. Hone; B. Dyack (2007). Drought and irrigation in Australia's Murray Darling Basin. Australian Commodities, Vol. 14 (2): 10.

Goldschmidt, A., et al. (2006). "Egypt." Microsoft® Encarta® [DVD].

Gordon, C. J. (1997). Behavioral thermoregulatory response to chlorpyrifos in the rat. Toxicology; Vol. 124 (3):165–71.

Grafton, R. Q.; K. Hussey (2011). Water Resources Planning and Management. Cambridge University Press, Cambridge, UK.

Green, J. (2009). The Kyoga catchment. In: H. J. Dumont (ed.), The Nile – origin, environments, limnology and human use. Monographiae Biologicae, Vol. 89: 205–214. Springer, Dordrecht.

Gregory, P. J., Johnson S. N., Newton, A. C.; J. S. I. Ingram (2009). Integrating pests and pathogens into the climate change/food security debate. Journal of Experimental Botany 60, 2827–2838.

Gruber, N.; J. N. Galloway (2008). An Earth-system perspective of the global nitrogen cycle. Nature, 451:293–296.

Guo, Q. F. (2000). Climate change and biodiversity conservation in Great Plains agroecosystems. Global Environmental Change-Human and Policy Dimensions, Vol. 10: 289–298.

Habib, O. A.; T. Ioriya; Y. Aruga (1987). The distribution of chlorophyll a as an index of primary productivity of phytoplankton in Khor El Ramla of the High Dam

Lake, Egypt. Journal of the Tokyo University of Fisheries, Vol. 74: 145–157.

Haffner, K.; H. j. Rosenfeld; G. Skrede; L. Wang (2002). Quality of red raspberry,

Rubus idaeus L. cultivars after storage in controlled and normal atmospheres. Postharvest Biology and Technology, Vol. 24 (3): 279–289.

Haines, A.; R. S. Kovats; D. Campbell-Lendrum; C. Corvalan (2006). Climate change and human health: impacts, vulnerability and public health. Public Health; Vol. 120 (7): 585–96.

Halpin, P. N. (1997). Global climate change and natural-area protection: management responses and research directions. Ecological Applications, Vol. 7: 828–843.

Hanjra, M. A.; M. E. Qureshi (2010). Global water crisis and future food security in an era of climate change. Food Policy, Vol. 35: 365–377.

Hanjra, M. A.; T. Ferede; D. G. Gutta (2009a). Pathways to breaking the poverty trap in Ethiopia: investments in agricultural water, education, and markets. Agricultural Water Management, Vol. 96 (11): 2–11.

Hanjra, M. A.; T. Ferede; D. G. Gutta (2009b). Reducing poverty in sub-Saharan Africa through investments in water and other priorities. Agricultural Water Management, Vol. 96 (7): 1062–1070.

Hansen, J. W. (2005). Integrating seasonal climate prediction and agricultural models for insights into agricultural practice. Philos T Roy Soc B 360:2037–2047.

ICARDA, Caravan. (2008). Review of agriculture in the dry areas. Issue No. 25 December 2008, pp. 1–61.

Idowu, O. J.; H. M. van Es; G. S. Abawi; D. W. Wolfe; R. R. Schindelbeck; B. N. Moebius-Clune; B. K. Gugino (2009). Use of an integrative soil health test for evaluation of soil management impacts. Renew Agric Food Sys 24:214–224.

Idso, S. B.; B. A. Kimball; P. E. Shaw; W. Widmer; J. T. Vanderslice; D. Higgs; et al. (2002). The effect of elevated atmospheric CO2 on the vitamin C concentration of (sour) orange juice. Agriculture, Ecosystems and Environment, Vol. 90: 1–7.

Iglesias, A.; L. Garrote; F. Flores; M. Moneo (2007). Challenges to manage the risk of water Scarcity and climate Change in the Mediterranean. Water Resour Manage, Vol. 21: 775–788.

Islam, M. S.; T. Matsui; Y. Yoshida (1996). Effect of carbon dioxide enrichment on physic-chemical and enzymatic changes in tomato fruits at various stages of maturity. Scientia Horticulturae, Vol. 65: 137–149.

Jacob, D. J.; D. A. Winner (2009). Effect of climate change on air quality. Atmospheric Environment, Vol. 43: 51–63.

Jackson, B. S., G. F. Arkin, and A. B. Hearn (1988). The cotton simulation model COTTAM: Fruiting model calibration and testing. Trans. ASAE 31: 846 - 854.

Jackson, P. B. N. (2000). Freshwater fisheries research in central and eastern Africa. A personal recollection. Transactions of the Royal Society of South Africa, Vol. 55: 1–81.

Jacobson, M. Z. (2007). Effects of ethanol (E85) versus gasoline vehicles on cancer and mortality in the United States. Environ. Sci. Technol. 41, 4150–4157.

Jain, V.; M. PAL; A. RAJ; S. Khetarpal (2007). Photosynthesis and nutrient composition of spinach and fenugreek grown under elevated carbon dioxide concentration. Biologia Plantarum, Vol. 51(3): 559–562.

Jetz, W.; D. S. Wilcove; A. P. Dobson (2007). Projected impacts of climate change and land-use change on the global diversity of birds. PLOS Biology, Vol. 5: 1211– 1219.

Jin, C.; S. Du; Y. Wang; J. Condon; X. Lin; Y. Zhang (2009). Carbon dioxide enrichment by composting in greenhouses and its effect on vegetable production. Journal of Plant Nutrition and Soil Science, Vol. 172: 418–424.

Jump, A. S.; J. Penuelas (2005). Running to stand still: adaptation and the response of plants to rapid climate change. Ecology Letters, Vol. 8: 1010–1020.

Kalnay, E.; M. Cai (2003). Impact of urbanization and land-use change on climate. Nature, Vol. 423: 528–531.

Kampunzu, A. B.; M. G. Bonhomme; M. Kanika (1998). Geochronology of volcanic rocks and evolution of the Cenozoic western branch of the East African Rift system. Journal of African Earth Sciences, Vol. 26: 441–461.

Kappelle, M.; M. M. I. Van Vuuren; P. Baas (1999). Effects of climate change on biodiversity: a review and identification of key research issues. Biodiversity and Conservation, Vol. 8: 1383–1397.

Karlen, D. L.; C. A. Ditzer; S. S. Andrews (2003). Soil quality: why and how? Geoderma 114:145–156.

Kasperson, R. E.; Archer E, Caceres D, Dow K, Downing T, Elmqvist T, Folke C, Han G, Iyengar K, Vogel C, Wilson K, Ziervogel G (2005). Vulnerable people and places. In: Scholes R, Rashid H (eds.) Millenium Ecosystem Assessment. Working group on conditions and trends, Island Press, Washington D.C.

Katnya, M. A. C.; G. Hoffmann-Thoma; A. A. Schriera; A. Fangmeierd; H. J. Jägerb; A. J. E. van Bel (2005). Increase of photosynthesis and starch in potato under elevated CO2 is dependent on leaf age. Journal of Plant Physiology, Vol. 162(4): 429–438.

Kays, S. J. (1997). Postharvest physiology of perishable plant products. Athens: AVI. 532 p.

Khan, S.; M. A. Hanjra (2009). Footprints of water and energy inputs in food production-global perspectives. Food Policy, Vol. 34 (2): 130–140.

Labat, D.; Y. Godderis; J. L. Probst et al (2004). Evidence for global runoff increase related to climate warming. Adv Water Pollut Res, Vol. 27: 631–642.

Ladenburger, C. G.; A. L. Hild; D. J. Kazmer; L. C. Munn (2006). Soil salinity patterns in Tamarix invasions in the Bighorn Basin,Wyoming, USA. J. Arid Environ., 65, 111-128.

Lal, R. (2005). Climate Change, Soil Carbon Dynamics, and Global Food Security. In: Climate change and global food security. (Eds.) R. Lal, B. A. Stewart, Norman

Uphoff and David O. Hansen, Books in soils, plants, and the environment, Vol. 96. Taylor & Francis Group, LLC, CRC Press. pp: 113 – 146.

Lal, R. (2006). Enhancing crop yield in developing countries through restoration of soil organic carbon pool in agricultural lands. Land Degrad Dev 17:197–209.

Lal, R. (2009). Tragedy of the Global Commons: Soil, Water and Air. In: (ed.) E. Lichtfouse, Climate Change, Intercropping, Pest Control and Beneficial Microorganisms, Sustainable Agriculture Reviews 2, Springer Science+Business Media B. V. pp: 9 – 11.

Lal, R. (2011). Soil Health and Climate Change: An Overview. In: Soil Health and Climate Change (Eds. Singh, B. P.; A. L. Cowie; K. Y. Chan). Soil Biology Series, Vol. 29. Springer-Verlag Berlin Heidelberg, Chapter 1, pp: 3 - 24.

Lunetta, R. S.; C. D. Elvidge (1998). Remote Sensing Change Detection: Environmental Monitoring Methods and Applications. Ann Arbor Press, MI.

McCarl, B. A.; R. M. Adams; B. H. Hurd (2001). Global climate change and its impact on agriculture. http://agecon2.tamu.edu/people/faculty/mccarlbruce/papers/879.pdf/28.12.2011.

McCarty, J. P. (2001). Ecological consequences of recent climate change. Conservation Biology, Vol. 15: 320–331.

MacDonald, R. W.; D. MacKay; B. Hickie (2002). Contaminant amplification in the environment. Environ Sci Technol; Vol. 36(23): 456A–62A.

Malcolm, J. R.; C. R. Liu; R. P. Neilson; L. Hansen; L. Hannah (2006). Global warming and extinctions of endemic species from biodiversity hotspots. Conservation Biology, Vol. 20 (2): 538–548.

Martin, S. (2004). An Introduction to Remote Sensing. Cambridge University Press, Cambridge, UK.

Mauzerall, D. L.; B. Sultan; N. Kim; D. F. Bradford (2005). NOx emissions from large point sources: variability in ozone production, resulting health damages and economic costs. Atmos Environ; Vol. 39 (16): 2851–66.

Mavi, H. S.; G. J. Tupper (2004). Agrometeorology: principles and applications of climate studies in agriculture. Food Products Press, Binghamton, NY.

Mearns, L. O. (2000). Climatic change and variability. In: K. R. Reddy & H. F. Hodges (Eds.), Climate change and global crop productivity (pp. 7–35). Wallingford, UK: CABI Publishing.

Nhemachena, C.; R. Hassan (2007). Micro-level analysis of farmers' adaptation to climate change in Southern Africa. IFPRI Discussion Paper No. 00714.

International Food Policy Research Institute, Washington, DC. Noss, R. F. (2001). Beyond Kyoto: forest management in a time of rapid climate change. Conservation Biology, Vol. 15: 578–590.

Noyes, D. P.; M. K. McElwee; H. D. Miller; B. W. Clark; L. A. Van Tiem; K. C. Walcott; K. N. Erwin; E. D. Levin (2009). The toxicology of climate change: Environmental contaminants in a warming world. Environment International, Vol. 35: 971–986.

NRC (National Research Council) (2010). Adapting to the Impacts of Climate Change America's Climate Choices: Panel on Adapting to the Impacts of Climate Change. National Academy of Sciences.

Nyssen, J.; J. Poesen; J. Moeyersons; J. Deckers; M. Haile; A. Lang (2004). Human impact on the environment in the Ethiopian and Eritrean highlands: a state of the art. Earth-Sci. Rev., 64, 273-320.

Oguto-Ohwayo, R. (1988). Reproductive potential of the Nile perch, Lates niloticus L. and the establishment of the species in Lakes Kioga and Victoria (East Africa). Hydrobiologia, Vol. 162: 193–200.

Phillips, D. L.; J. L. Lee; R. F. Dodson (1996). Sensitivity of the U.S. Corn Belt to climate change and elevated CO2. I: Corn and soybean yields. Agric. Syst., 52:481–502.

Piesse, J.; C. Thirtle (2009). Three bubbles and a panic: an explanatory review of recent food commodity price events. Food Policy, doi:10.1016/j.foodpol.2009.1001.1001.

Piper, B.S.; D. T. Plinston; J. V. Sutcliffe (1986). The water balance of Lake Victoria. Hydrol. Sci. J., Vol. 31(1): 25-37.

Pittock, B. (2009). Climate change: the science – impacts and solutions. 2nd Ed. CSIRO PUBLISHING, Australia, Earthscan, Dunstan House, UK.

Polley, H. W. (2002). Implications of atmospheric and climate change for crop yield. Crop Science, Vol. 42: 131–140.

Poloczanska, E. S.; A. J. Richardson (2009). Climate Change. Marine climate change in Australia, Impacts and adaptation responses, 2009, Report Card. www.oceanclimatechange.org.au/ 25.12.2011).

Porro, B. (2002). Preface to Opportunities and Risks of Climate Change. Swiss Reinsurance, p. 4, available at http://www.swissre.com.

Porter, J. R.; M. A. Semenov (2005). Crop responses to climatic variation. Philosophical Transactions of the Royal Society B: Biological Sciences, Vol. 360: 2021–2035.

Pounds, J. A.; M. R. Bustamante; L. A. Coloma; J. A. Consuegra; M. P. L. Fogden; P. N. Foster; E. La Marca; K. L. Masters; A. Merino-Viteri; R. Puschendorf; S. R. Ron; G. Sanchez; A. Azofeifa; C. J. Still; B. E. Young (2006).Widespread amphibian extinctions from epidemic disease driven by global warming. Nature, Vol. 439: 161–167.

Qaderi, M. M.; D. M. Reid (2009). Crop Responses to Elevated Carbon Dioxide and Temperature. In: S. N. Singh (ed.), Climate Change and Crops, Environmental Science and Engineering, Springer-Verlag Berlin Heidelberg.

Vandermeiren, K.; H. Harmens, G. Mills and L. De Temmerman (2009). Impacts of Ground-Level Ozone on Crop Production in a Changing Climate. In: S. N. Singh (ed.), Climate Change and Crops, Environmental Science and Engineering, Springer-Verlag Berlin Heidelberg.

Råberg, T. (2008). Agro ecosystems in a changing climate: Adaptation through crop rotations. Faculty of Landscape Planning, Horticulture and Agricultural Science, SLU-Alnarp. Master project in the Horticultural Science programme 2008:9 20 p.

Ramankutty, N.; J. A. Foley; N. J. Olejniczak (2002). People on the land: Changes in global population and croplands during the 20th century. Ambio 31(3):251–257.

Ruddiman, W. F. (2005). How did human first alter global climate. Sci Am 292:429– 436.

Rzóska, J. (1976). The Nile: Biology of an ancient river. Monographiae Biologicae, Vol. 29: 417 pp. Junk, The Haguc.

Saad, M. A. H.; M. H. Abbas (1985). Limnological investigations on the Rosetta branch of the Nile. 11. Seasonal variations of nutrients. Freshwater Biology, Vol. 15: 655–660.

Sabater, S.; D. Barceló (2010). Water scarcity in the Mediterranean: perspectives under global change. The Handbook of Environmental Chemistry Series (Vol. 8), Springer-Verlag Berlin, Heidelberg, Germany.

Sabater, S.; K. Tockner (2010). Effects of hydrologic alterations on the ecological quality of river ecosystems. In: Sabater, S.; D. Barceló (Eds.). Water scarcity in the

Mediterranean: perspectives under global change. The Handbook of Environmental Chemistry Series (Vol. 8), Springer-Verlag Berlin, Heidelberg, Germany.

Sage, R. F.; D. A. Way; D. S. Kubien (2008). Rubisco, Rubisco activase, and global climate change. Journal of Experimental Botany, Vol. 59: 1581–1595.

Said, R. (1981). The Geological Evolution of the River Nile. Springer, New York, 151 pp.

Sampath, R. K. (1992). Issues in irrigation pricing in developing countries. World Development, 967–977.

Sala, O. E.; D. D. van Vuuren; H. M. Pereira; D. Lodge; J. Alder; G. Cumming; A. Dobson; V. Wolters; M. A. Xenopoulos (2005). Chapter 10: biodiversity across scenarios. In: Ecosystems and Human Well-being, Vol. 2: scenarios. Island Press, Washington, DC, pp. 375–408.

Schaefer-Elinder, L. (2005). Obesity, hunger, and agriculture: the damaging role of Zubsidies. British Medical Journal, Vol. 331: 1333–1336.

Swetnam, T. W.; C. D. Allen; J. L. Betancourt (1999). Applied historical ecology: using the past to manage for the future. Ecological Applications, Vol. 9: 1189–1206.

Tagaris, E., Manomaiphiboon, K., Liao, K.-J., Leung, L.R., Woo, J.-H., He, S., Amar, P., Russell, A. G. (2007). Impacts of global climate change and emissions on regional ozone and fine particulate matter concentrations over the United States. J. Geophys. Res. 112, D14312.

Talling, J. F. (2006). A brief history of the scientific study of tropical African inland waters. Freshwater Forum, Vol. 26: 3–37.

Talling, J. F. (2009). The Nile: History of scientific research. In: H. J. Dumont (ed.), The Nile – origin, environments, limnology and human use. Monographiae Biologicae, Vol. 89: 205–214. Springer, Dordrecht.

Talbot, M. R.; M. A. J. Williams (2009). Cenozoic evolution of the Nile basin. In: H. J. Dumont (ed.), The Nile – origin, environments, limnology and human use. Monographiae Biologicae, Vol. 89: 37 – 60. Springer, Dordrecht.

Taub, D.; B. Miller; H. Allen (2008). Effects of elevated CO2 on the protein concentration of food crops: A meta-analysis. Global Change Biology, Vol. 14: 565–575.

Tawfik, M. (2009). Practices for wastewater treatment and re-use in the Mediterranean: case of Egypt. INNOVA-MED: innovative processes and practices for wastewater treatment and re-use in the Mediterranean region. FP 6.

Tharme, R. E. (2003). A global perspective on environmental flow assessment: emerging trends in the development and application of environmental flow methodologies for rivers. River Res Appl, Vol. 19: 397–441.

Thomas, C. D.; A. Cameron; R. E. Green; M. Bakkenes; L. J. Beaumont; Y. C. Collingham;

B. F. N. Erasmus; M. F. de Siqueira; A. Grainger; L. Hannah; L. Hughes; B. Huntley; A. S. van Jaarsveld; G. F. Midgley; L. Miles; M. A. Ortega-Huerta; A. T. Peterson; O. L. Phillips; S. E. William. (2004). Extinction risk from climate change. Nature, Vol. 427: 145–148.

Thomas, C. D.; A. M. A. Franco; J. K. Hill (2006). Range retractions and extinction in the face of climate change. Trends in Ecology and Evolution, Vol. 21 (8).

Turner II, B. L.; D. Skole; S. Sanderson; G. Fischer; L. Fresco; R. Leemans (1995). Land use and Land-cover Change Science/Research Plan. IGBP Global Change Report No. 35 and HDP Report No. 7. International Geosphere–Biosphere Program and the Human Dimensions of Global Environmental Change Program: Stockholm and Geneva. UNEP, 2002: Global Environmental Outlook 3: past, present and future perspectives. A report from the GEO Project, United Nations Environment Program, Nairobi, Kenya, 416 pp.

Turner, B. L. II; S. McCandless (2004). How humankind came to rival nature: A brief history of the human-environment condition and the lessons learned. In: Clark

WC, Crutzen P, Schellnhuber H-J (eds.) Earth system analysis for sustainability. Dahlem Workshop report No. 91, MIT Press, Cambridge, pp 227–243.

Tvedt (2004). The River Nile in the Age of the British: Political Ecology & the Quest for Economic Power. IB Tauris, London.

Uchida, H. and A. Nelson (2009). Agglomeration index: towards a new measure of urban concentration. Background paper for the World Bank's World Development Report 2009. Available at: http://bioval.jrc.ec.europa.eu/ products/gam/index.htm (accessed 10 August 2009).

UNECE (United Nations Economic Commission for Europe) (2007). Hemispheric Transport of Air Pollution 2007. Air Pollution Studies No. 16. United Nations, New York and Geneva.

UNEP (1998). The UNEP handbook on adaptation is by Feenstra and others, Handbook on Methods for Climate Change – Impacts Assessment and Adaptation Strategies, United Nations Environment Program, Nairobi, Kenya.

UNEP (2002). Global Environment Outlook GEO 3. Past, Present and Future Perspectives. UNEP, Nairobi and Earthscan Publications, London.

UNFCCC (2005). Feeling the heat. Electronic background document - United Nations Framework Convention on Climate Change, (Available at http://unfccc.int/essential_background/feeling_the_heat/items/2918.php)

VijayaVenkataRamana, S.; S. Iniyanb; R. Goic (2012). A review of climate change, mitigation and adaptation. Renewable and Sustainable Energy Reviews, Vol. 16: 878– 897.

Vivian, G. (2005). "Natural Hazards", In Encyclopedia of Coastal Science, by Maurice L. Schwartz, 678-684. Dordrecht, The Netherlands: Springer.

Voeroesmarty, C. J.; D Sahagian (2000). Anthropogenic disturbance of the terrestrial water cycle. Bioscience, Vol. 50: 753–765.

Voeroesmarty, C. J.; P. Green; J. Salisbury; R. B. Lammers (2000). Global water resources: vulnerability from climate change and population growth. Science, Vol. 289: 284–288.

Vorne, V.; K. Ojanperä; L. De Temmerman; M. Bindi; P. Hoegy; M. Jones; et al. (2002). Effects of elevated carbon dioxide and ozone on potato tuber quality in the European multiple-site experiment 'CHIP project'. European Journal of Agronomy, Vol. 17(4): 369–381.

Walther, G. R.; E. Post; P. Convey; A. Menzel; C. Parmesan; T. J. C. Beebee; J. M. Fromentin; O. Hoegh-Guldberg; F. Bairlein (2002). Ecological responses to recent climate change. Nature, Vol. 416: 389–395.

Wang, S. Y.; J. A. Bunce; J. Maas (2003). Elevated carbon dioxide increases contents of antioxidant compounds in field-grown strawberries. Journal of Agricultural and Food Chemistry, Vol. 51: 4315–4320.

Wassmann, R. S.;V. K. Jagdish; S. Hauer.; A. Ismail; E. Redona; R. Serraj; R. K. Singh; G. Howell; H. Pathak; K. Sumfleth (2009). Climate change affecting rice production: The physiological and agronomic basis for possible adaptation. Adv. Agron. 101:59–122.

Watkinson, W. P.; M. J. Campen; L. B. Wichers; J. P. Nolan; D. L. Costa (2003). Cardiac and thermoregulatory responses to inhaled pollutants in healthy and compromised rodents: modulation via interaction with environmental factors. Environ Res; Vol. 92(1): 35–47.

Watson, R. T. (2001). Climate Change 2001:Synthesis Report, Third Assessment Report of the Intergovernmental Panel on Climate Change, Cambridge: Cambridge University Press.

Waughray, D. (2011). Water security: the water-food-energy-climate nexus: the World Economic Forum water initiative. World Economic Forum, ISLAND PRESS.

Weissflog, L.; N. Elansky; E. Putz; G. Krueger; C. A. Lange; L. Lisitzina; A.

Pfennigsdorff (2004): Trichloroacetic acid in the vegetation of polluted and remote areas of both hemispheres. Part II. Salt lakes as novel sources of natural chlorohydrocarbons. Atmos, Environ,, 38, 4197-4204.

Welch, D. (2005). What should protected area managers do in the face of climate change. The George Wright Forum 22: 75–93.

Weng, Q. (2002). Land use change analysis in the Zhujiang Delta of China using satellite remote sensing, GIS, and stochastic modeling. Journal of Environmental Management 64, 273–284.

Weng, Q.; D. Lu (2009). Landscape as a continuum: An examination of the urban landscape structures and dynamics of Indianapolis city, 1991–2000. International Journal of Remote Sensing, 30: 2547–2577.

Xingli, G. et al. (2010). Future habitat loss and the conservation of plant biodiversity. Biol. Conserv., Vol. 143: 1594–1602.

Yadav, S. S.; B. Redden; D. L. McNeil; Y. Gan; A. Rizvi, A. K. Vrema; P. N. Bahl (2010). Climate Change and Management of Cool Season Grain. In: Yadav, S. S.; D. L. McNeil; R. Redden; S. A. Patil (Eds.) Climate Change and Management of Cool Season Grain Legume Crops. Springer Science+Business Media B.V.

Yadav, S. S.; R. Redden; J. L. Hatfield, H. Lotze-Campen; A. Hall (2011). Crop Adaptation to Climate Change. Inc., New Delhi, India, John Wiley & Sons, Inc.

Yates, D. N.; K. M. Strzepek (1998). Modeling the Nile basin under climatic change. Journal of Hydrologic Engineering, Vol. 3: 98–108.

You, L., Wood, S. and Wood-Sichra, U. (2009). Generating plausible crop distribution maps for sub-Saharan Africa using a spatially disaggregated data fusion and optimization approach. Agricultural Systems 99, 126–140.

Zereini, F.; H. Hoetzl (2008). Climatic Changes and Water Resources in the Middle East and North Africa. Environmental Science and Engineering Series. Springer-Verlag Berlin Heidelberg.

Zeitoun, M.; J. A. Allan; Y. Mohieldeen (2010). Virtual water 'flows' of the Nile Basin, 1998 2004: A first approximation and implications for water security. Global Environmental Change, Vol. 20: 229–242.

INDEX